Life Annuities and Life Insurance

MATHEMATICS
OF FINANCE

Fifth Edition

Petr Zima
University of Waterloo

Robert L. Brown
University of Waterloo

 **McGraw-Hill
Ryerson**

Toronto   Montréal   Boston   Burr Ridge, IL   Dubuque, IA
Madison, WI   New York   San Francisco   St. Louis
Bangkok   Bogotá   Caracas   Kuala Lumpur   Lisbon   London
Madrid   Mexico City   Milan   New Delhi   Santiago   Seoul
Singapore   Sydney   Taipei

Mathematics of Finance
Fifth Edition

ISBN: 0-07-087135-3

3 4 5 6 7 8 9 10 TRI 0 9 8 7 6 5 4 3

Printed and bound in Canada.

Care has been taken to trace ownership of copyright material contained
in this text; however, the publisher will welcome any information that
enables them to rectify any reference or credit for subsequent editions.

Vice-President and Editorial Director: Pat Ferrier
Senior Sponsoring Editor: Lynn Fisher
Associate Sponsoring Editor: Jason Stanley
Developmental Editor: Maria Chu
Supervising Editor: Alissa Messner
Production Editor: Dianne Broad
Senior Marketing Manager: Jeff MacLean
Production Coordinator: Brad Madill
Composition: Pronk&Associates
Cover Design: Greg Devitt
Cover Image: © Jim Krantz/Stone
Printer: Tri-Graphic Printing

Canadian Cataloguing in Publication Data

Zima, Petr, 1941 –
   Mathematics of finance

5th ed.
Includes index.
ISBN 0-07-087135-3

1. Business mathematics. I. Brown, Robert L., 1949- . II. Title.

HF5691.Z55 2000   650'.01'513   C00-932174-8

Contents

Preface

It is more than twenty years now since the publication of the first edition of *Mathematics of Finance*, and over thirty years since we started teaching this material. In that time, we have learned how volatile interest rates can be. The effects are everywhere, including the interest earned by a school child in his or her first bank account, the investment earnings of a pension fund, the interest rate charged on a credit card or mortgage loan, or rates in the short-term money market used by multinational corporations.

This text is designed to provide readers with a generic approach to understanding using interest rates in respect to a wide range of financial transactions, including annuities, home mortgage and personal loans, bonds and the assessment of future investment projects. A wide variety of interest rates are used in the solved examples and in the hundreds of exercises available to the student.

While finance and business students at colleges or universities can readily use the text, it is also appropriate for others who wish to gain a better understanding of these issues. A limited level of mathematical knowledge is assumed at the start of the text, thus permitting introductory chapters on algebra to be omitted. Nevertheless, the appendices include detailed explanations of logarithms, progressions and linear interpolation, which cover the major requirements.

The text also includes a number of spreadsheet examples. Spreadsheets are a very powerful tool in financial calculations as they permit complicated models to be developed and many calculations to be completed very quickly. As well, we like to emphasize the use of pocket calculators in problem-solving techniques. However, it is essential that the fundamental principles of financial mathematics be appreciated before any spreadsheet is developed or calculator is used.

Despite scrutiny and review, it is possible that some errors will remain. For these, we accept full responsibility and welcome any correspondence and suggestions. Correspondence should be sent to the Finance Editor at McGraw-Hill Ryerson.

Our interest in the mathematics of finance remains undiminished. It is our hope that this text provides the reader with improved understanding of the calculations that underlie most financial transactions.

New Features to the Fifth Edition

The most obvious new feature is the **new design**. It emphasizes key formulas and other pedagogical features and makes them more readily identifiable and easier to remember for students. **Calculation Tip** and **Observation** boxes are located throughout the text and draw students' attention to pitfalls, shortcuts, helpful suggestions for using calculators and remind them of important concepts and formulas. **Summaries** that list key points, concepts and formulas presented in the chapter are found at the end of each chapter in the form of a bulleted list that serves as a useful review feature for students. Many **new and revised exercises** can be found at the end of each chapter. There are also **Case Studies** in selected Review sections that present more complex real-world problems. **Web addresses** for interesting useful Internet sites for finance are incorporated into each chapter.

Other Key Features

Financial mathematics is easier to explain and understand through the use of examples, and, for this reason, most sections include a number of solved problems. **Part A exercises** are designed to help students understand the major concepts while **Part B exercises** provide more difficult problems and include advanced material. Also, each chapter concludes with a set of review exercises. The authors emphasize that one of the most important aspects for the student is visualizing the problem with the help of a **timeline** diagram to assist the student in setting up the solution. A **glossary** at the end of the text contains definitions for key financial terminology. **Answers to all exercises problems** are located in back of the text. For easy reference, the book contains **inside cover summaries** of formulas.

Supplements

A **Solutions Manual** presents full solutions to all of the problems in the text. Also, a **Problems Diskette** in Word format contains exercises from the text that instructors can use to build tests.

PageOut is the McGraw-Hill Ryerson Web site development centre. This web page generation software is free to adoptors and is designed to help faculty create and deliver web-based course content. PageOut features include: Assignments/Quizzes, Course Calendar, Discussion Area, Student Homepages, Gradebook, and more in a matter of minutes. To find out more about PageOut, contact your McGraw-Hill Ryerson representative, visit the Web site at http://www.pageout.net or send an e-mail to mail@mcgrawhill.ca.

Acknowledgements

The fifth edition is a reflection of the contributions from many dedicated teachers and users of this text. For the fifth edition we are particularly grateful for the input of the following people: Helen Dakin, Mohawk College; Richard Brown, York University; Josephine Evans, Langara College; Barry Garner, University College of Fraser Valley; Steve Kopp, University of Western Ontario; Diane Huysmans, Fanshawe College.

We also wish to thank the staff at McGraw-Hill Ryerson. This includes: Lynn Fisher, Senior Sponsoring Editor; Maria Chu, Developmental Editor; Alissa Messner, Supervising Editor; Jeff MacLean, Senior Marketing Manager and others who we do not know personally, but made important contributions. Throughout the production process, Dianne Broad did an excellent job of editing. We would also like to extend special acknowledgement to Mary Lou Dufton who input the manuscript changes and the solutions manual.

Petr Zima
Robert L. Brown

Chapter One

Simple Interest and Simple Discount

Section 1.1 ## Simple Interest

Suppose that an investor lends money to a debtor. The debtor must pay back the money originally borrowed, and also the fee charged for the use of the money, called **interest**. From the investor's point of view, interest is income from invested capital. The capital originally invested in an interest transaction is called **the principal**. The sum of the principal and interest due is called the **amount** or **accumulated value**. Any interest transaction can be described by the **rate of interest**, which is the ratio of the interest earned in one time unit to the principal.

In early times, the principal lent and the interest paid might be tangible goods (e.g., grain). Now, they are most commonly in the form of money. The practice of charging interest is as old as the earliest written records of humanity. Four thousand years ago, the laws of Babylon referred to interest payments on debts.

At **simple interest**, the interest is computed on the original principal during the whole time, or term of the loan, at the stated annual rate of interest.

We shall use the following notation:

P = the principal, or the present value of S, or the discounted value of S, or the proceeds.

I = simple interest.

S = the amount, or the accumulated value of P, or the maturity value of P.

r = annual rate of simple interest.

t = time in years.

Simple interest is calculated by means of the formula

$$\boxed{I = Prt} \tag{1}$$

From the definition of the amount S we have

$$S = P + I$$

By substituting for $I = Prt$ we obtain S in terms of P, r, and t

$$S = P + Prt$$

$$\boxed{S = P(1 + rt)} \tag{2}$$

> The factor $(1 + rt)$ in formula **(2)** is called an **accumulation factor at a simple interest rate r** and the process of calculating S from P by formula **(2)** is called **accumulation at a simple interest rate r**.

The time t must be in years. When the time is given in months, then

$$t = \frac{\text{number of months}}{12}$$

When the time is given in days, there are two different varieties of simple interest in use:

1. **Exact interest**, where $t = \dfrac{\text{number of days}}{365}$

   i.e., the year is taken as 365 days (leap year or not).

2. **Ordinary interest**, where $t = \dfrac{\text{number of days}}{360}$

   i.e., the year is taken as 360 days.

CALCULATION TIP: The general practice in Canada is to use exact interest, whereas the general practice in the United States and in international business transactions is to use ordinary interest (also referred to as the Banker's Rule). In this textbook, exact interest is used all the time unless specified otherwise. When the time is given by two dates we calculate the exact number of days between the two dates from a table listing the serial number of each day of the year (see the table on the inside back cover). The exact time is obtained as the difference between serial numbers of the given dates. In leap years (years divisible by 4) an extra day is added to February.

EXAMPLE 1 A loan taken out on April 7 is to be repaid on September 25. Find the exact number of days between the two dates.

Solution Using the table on the inside back cover we find that April 7 is day 97; September 25 is day 268, and the difference $268 - 97 = 171$ represents the exact number of days from April 7 to September 25.

∎

EXAMPLE 2   Find the exact and ordinary simple interest on a 90-day loan of $500 at $8\frac{1}{2}\%$.

Solution   We have $P = 500$, $r = 0.085$, $t = 90$ days

$$\text{Exact interest} = 500 \times 0.085 \times \tfrac{90}{365} = \$10.48$$

$$\text{Ordinary interest} = 500 \times 0.085 \times \tfrac{90}{360} = \$10.63$$

■

OBSERVATION: Notice that ordinary interest is always greater than the exact interest and thus it brings increased revenue to the lender.

EXAMPLE 3   A couple borrows $10 000. The annual interest rate is $10\frac{1}{2}\%$, payable monthly, and the monthly payment is $200. How much of the first payment goes to interest and how much to principal?

Solution   We have $P = 10\ 000$, $r = 0.105$, $t = \frac{1}{12}$, and

$$I = 10\ 000 \times 0.105 \times \tfrac{1}{12} = \$87.50$$

The interest for the first month is $87.50 and $112.50 is applied to principal reduction.

■

EXAMPLE 4   A loan shark made a loan of $100 to be repaid with $120 at the end of one month. What was the annual interest rate?

Solution   We have $P = 100$, $I = 20$, $t = \frac{1}{12}$, and

$$r = \frac{I}{Pt} = \frac{20}{100 \times \frac{1}{12}} = 240\%$$

■

EXAMPLE 5   How long will it take $3000 to earn $60 interest at 6%?

Solution   We have $P = 3000$, $I = 60$, $r = 0.06$, and

$$t = \frac{I}{Pr} = \frac{60}{3000 \times 0.06} = \frac{1}{3} = 4 \text{ months}$$

■

Demand Loans

On a **demand loan**, the lender may demand full or partial payment of the loan at any time and the borrower may repay all of the loan or any part at any time without notice and without interest penalty. Interest on demand loans is based on the unpaid balance and is usually payable monthly. The interest rate on demand loans is not usually fixed but fluctuates with market conditions.

EXAMPLE 6 Jessica borrowed $1500 from her credit union on a demand loan on August 16. Interest on the loan, calculated on the unpaid balance, is charged to her account on the 1st of each month. Jessica made a payment of $300 on September 17, a payment of $500 on October 7, a payment of $400 on November 12 and repaid the balance on December 15. The rate of interest on the loan on August 16 was 12% per annum. The rate was changed to 11.5% on September 25 and 12.5% on November 20. Calculate the interest payments required and the total interest paid.

Solution

Interest period	# of days	Balance	Rate	Interest
Aug. 16 - Sep. 1	16	$1500	12%	$1500(0.12)(\frac{16}{365}) = \underline{\quad 7.89}$ Sep. 1 interest = $ 7.89
Sep. 1 - Sep. 17	16	$1500	12%	$1500(0.12)(\frac{16}{365}) = \quad 7.89$
Sep. 17 - Sep. 25	8	$1200	12%	$1200(0.12)(\frac{8}{365}) = \quad 3.16$
Sep. 25 - Oct. 1	6	$1200	11.5%	$1200(0.115)(\frac{6}{365}) = \underline{\quad 2.27}$ Oct. 1 interest = $13.32
Oct. 1 - Oct. 7	6	$1200	11.5%	$1200(0.115)(\frac{6}{365}) = \quad 2.27$
Oct. 7 - Nov. 1	25	$ 700	11.5%	$700(0.115)(\frac{25}{365}) = \underline{\quad 5.51}$ Nov. 1 interest = $ 7.78
Nov. 1 - Nov. 12	11	$ 700	11.5%	$700(0.115)(\frac{11}{365}) = \quad 2.43$
Nov. 12 - Nov. 20	8	$ 300	11.5%	$300(0.115)(\frac{8}{365}) = \quad 0.76$
Nov. 20 - Dec. 1	11	$ 300	12.5%	$300(0.125)(\frac{11}{365}) = \underline{\quad 1.13}$ Dec. 1 interest = $ 4.32
Dec. 1 - Dec. 15	14	$ 300	12.5%	$300(0.125)(\frac{14}{365}) = \underline{\quad 1.44}$ Dec. 15 interest = $ 1.44

Total interest paid = 7.89 + 13.32 + 7.78 + 4.32 + 1.44 = $34.75

■

Invoice Cash Discounts

To encourage prompt payments of invoices many manufacturers and whole-salers offer cash discounts for payments in advance of the final due date. The following typical credit terms may be printed on sales invoices:

2/10, *n*/30 — Goods billed on this basis are subject to a cash discount of 2% if paid within ten days. Otherwise, the full amount must be paid not later than thirty days from the date of the invoice.

A merchant may consider borrowing the money to pay the invoice in time to receive the cash discount. Assuming that the loan would be repaid on the day the invoice is due, the interest the merchant should be willing to pay on the loan should not exceed the cash discount.

EXAMPLE 7 A merchant receives an invoice for a motor boat for $4000 with terms 4/30, n/100. What is the highest simple interest rate at which he can afford to borrow money in order to take advantage of the discount?

Solution Suppose the merchant will take advantage of the cash discount of 4% of 4000 = $160 by paying the bill within 30 days from the date of invoice. He needs to borrow 4000 − 160 = $3840 for 70 days. The interest he should be willing to pay on borrowed money should not exceed the cash discount $160.

We have $P = 3840$, $I = 160$, $t = \frac{70}{365}$, and we calculate

$$r = \frac{I}{Pt} = \frac{160}{3840 \times \frac{70}{365}} = 21.73\%$$

The highest simple interest rate at which the merchant can afford to borrow money is 21.73%. This is a break-even rate. If he can borrow money, say at a rate of 15%, he should do so. He would borrow $3840 for 70 days at 15%. Maturity value of the loan is $3840[1 + (.15)(\frac{70}{365})] = \$3950.47$. Thus his savings would be $4000 − 3950.47 = \$49.53$. ■

Exercise 1.1

1. Find the maturity value of
   a) a $2500 loan for 18 months at 12% simple interest,
   b) a $1200 loan for 120 days at 8.5% ordinary simple interest, and
   c) a $10 000 loan for 64 days at 15% exact simple interest.

2. At what rate of simple interest will
   a) $1000 accumulate to $1420 in $2\frac{1}{2}$ years,
   b) money double itself in 7 years, and
   c) $500 accumulate $10 interest in 2 months?

3. How many days will it take $1000 to accumulate to at least $1200 at 5.5% simple interest?

4. Find the ordinary and exact simple interest on $5000 for 90 days at $10\frac{1}{2}\%$.

5. A student lends his friend $10 for one month. At the end of the month he asks for repayment of the $10 plus purchase of a chocolate bar worth 50¢. What simple interest rate is implied?

6. What principal will accumulate to $5100 in 6 months if the simple interest rate is 9%?

7. What principal will accumulate to $580 in 120 days at 18% simple interest?

8. Find the accumulated value of $1000 over 65 days at $11\frac{1}{2}\%$ using both ordinary and exact simple interest.

9. A man borrows $1000 for 220 days at 17% simple interest. What amount must he repay?

10. A sum of 2000 is invested from May 18, 2001, to April 8, 2002, at 4.5% simple interest. Find the amount of interest earned.

11. On May 13, 2001, Jacob invested $4000. On February 1, 2002, he intends to pay Fred $4300 for a used car. The bank assured Jacob that his investment would be adequate to cover the purchase. Determine the minimum simple interest rate that Jacob's money must be earning.

12. On January 1, Mustafa borrows $1000 on a demand loan from his bank. Interest is paid at the end of each quarter (March 31, June 30, September 30, December 31) and at the time of the last payment. Interest is calculated at the rate of 12% on the balance of the loan outstanding. Mustafa repaid the loan with the following payments:

March 1	$100
April 17	$300
July 12	$200
August 20	$100
October 18	$300
	$1000

Calculate the interest payments required and the total interest paid.

13. On February 3, a company borrowed $50 000 on a demand loan from a bank. Interest on the loan is charged to the company's current account on the 11th of each month. The company repaid the loan with the following payments:

February 20	$10 000
March 20	$10 000
April 20	$15 000
May 20	$15 000
	$50 000

The rate of interest on the loan was originally 15%. The rate was changed to 16% on April 1, and to 15.5% on May 1. Calculate the interest payments required and the total interest paid.

14. A cash discount of 2% is given if a bill is paid 20 days in advance of its due date. At what interest rate could you afford to borrow money to take advantage of this discount?

15. A merchant receives an invoice for $2000 with terms 2/20, $n/60$. What is the highest simple interest rate at which he can afford to borrow money in order to take advantage of the discount?

16. The ABC general store receives an invoice for goods totalling $500. The terms were 3/10, $n/30$. If the store were to borrow the money to pay the bill in 10 days, what is the highest interest rate at which the store could afford to borrow?

17. I.C.U. Optical receives an invoice for $2500 with terms 2/10, $n/50$.
    a) If the company is to take advantage of the discount, what is the highest simple interest rate at which it can afford to borrow money?
    b) If money can be borrowed at 10% simple interest, how much does I.C.U. Optical actually save by using the cash discount?

Section 1.2   Discounted Value at Simple Interest

From formula (2) we can express P in terms of S, r and t and obtain

$$P = \frac{S}{1 + rt} = S(1 + rt)^{-1}$$

(3)

> The factor $(1 + rt)^{-1}$ in formula **(3)** is called a **discount factor at a simple interest rate** r and the process of calculating P from S is called **discounting at a simple interest rate** r, or simple discount at an interest rate r.

When we calculate P from S, we call P the present value of S or the discounted value of S. The difference $D = S - P$ is called the simple discount on S at an interest rate.

For a given interest rate r, the difference $S - P$ has two interpretations.

1. The interest I on P which when added to P gives S.

2. The discount D on S which when subtracted from S gives P.

EXAMPLE 1 Sixty days after borrowing money, a person pays back exactly $200. How much was borrowed if the $200 payment includes the principal and simple interest at 9%?

Solution We have $S = 200$, $r = 0.09$, and $t = \frac{60}{365}$. Substituting in formula **(3)** gives

$$P = \frac{200}{1 + 0.09(\frac{60}{365})} = \$197.08$$

∎

EXAMPLE 2 What is the discounted value of $800 due at the end of 8 months if the simple interest rate is 10%? What is the simple discount?

Solution We have $S = 800$, $r = 10\%$, $t = \frac{8}{12}$ and calculate

$$P = \frac{S}{1 + rt} = \frac{800}{1 + (0.10)(\frac{8}{12})} = \$750$$

The simple discount

$$D = S - P = 800 - 750 = \$50$$

∎

Promissory Notes

A **promissory note** is a written promise by a debtor, called the **maker** of the note, to pay to, or to the order of, the creditor, called the **payee** of the note, a sum of money, with or without interest, on a specified date.

The following is an example of an interest-bearing note.

> $2000.00   Toronto, September 1, 2001
>
> Sixty days after date, I promise to pay to the order of Mr. A
>
> Two thousand and $\frac{00}{100}$ dollars
>
> for value received with interest at 11% per annum.
>
>   (Signed) Mr. B

The **face value** of the note is $2000. The **term** of the note is 60 days. The **due date** is 60 days after September 1, 2001, that is October 31, 2001.

In Canada, *three days of grace**are added to the 60 days to arrive at the **legal due date**, or the "**maturity date**," that is November 3, 2001. By a **maturity value** of a note we shall understand the value of the note at the maturity date. In our example, the maturity value of the note is the accumulated value of $2000 for 63 days at 11%, i.e., $2000[1 + (0.11)(\frac{63}{365})] = \$2037.97$.

If Mr. B chooses to pay on October 31, he will pay interest for 60 days, not 63 days. However, no legal action can be taken against him until the expiry of the three days of grace. Thus, Mr. B would be within his legal rights to repay as late as November 3, and in that case he would pay interest for 63 days.

A promissory note may be sold one or several times before its maturity. Each buyer discounts the maturity value of the note for the time from the date of sale to the maturity date at his interest rate and the seller receives the proceeds of the sale. The proceeds are determined by formula **(3)**.

The procedure for discounting of promissory notes can be summarized in 2 steps:

Step 1 Get the maturity value, S, of the note.
   The maturity value of a noninterest-bearing note is the face value of the note. The maturity value of an interest-bearing note is the accumulated value of the face value of the note at the maturity date.

Step 2 Get the proceeds, P, by discounting the maturity value, S, at a specified simple interest rate from the maturity date back to the date of sale.

EXAMPLE 3 The note described in the beginning of this section is sold by Mr. A on October 1, 2001 to a bank that discounts notes at a $9\frac{1}{2}$% simple interest rate.

a) How much money would Mr. A receive for the note?
b) What rate of interest will the bank realize on its investment, if it holds the note till maturity?
c) What rate of interest will Mr. A realize on his investment, when he sells the note on October 1, 2001?

---

*Three days of grace. In Canada, "three days of grace" are allowed for the payment of a promissory note. This means that the payment becomes due on the third day after the due date of the note.

If interest is being charged, the three days of grace must be added to the time stated in the note, to compute it. If the third day of grace falls on a holiday, payment will become due the next day that is itself not a holiday.

If the time is stated in months, these must be calendar months and not months of 30 days. For example, 2 months after July 5 is September 5. Thus the legal due date is September 8, and the number of days to be used in calculating the interest (if any) will be 65 days.

In the month in which the payment falls due, there may be no corresponding date to that from which the time is computed. In such a case, the last day of the month is taken as the corresponding date. For example, two months after December 31 is February 28 (or February 29 in leap years) and March 3 would be the legal due date.

Solution a We arrange the dated values on a time diagram below.

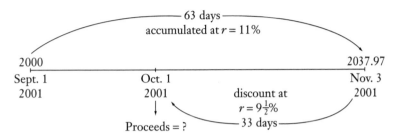

Maturity value of the note is $S = 2000[1 + (0.11)(\frac{63}{365})] = \$2037.97$.

Proceeds on October 1, 2001, are $P = 2037.97[1 + (0.095)(\frac{33}{365})]^{-1} = \$2020.61$

Thus Mr. A would receive \$2020.61 for the note.

Solution b The bank will realize a profit of $2037.97 - 2020.61 = \$17.36$ for 33 days. Thus we have $P = 2020.61$, $I = 17.36$, and $t = \frac{33}{365}$ and calculate

$$r = \frac{I}{Pt} = \frac{17.36}{2020.61 \times \frac{33}{365}} = 9.50\%$$

Solution c Mr. A will realize a profit of $2020.61 - 2000 = \$20.61$ on his investment of \$2000 for the 30 days he held the note. The rate of interest Mr. A will realize is

$$r = \frac{I}{Pt} = \frac{20.61}{2000 \times \frac{30}{365}} = 12.54\%$$

■

EXAMPLE 4 On April 21, a retailer buys goods amounting to \$5000. If he pays cash he will get a 4% cash discount. To take advantage of this cash discount, he signs a 90-day noninterest-bearing note at his bank that discounts notes at an interest rate of 9%. What should be the face value of this note to give him the exact amount needed to pay cash for the goods?

Solution The cash discount 4% of \$5000 is \$200. The retailer needs \$4800 in cash. He will sign a noninterest-bearing note with maturity value S calculated by formula **(2)**, given $P = 4800$, $r = 9\%$, and $t = \frac{93}{365}$. Thus

$$S = P(1 + rt) = 4800[1 + (.09)(\tfrac{93}{365})] = \$4910.07$$

The face value of the noninterest-bearing note should be \$4910.07.

Note: The retailer could also sign a 90-day 9% interest-bearing note with a face value of \$4800 and a maturity value of \$4910.07 in 93 days. He would receive cash of \$4800 immediately as the proceeds $P = 4910.07[1 + (.09)(\frac{93}{365})]^{-1} = \$4800$.

■

Exercise 1.2

For each promissory note in problems 1 to 4 find the maturity date, maturity value, discount period, and the proceeds:

No.	Date of Note	Face Value	Interest Rate on Note	Term	Date of Discount	Interest Rate for Discounting
1.	Nov. 3	$3000	None	3 months	Dec. 1	10.25%
2.	Sept. 1	$1200	5.5%	60 days	Oct. 7	4.75%
3.	Dec. 14	$ 500	7.0%	192 days	Dec. 24	7.00%
4.	Feb. 4	$4000	None	2 months	Mar. 1	8.50%

5. A 90-day note promises to pay Ms. Chiu $2000 plus simple interest at 13%. After 51 days it is sold to a bank that discounts notes at a 12% simple interest rate.
   a) How much money does Ms. Chiu receive?
   b) What rate of interest does Ms. Chiu realize on her investment?
   c) What rate of interest will the bank realize if the note is paid in full in exactly 90 days?

6. An investor lends $5000 and receives a promissory note promising repayment of the loan in 90 days with 12% simple interest. This note is immediately sold to a bank that charges 10% simple interest. How much does the bank pay for the note? What is the investor's profit? What is the bank's profit on this investment when the note matures?

7. A 90-day note for $800 bears interest at 10% and is sold 60 days before maturity to a bank that uses a 12% simple interest rate. What are the proceeds?

8. Jacob owes Kieran $1000. Kieran agrees to accept as payment a noninterest-bearing note for 90 days that can be discounted immediately at a local bank that charges a simple interest rate of 10%. What should be the face value of the note so that Kieran will receive $1000 as proceeds?

9. Justin has a note for $5000 dated October 17, 2001. The note is due in 120 days with interest at 10%. If Justin discounts the note on January 15, 2002, at a bank charging a simple interest rate of 9%, what will be the proceeds?

10. A merchant buys goods worth $2000 and signs a 90-day noninterest-bearing promissory note. Find the proceeds if the supplier sells the note to a bank that uses a 13% simple interest rate. How much profit did the supplier make if the goods cost $1500?

11. On August 16 a retailer buys goods worth $2000. If he pays cash he will get a 3% cash discount. To take advantage of this he signs a 60-day noninterest-bearing note at a bank that discounts notes at a 6% simple interest rate. What should be the face value of this note to give the retailer the exact amount needed to pay cash for the goods?

12. Monique has a note for $1500 dated June 8, 2002. The note is due in 120 days with interest at 12%.
    a) If Monique discounts the note on August 1, 2002, at a bank charging an interest rate of 15%, what will the proceeds be?
    b) What rate of interest will Monique realize on her investment?
    c) What rate of interest will the bank realize if the note is paid off in full in exactly 120 days?

Equations of Value

All financial decisions must take into account the basic idea that **money has time value**. In a financial transaction involving money due on different dates, every sum of money should have an attached date, the date on which it falls due. That is, the mathematics of finance deals with **dated values**. This is one of the most important facts in the mathematics of finance.

Illustration: At a simple interest rate of 8%, $100 due in 1 year is considered to be equivalent to $108 in 2 years since $100 would accumulate to $108 in 1 year. In the same way

$$100(1 + 0.08)^{-1} = \$92.59$$

would be considered an equivalent sum at present.

In general, we compare dated values by the following **definition of equivalence**:

> X due on a given date is equivalent at a given simple interest rate r to Y due t years later if
>
> $$Y = X(1 + rt) \text{ or } X = \frac{Y}{1 + rt} = Y(1 + rt)^{-1}$$

The following time diagram illustrates dated values equivalent to a given dated value X.

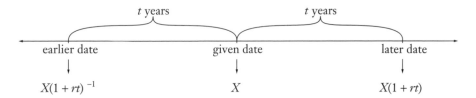

Based on the time diagram above we can state the following simple rules:

> 1. When we move money forward, we accumulate, i.e., multiply the sum by an accumulation factor $(1 + rt)$
> 2. When we move money backward, we discount, i.e., multiply the sum by a discount factor $(1 + rt)^{-1}$

EXAMPLE 1   A debt of $1000 is due at the end of 9 months. Find an equivalent debt at a simple interest rate of 9% at the end of 4 months and at the end of 1 year.

Solution Let us arrange the data on a time diagram below.

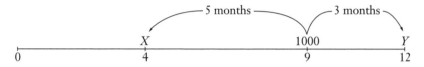

According to the definition of equivalence

$$X = 1000[1 + (0.09)(\tfrac{5}{12})]^{-1} = \$963.86$$

$$Y = 1000[1 + (0.09)(\tfrac{3}{12})] = \$1022.50$$

∎

The sum of a set of dated values, due on different dates, has no meaning. We have to replace all the dated values by **equivalent dated values**, due on the same date. The sum of the equivalent values is called the **dated value of the set**.

EXAMPLE 2 A person owes $300 due in 3 months and $500 due in 8 months. What single payment a) now; b) in 6 months; c) in 1 year, will liquidate these obligations if money is worth 8%?

Solution

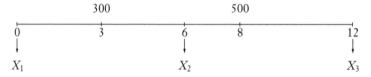

We calculate equivalent dated values of both obligations at the three different times and arrange them in the table below.

Obligations	Now	In 6 Months	In 1 Year
First	$300[1 + (.08)(\tfrac{3}{12})]^{-1} = 294.12$	$300[1 + (.08)(\tfrac{3}{12})] = 306.00$	$300[1 + (.08)(\tfrac{9}{12})] = 318.00$
Second	$500[1 + (.08)(\tfrac{8}{12})]^{-1} = 474.68$	$500[1 + (.08)(\tfrac{2}{12})]^{-1} = 493.42$	$500[1 + (.08)(\tfrac{4}{12})] = 513.33$
Sum	$X_1 = 768.80$	$X_2 = 799.42$	$X_3 = 831.33$

∎

One of the most important problems in the mathematics of finance is the replacing of a given set of payments by an equivalent set.

We say that two sets of payments are equivalent at a given simple interest rate if the dated values of the sets, on any common date, are equal. An equation stating that the dated values, on a common date, of two sets of payments are equal is called an **equation of value** or an **equation of equivalence**. The date used is called the **focal date** or the **comparison date** or the **valuation date**.

A very effective way to solve many problems in mathematics of finance is to use the equation of value. The procedure is carried out in the following steps:

Step 1   Make a good time diagram showing the dated values of obligations on one side of the time line and the dated values of payments on the other side. A good time diagram is of great help in the analysis and solution of problems.

Step 2   Select a focal date and bring all the dated values to the focal date using the specified interest rate.

Step 3   Set up an equation of value at the focal date.

Step 4   Solve the equation of value using methods of algebra.

EXAMPLE 3   Debts of $500 due 20 days ago and $400 due in 50 days are to be settled by a payment of $600 now and a final payment 90 days from now. Find the value of the final payment at a simple interest rate of 11% with a focal date at the present.

Solution   We arrange the dated values on a time diagram.

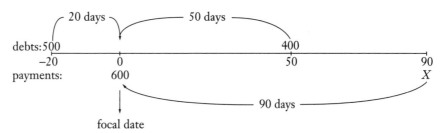

Equation of value at the present time:

dated value of the payments = dated value of the debts

$$X[1 + (0.11)(\tfrac{90}{365})]^{-1} + 600 = 500[1 + (0.11)(\tfrac{20}{365})] + 400[1 + (0.11)(\tfrac{50}{365})]^{-1}$$

$$0.973592958X + 600 = 503.01 + 394.06$$

$$0.973592958X = 297.07$$

$$X = \$305.13$$

The final payment to be made in 90 days is $305.13.

∎

EXAMPLE 4   Dated payments X, Y, and Z are as follows: X is $8557.92 on January 12, 2001; Y is $8739.86 on April 19, 2001; and Z is $9159.37 on November 24, 2001. If the annual rate of simple interest is 8% show that X is equivalent to Y and Y is equivalent to Z but that X is not equivalent to Z.

Solution   We arrange the dated values X, Y, and Z on a time diagram, listing for each date the day number from the table on the inside back cover.

$X = \$8557.92$	$Y = \$8739.86$	$Z = \$9159.37$
Jan. 12	Apr. 19	Nov. 24
2001 97 days	2001 219 days	2001
(12)	(109)	(328)

At a simple rate of interest $r = .08$

X is equivalent to Y since $Y = 8557.92[1 + (.08)(\frac{97}{365})] = \$8739.86$

and Y is equivalent to Z since $Z = 8739.86[1 + (.08)(\frac{219}{365})] = \$9159.37$

but X **is not** equivalent to Z since $Z \neq 8557.92[1 + (.08)(\frac{97+219}{365})] = \$9150.64$

∎

In mathematics, an equivalence relationship must satisfy the so-called **property of transitivity**, that is, if X is equivalent to Y and Y is equivalent to Z, then X is equivalent to Z.

> **OBSERVATION:**   Example 4 illustrates that the definition of equivalence of dated values at simple interest r does not satisfy the property of transitivity. As a result, the solutions to the problems of equations of value at simple interest do depend on the selection of the focal date. The following Example 5 illustrates that in problems involving equations of value at simple interest the answer will vary slightly with the location of the focal date. It is therefore important that the parties involved in the financial transaction agree on the location of the focal date.

EXAMPLE 5   A person borrows \$1000 at 11% simple interest. She is to repay the debt with 3 equal payments, the first at the end of 3 months, the second at the end of 6 months and the third at the end of 9 months. Find the size of the payments. Put the focal date a) at the present time; b) at the end of 9 months.

Solution a   Arrange all the dated values on a time diagram.

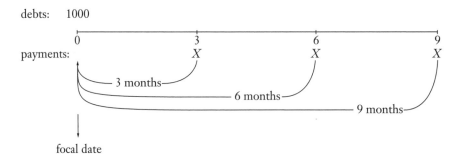

Equation of value at the present time:

$$X[1 + (0.11)(\tfrac{3}{12})]^{-1} + X[1 + (0.11)(\tfrac{6}{12})]^{-1} + X[1 + (0.11)(\tfrac{9}{12})]^{-1} = 1000.00$$
$$0.97323601X + 0.9478673X + 0.92378753X = 1000.00$$
$$2.8448908X = 1000.00$$
$$X = \$351.51$$

Solution b   Equation of value at the end of 9 months:

$$X[1 + (0.11)(\tfrac{6}{12})] + X[1 + (0.11)(\tfrac{3}{12})] + X = 1000[1 + (0.11)(\tfrac{9}{12})]$$
$$1.055X + 1.0275X + X = 1082.50$$
$$3.0825X = 1082.50$$
$$X = \$351.18$$

Notice the slight difference in the answer for the different focal dates.

■

Exercise 1.3

For problems 1 to 6 make use of the following table. In each case find the equivalent payments for each of the debts.

No.	Debts	Equivalent Payments	Focal Date	Rate
1.	$1200 due in 80 days	In full	Today	12%
2.	$100 due in 6 months, $150 due in 1 year	In full	Today	6%
3.	$200 due in 3 months, $800 due in 9 months	In full	6 months hence	8%
4.	$600 due 2 months ago, $400 due in 3 months	$500 today and the balance in 6 months	Today	13%
5.	$800 due today	Two equal payments due in 4 and 7 months	7 months hence	12%
6.	$2000 due 300 days ago	Three equal payments due today, in 60 days and in 120 days	Today	11%

7. Thérèse owes $500 due in 4 months and $700 due in 9 months. If money is worth 11%, what single payment a) now; b) in 6 months; c) in 1 year, will liquidate these obligations?

8. Andrew owes Nicola $500 due in 3 months and $200 due in 6 months. If Nicola accepts $300 now, how much will Andrew be required to repay at the end of 1 year, provided they agree to use an interest rate of 10% and a focal date at the end of 1 year?

9. A person borrows $1000 to be repaid with two equal instalments, one in six months, the other at the end of 1 year. What will be the size of these payments if the interest rate is 8% and the focal date is 1 year hence? What if the focal date is today?

10. Frank borrows $5000 on January 8, 2003. He pays $2000 on April 30, 2003, and $2000 on August 31, 2003. The last payment is to be on January 2, 2004. If interest is at 6% and the focal date is January 2, 2004, find the size of the final payment.

11. Mrs. Adams has two options available in repaying a loan. She can pay $200 at the end of 5 months and $300 at the end of 10 months, or she can pay $X at the end of 3 months and $2X at the end of 6 months. Find X if interest is at 12% and the focal date is 6 months hence and the options are equivalent. What is the answer if the focal date is 3 months hence and the options are equivalent?

12. Mr. A will pay Mr. B $2000 at the end of 5 years and $8000 at the end of 10 years if Mr. B will give him $3000 today plus an additional sum of money $X at the end of 2 years. Find X if interest is at 13% and the comparison date is today. Find X if the comparison date is 2 years hence (i.e., at the time $X is paid).

13. Mr. Malczyk borrows $2000 at 14%. He is to repay the debt with 4 equal payments, one at the end of each 3-month period for 1 year. Find the size of the payments given a focal date a) at the present time; b) at the end of 1 year.

14. A person borrows $800 at 16%. He agrees to pay off the debt with payments of size $X, $2X and $4X in 3 months, 6 months and 9 months respectively. Find X using all four transaction dates as possible focal dates.

Partial Payments

Financial obligations are sometimes liquidated by a series of partial payments during the term of obligation. Then it is necessary to determine the balance due on the final due date. There are two common ways to allow interest credit on short-term transactions.

Method 1 (also known as **Merchant's Rule**) The entire debt and each partial payment earn interest to the final settlement date. The balance due on the final due date is simply the difference between the accumulated value of the debt and the accumulated value of the partial payments.

EXAMPLE 1 On February 4 David borrowed $3000 at 11%. He paid $1000 on April 21, $600 on May 12 and $700 on June 11. What is the balance due on August 15 using the Merchant's Rule?

Solution We arrange all dated values on a time diagram.

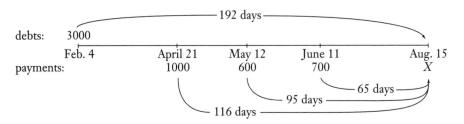

Simple interest is computed at 11% on the original debt of $3000 for 192 days, on the first partial payment of $1000 for 116 days, on the second partial payment of $600 for 95 days, and on the third partial payment of $700 for 65 days.

Calculations are given below.

Original debt	3000.00	1st partial payment	1000.00
Interest for 192 days	173.59	Interest for 116 days	34.96
		2nd partial payment	600.00
Accumulated value	$3173.59	Interest for 95 days	17.18
of the debt		3rd partial payment	700.00
		Interest for 65 days	13.71
		Accumulated value	$2365.85
		of the partial	
		payments	

Balance due on August 15: $3173.59 - 2365.85 = \$807.74$

Alternative Solution   We can write an equation of value with August 15 as the focal date.

On August 15: value of the payments = value of the debts

$$X + 1000[1 + (0.11)(\tfrac{116}{365})] + 600[1 + (0.11)(\tfrac{95}{365})] + 700[1 + (0.11)(\tfrac{65}{365})] = 3000[1 + (0.11)(\tfrac{192}{365})]$$

$$X + 1034.96 + 617.18 + 713.71 = 3173.59$$

$$X = \$807.74$$

∎

Method 2 (also known as **Declining Balance Method**) The interest on the unpaid balance of the debt is computed each time a partial payment is made. If the payment is greater than the interest due, the difference is used to reduce the debt. If the payment is less than the interest due, it is held without interest until other partial payments are made whose sum exceeds the interest due at the time of the last of these partial payments. (This point is illustrated in problems 4 and 5 of the exercise that follows.) The balance due on the final date is the outstanding balance after the last partial payment carried to the final due date.

EXAMPLE 2   Solve Example 1 using the Declining Balance Method.

Solution   We arrange all dated values on a time diagram.

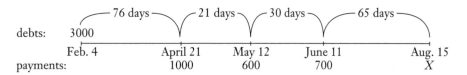

Instead of having one comparison date we have 4 comparison dates. Each time a payment is made, the preceding balance is accumulated at a simple interest rate of 11% to this point and a new balance is obtained.

The calculations are given below.

Original debt	3000.00
Interest for 76 days	68.71
Amount due on April 21	3068.71
First partial payment	1000.00
Balance due on April 21	2068.71
Interest for 21 days	13.09
Amount due on May 12	2081.80
Second partial payment	600.00
Balane due on May 12	1481.80
Interest for 30 days	13.40
Amount due on June 11	1495.20
Third partial payment	700.00
Balance due on June 11	795.20
Interest for 65 days	15.58
Balance due on August 15	$810.78

The above calculations also may be carried out in a shorter form, as shown below.

Balance due on April 21	$3000[1 + (.11)(\frac{76}{365})] - 1000 = \$2068.71$
Balance due on May 12	$2068.71[1 + (.11)(\frac{21}{365})] - 600 = \$1481.80$
Balance due on June 11	$1481.80[1 + (.11)(\frac{30}{365})] - 700 = \$795.20$
Balance due on August 15	$795.20[1 + (.11)(\frac{65}{365})] = \$810.78$

Note: The methods result in two different concluding payments. It is important that the two parties to a business transaction agree on the method to be used. Common business practice is the Declining Balance Method.

■

Exercise 1.4

1. Jean-Luc borrowed $1000, repayable in one year, with interest at $14\frac{1}{4}\%$. He pays $200 in 3 months and $400 in 7 months. Find the balance in one year using the Merchant's Rule and the Declining Balance Method.

2. On June 1, 2001, Sheila borrows $2000 at 12%. She pays $800 on August 17, 2001; $400 on November 20, 2001; and $500 on February 2, 2002. What is the balance due on April 18, 2002, by the Merchant's Rule? By the Declining Balance Method?

3. A debt of $5000 is due in six months with interest at 10%. Partial payments of $3000 and $1000 are made in 2 and 4 months respectively. What is the balance due on the final statement date by the Merchant's Rule? By the Declining Balance Method?

4. Alexandra borrows $1000 on January 8 at 16%. She pays $350 on April 12, $20 on August 10, and $400 on October 3. What is the balance due on December 15 using the Merchant's Rule? The Declining Balance Method?

5. A loan of $1400 is due in one year with simple interest at 12%. Partial payments of $400 in 2 months, $30 in 6 months and $600 in 8 months are made. Find the balance due in one year using the Declining Balance Method.

Simple Discount at a Discount Rate

The calculation of present (or discounted) value over durations of less than one year is sometimes based on a simple discount rate. The **annual simple discount rate** d is the ratio of the discount D for the year to the amount S on which the discount is given. The simple discount D on an amount S for t years at the discount rate d is calculated by means of the formula:

$$\boxed{D = Sdt}$$ (4)

and the discounted value P of S, or the proceeds P, is given by

$$P = S - D$$

By substituting for $D = Sdt$ we obtain P in terms of S, d, and t

$$P = S - Sdt$$

$$\boxed{P = S(1 - dt)}$$ (5)

The factor $(1 - dt)$ in formula **(5)** is called **a discount factor at a simple discount rate d**.

The considerations regarding the measurement of t (day count, etc.) are the same as for simple interest. The charge for some short-term loans, called discounted loans, is based on the final amount rather than on the present value. The lender calculates the discount D on the final amount S that must be paid on the due date, deducts it from S and lends the borrower the proceeds P. For this reason it is sometimes referred to as **interest in advance**.

From formula **(5)** we can express S in terms of P, d and t and obtain

$$\boxed{S = \frac{P}{1 - dt} = P(1 - dt)^{-1}}$$ (6)

The factor $(1 - dt)^{-1}$ in formula **(6)** is called **an accumulation factor at a simple discount rate d**.

Formula **(6)** is used to calculate the maturity value of a loan for a specified proceeds.

EXAMPLE 1   A person borrows $500 for 6 months from a lender who charges a $9\frac{1}{2}\%$ discount rate. a) What is the discount, and how much money does the borrower receive? b) What size loan should the borrower ask for if he wants to receive $500 cash?

Solution a   We have $S = 500$, $d = 9\frac{1}{2}\%$, $t = \frac{1}{2}$ and calculate

$$\text{discount } D = Sdt = 500 \times 0.095 \times \tfrac{1}{2} = \$23.75$$

$$\text{proceeds } P = S - D = 500 - 23.75 = \$476.25$$

We could calculate the proceeds by formula **(5)**

$$P = S(1 - dt) = 500[1 - (0.095)(\tfrac{1}{2})] = \$476.25$$

Solution b   We have $P = 500$, $d = 9\frac{1}{2}\%$, $t = \frac{1}{2}$ and calculate the maturity value of the loan

$$S = \frac{P}{1 - dt} = \frac{500}{1 - (0.095)(\tfrac{1}{2})} = \$524.93$$

The borrower should ask for a loan of $524.93 to receive proceeds of $500.

∎

EXAMPLE 2   Calculate the discounted value of $1000 due in 1 year: a) at a simple interest rate of 10%; b) at a simple discount rate of 10%.

Solution a   We have $S = 1000$, $r = 10\%$, $t = 1$ and calculate the discounted value P by formula **(2)**

$$P = \frac{S}{1 + rt} = \frac{1000}{1 + (0.1)(1)} = \$909.09$$

Solution b   We have $S = 1000$, $d = 10\%$, $t = 1$ and calculate the discounted value P by formula **(5)**

$$P = S(1 - dt) = 1000[1 - (0.1)(1)] = 1000(0.9) = \$900.00$$

Note the difference of $9.09 between the discounted value at a simple interest rate and the discounted value at a simple discount rate. We can conclude that a given simple discount rate results in a larger money return to a lender than the same simple interest rate.

∎

EXAMPLE 3:   If $1200 is the present value of $1260 due at the end of 9 months, find a) the annual simple interest rate, and b) the annual simple discount rate.

Solution a   We have $P = 1200$, $I = 60$, $t = \frac{9}{12}$, and calculate

$$r = \frac{I}{Pt} = \frac{60}{1200(\frac{9}{12})} \doteq 6.67\%$$

Solution b   We have $S = 1260$, $D = 60$, $t = \frac{9}{12}$, and calculate

$$d = \frac{D}{St} = \frac{60}{1260(\frac{9}{12})} \doteq 6.35\%$$

∎

In general, we can calculate **equivalent rates** of interest or discount by using the following definition:

> Two rates are equivalent if they have the same effect on money over the same period of time.

The above definition can be used to find simple (or compound) equivalent rates of interest or discount.

EXAMPLE 4   What simple rate of discount d is equivalent to a simple rate of interest $r = 12\%$ if money is invested for a) 1 year; b) 2 years?

Solution a   Comparing the discounted values of $1 due at the end of 1 year:

$$[1 - d(1)] = [1 + (.12)(1)]^{-1}$$
$$1 - d = (1.12)^{-1}$$
$$d \doteq 10.71\%$$

Solution b   Comparing the discounted values of $1 due at the end of 2 years:

$$[1 - d(2)] = [1 + (.12)(2)]^{-1}$$
$$1 - 2d = (1.24)^{-1}$$
$$d \doteq 9.68\%$$

∎

OBSERVATION: Notice that at simple rates of interest and discount, the equivalent rates are dependent on the period of time.

Promissory notes are occasionally sold based on simple discount from the maturity value, as illustrated in the following example.

EXAMPLE 5   On August 10 Smith borrows $4000 from Jones and gives Jones a promissory note at a simple interest rate of 10% with a maturity date of February 4 in the following year. Brown buys the note from Jones on November 3, based on a simple discount rate of 11%. Determine Brown's purchase price.

Solution   The maturity value of the note is $4000[1 + (.10)(\frac{178}{365})] = \$4195.07$. The discount period is from November 3 to February 4, i.e., 93 days. The amount paid by Brown is determined by equation **(5)**

$$4195.07[1 - (.11)(\tfrac{93}{365})] = \$4077.49$$

∎

Exercise 1.5

1. Find the discounted value of $2000 due in 130 days
   a) at a simple interest rate of 8.5%;
   b) at a simple discount rate of 8.5%.

2. What simple discount rate d is equivalent over a 300-day period to a simple interest rate of 9% for the first 120 days and 12% thereafter, if
   a) interest from the first 120 days does not earn interest during the last 180 days
   b) interest from the first 120 days earns interest during the last 180 days?

3. A note with a maturity value of $700 is sold at a discount rate of 13%, 45 days before maturity. Find the discount and the proceeds.

4. A promissory note with face value $2000 is due in 175 days and bears 15% simple interest. After 60 days it is sold for $2030. What simple discount rate can the purchaser expect to earn?

5. A storekeeper buys goods costing $800. He asks his creditor to accept a 60-day noninterest-bearing note, which, if his creditor discounts immediately at a 10% discount rate, will result in proceeds of $800. For what amount should he make the note?

6. A company borrowed $50 000 on May 1, 2002, and signed a promissory note bearing interest at 11% for 3 months. On the maturity date, the company paid the interest in full and gave a second note for 3 months without interest and for such an amount that when it was discounted at a 12% discount rate on the day it was signed, the proceeds were just sufficient to pay the debt. Find the amount of interest paid on the first note and the face value of the second note.

Section 1.6	**Summary and Review Exercises**

- Accumulated value of X at the end of t years at a simple interest rate r is given by $X(1 + rt)$ where $(1 + rt)$ is called an accumulation factor at simple interest rate r.

- Discounted value of X due in t years at a simple interest rate r is given by $X(1 + rt)^{-1}$ where $(1 + rt)^{-1}$ is called a discount factor at a simple interest rate r.

- Discounted value of X due in t years at a simple discount rate d is given by $X(1 - dt)$ where $(1 - dt)$ is called a discount factor at a simple discount rate d.

- Accumulated value of X at the end of t years at a simple discount rate d is given by $X(1 - dt)^{-1}$ where $(1 - dt)^{-1}$ is called an accumulation factor at a simple discount rate d.

- Equivalence of dated values at a simple interest rate r:
  X due on a given date is equivalent to Y due t years later,
  if $Y = X(1 + rt)$ or $X = Y(1 + rt)^{-1}$

Note: The above definition of equivalence of dated values at a simple interest rate r does not satisfy the property of transitivity. The lack of transitivity leads to different answers when different comparison dates are used for equations of value.

- Equivalent rates:
  Two rates are equivalent if they have the same effect on money over the same period of time.

Note: At simple rates of interest and discount, the equivalent rates are dependent on the period of time.

Review Exercises 1.6

1. How long will it take $1000
   a) to earn $100 at 6% simple interest?
   b) to accumulate to $1200 at $13\frac{1}{2}$% simple interest?

2. Find the accumulated and the discounted value of $1000 over 55 days at 15% using both ordinary and exact simple interest.

3. A taxpayer expects an income tax refund of $380 on May 1. On March 10, a tax discounter offers 85% of the full refund in cash. What rate of simple interest will the tax discounter earn?

4. A retailer receives an invoice for $8000 for a shipment of furniture with terms 3/10, *n*/40.
   a) What is the highest simple interest rate at which he can afford to borrow money in order to take advantage of the cash discount?
   b) If the retailer can borrow at a simple interest rate of 12%, find his savings resulting from the cash discount, when he pays the invoice within 10 days.

5. A cash discount of 3% is given if a bill is paid 40 days in advance of its due date.
   a) What is the highest simple interest rate at which you can afford to borrow money if you wanted to take advantage of this discount?
   b) If you can borrow money at a simple interest rate 8%, how much can you save by paying an invoice for $5000 forty days in advance of its due date?

6. A loan of $2500 taken out on April 2 requires equal payments on May 25, July 20 and September 10 and a final payment of $500 on October 15. If the focal date is October 15, what is the size of the equal payments at 9% simple interest?

7. Last night Marion won $5000 in a lottery. She was given two options. She can take $5000 today or *X* every six months (beginning six months from now) for 2 years. If the options are equivalent and the simple interest rate is 10%, find *X* using a focal date of 2 years.

8. Today Mr. Mueller borrowed $2400 and arranged to repay the loan with three equal payments of $*X* at the end of 4, 8 and 12 months respectively. If the lender charges a simple interest rate of 6%, find *X* using today as a focal date.

9. Roger borrows $4500 at 9% simple interest on July 3, 2001. He pays $1250 on October 27, 2001, and $2500 on January 7, 2002. Find the balance due on May 1, 2002, using the Merchant's Rule and the Declining Balance Method.

10. Paul borrows $4000 at an 18% simple interest rate. He is to repay the loan by paying $1000 at the end of 3 months and two equal payments at the end of 6 months and 9 months. Find the size of the equal payments using
    a) the end of 6 months as a focal date.
    b) the present time as a focal date.

11. Melissa borrows $1000 on May 8, 2001 at a $10\frac{1}{2}$% simple interest rate. She pays $500 on July 17, 2001 and $400 on September 29, 2001. What is the balance due on October 31, 2001 using the Merchant's Rule and the Declining Balance Method?

12. Robert borrows $1000 for 8 months from a lender who charges an 11% discount rate.
a) How much money does Robert receive?
b) What size loan should Robert ask for in order to receive $1000 cash?

13. Consider the following promissory note:

> $1500.00   Vancouver, May 11, 2001
>
> Ninety days after date, I promise to pay to the order of J.D. Prasad
>
>   Fifteen hundred and $\frac{00}{100}$ dollars
>
> for value received with interest at 12% per annum.
>
>   (Signed) J.B. Lau

Mr. Prasad sells the note on July 2, 2001, to a bank that discounts notes at a 13% discount rate.
a) How much money does Mr. Prasad receive?
b) What rate of interest does Mr. Prasad realize on his investment?
c) What rate of interest does the bank realize on its investment if it holds the note till maturity?
d) What rate of interest does the bank realize if Mr. Lau pays off the loan as due in exactly 90 days?

14. A 180-day promissory note for $2000 bears 14% simple interest. After 60 days it is sold to a bank that charges a discount rate of 14%.
a) Find the price paid by the bank for the note.
b) What simple interest rate did the original owner of the note actually earn?
c) If the note is actually paid on the **due** date, what discount rate will the bank have realized?

15. A note for $800 is due in 90 days with simple interest at 11%. On the maturity date, the maker of the note paid the interest in full and gave a second note for 60 days without interest and for such an amount that when it was discounted at a 10% discount rate on the day it was signed, the proceeds were just sufficient to pay the debt. Find the interest paid on the first note and the face value of the second note.

16. A debt of $1200 is to be paid off by payments of $500 in 45 days, $300 in 100 days, and a final payment of $436.92. Interest is at $r = 11\%$ and the Merchant's Rule was used to calculate the final payment. In how many days should the final payment be made?

17. Find the difference in simple interest on a $20 000 loan at a simple interest rate of 12% from April 15 to the next February 4, using the Banker's Rule and Canadian practice.

18. A person borrows $3000 at a 14% simple interest rate. He is to repay the debt with 3 equal payments, the first at the end of 3 months, the second at the end of 7 months and the third at the end of 12 months. Find the difference between payments resulting from the selection of the focal date at the present time and at the end of 12 months.

19. A promissory note for $1500 is due in 170 days with simple interest at 9.5%. After 40 days it is sold to a finance company that charges a simple discount rate of 14%. How much does the finance company pay for the note?

20. You borrow $1000 now and $2000 in 4 months. You agree to pay $X in 6 months and $2X in 8 months (from now). Find X using a focal date 8 months from now
a) at simple interest rate $r = 11\%$;
b) at simple discount rate $d = 11\%$.

Chapter Two

Compound Interest and Compound Discount

Section 2.1 — Fundamental Compound Interest Formula

If the interest due is added to the principal at the end of each interest period and thereafter earns interest, the interest is said to be **compounded**. The sum of the original principal and total interest is called the **compound amount** or **accumulated value**. The difference between the accumulated value and the original principal is called the **compound interest**. The time between two successive interest computations is called the **interest period** or **conversion period**. This time unit need not be a year. Most of you will already be familiar with situations where interest is "payable quarterly," or "compounded semi-annually" or "convertible monthly."

EXAMPLE 1   Find the compound interest earned on $1000 for 2 years at 10% compounded semi-annually and compare it with the simple interest earned on $1000 for 2 years at 10% per annum.

Solution   Since the conversion period is 6 months, interest is earned at the rate of 5% per period, and there are 4 interest periods in 2 years.

At the End of	Compound Interest	Accumulated Value
the 1st interest period	$1000 \times 0.05 = 50$	$1050.00
the 2nd interest period	$1050 \times 0.05 = 52.50$	$1102.50
the 3rd interest period	$1102.50 \times 0.05 = 55.13$	$1157.63
the 4th interest period	$1157.63 \times 0.05 = 57.88$	$1215.51

The compound interest earned on $1000 for 2 years at 10% compounded semi-annually is $215.51; whereas the simple interest on $1000 for 2 years at 10% is $I = 1000 \times 0.10 \times 2 = \$200$.

■

In the following we shall develop quicker methods for calculating the compound interest.

The following notation will be used:

P = the original principal, or the present value of S, or the discounted value of S.

S = the compound amount of P, or the accumulated value of P.

n = the number of interest (or conversion) periods involved.

m = the number of interest periods per year, or the frequency of compounding.

j_m = the nominal (yearly) interest rate that is compounded (payable, convertible) m times per year. Actuaries use the symbol $i^{(m)}$ instead of j_m.

i = the interest rate per conversion period.

Note 1: Commonly used frequencies of compounding are $m = 1$ for annual compounding, $m = 2$ for semi-annual compounding, $m = 4$ for quarterly compounding, $m = 12$ for monthly compounding, $m = 52$ for weekly compounding, and $m = 365$ for daily compounding (leap year or not). Continuous compounding that uses compounding intervals shorter than one day is covered in Section 2.9.

Note 2: The rate i equals $\frac{j_m}{m}$ and is always used in the compound interest calculation. For example, $j_{12} = 9\%$ means that a yearly rate of 9% is converted (compounded, payable) 12 times per year and that $i = \frac{9}{12}\% = \frac{3}{4}\% = 0.0075$ is the interest rate per month.

Let P represent the principal at the beginning of the first interest period and i the interest rate per conversion period. We shall calculate the accumulated values at the ends of the successive interest periods for n periods.

At the end of the 1st period:
the interest due $P i$
the accumulated value $P + P i = P(1 + i)$

At the end of the 2nd period:
the interest due $[P(1 + i)]i$
the accumulated value $P(1 + i) + [P(1 + i)]i$
                                     $= P(1 + i)(1 + i) = P(1 + i)^2$

At the end of the 3rd period:
the interest due $[P(1 + i)^2]i$
the accumulated value $P(1 + i)^2 + [P(1 + i)^2]i$
                                     $= P(1 + i)^2(1 + i) = P(1 + i)^3$

Continuing in this manner for n periods the accumulated value S at the end of n periods is given by the **fundamental compound interest formula**

$$\boxed{S = P(1 + i)^n} \tag{7}$$

> The factor $(1 + i)^n$ is called the **accumulation factor** or the **accumulated value of \$1**. The process of calculating S from P is called **accumulation**. To obtain the accumulated value S of P for n periods at rate i, we multiply P by the corresponding accumulation factor $(1 + i)^n$.

EXAMPLE 2   Accumulate \$100 at $j_{12} = 12\%$ for a) 5 years; b) 25 years.

Solution a   We have $P = 100$, $i = .01$, $n = 60$, and calculate

$$S = 100(1.01)^{60} = \$181.67$$

The compound interest on \$100 at $j_{12} = 12\%$ for 5 years is \$81.67.

Solution b   We have $P = 100$, $i = .01$, $n = 300$, and calculate

$$S = 100(1.01)^{300} = \$1978.85.$$

The compound interest on \$100 at $j_{12} = 12\%$ for 25 years is \$1878.85, which is more than 18 times the original investment of \$100. If the investment had been at 1% simple interest per month, the interest earned would have been only \$300. This illustrates the power of compound interest at a high rate of interest for a long period of time.

■

CALCULATION TIP: It is assumed that students will be using pocket calculators equipped with the functions y^x and $\log x$ to solve the problems in *Mathematics of Finance*. In the examples in this textbook, we have used all digits of the factors provided by a pocket calculator and rounded off to the nearest cent only in the final answer.

The table and graph below show the effect of time and rate on the growth of money at compound interest.

Growth of \$100 at Compound Interest Rate j_{12}

Years	6%	8%	10%	12%
5	134.89	148.98	164.53	181.67
10	181.94	221.96	270.70	330.04
15	245.41	330.69	445.39	599.58
20	331.02	492.68	732.81	1 089.26
25	446.50	734.02	1 205.69	1 978.85
30	602.26	1 093.57	1 983.74	3 594.96
35	812.36	1 629.26	3 263.87	6 530.96
40	1 095.75	2 427.34	5 370.07	11 864.77
45	1 478.00	3 616.36	8 835.42	21 554.69
50	1 993.60	5 387.82	14 536.99	39 158.34

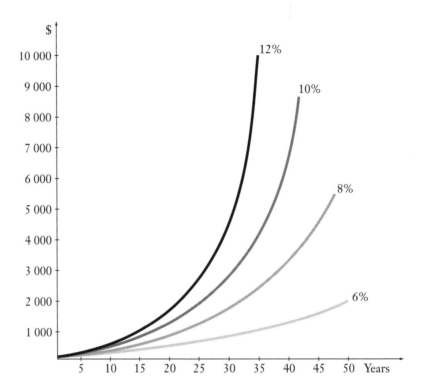

EXAMPLE 3   A person deposits $1000 into a savings account that earns interest at 4.25% compounded daily. How much interest will be earned: a) during the first year? b) during the second year?

Solution a   We have $P = 1000$, $i = \frac{.0425}{365}$, $n = 365$ and calculate the accumulated value at the end of 1 year

$$S = 1000\left(1 + \tfrac{.0425}{365}\right)^{365} = \$1043.41$$

The compound interest earned during the first year is $43.41.

Solution b   We calculate the accumulated value at the end of 2 years

$$S = 1000\left(1 + \tfrac{.0425}{365}\right)^{730} = \$1088.71$$

The compound interest earned during the second year is

$$1088.71 - 1043.41 = \$45.30$$

∎

CALCULATION TIP: When using the fundamental compound interest formula **(7)** in calculations, do not round off the value of i. Use all digits provided by your calculator.

Exercise 2.1

Part A

Problems 1 to 8 make use of the following table. In each case find the accumulated value and the compound interest earned.

No.	Principal	Nominal Rate	Conversion Frequency	Time
1.	$ 100	$5\frac{1}{2}\%$	Annually	5 years
2.	$ 500	$11\frac{1}{4}\%$	Monthly	2 years
3.	$ 220	8.8%	Quarterly	3 years
4.	$1000	9%	Semi-annually	6 years
5.	$ 50	12%	Monthly	4 years
6.	$ 800	$7\frac{3}{4}\%$	Annually	10 years
7.	$ 300	8%	Weekly	3 years
8.	$1000	10%	Daily	2 years

9. Accumulate $500 for one year at a) $j_{12} = 8\%$, b) $j_{12} = 12\%$, c) $j_{12} = 16\%$.

10. How much money will be required on December 31, 2006, to repay a loan of $2000 made December 31, 2003, if $j_4 = 12\%$?

11. Find the accumulated value of $100 over 5 years at an 8% nominal rate compounded: a) annually; b) semi-annually; c) quarterly; d) monthly; e) daily.

12. Parents put $1000 into a savings account at the birth of their daughter. If the account earns interest at 12% compounded monthly, how much money will be in the account when their daughter is 18 years old?

13. In 1492, Queen Isabella sponsored Christopher Columbus' journey by giving him $10 000. If she had placed this money in a bank account at $j_1 = 3\%$, how much money would be in the account in 2002? If the bank account earned a simple interest rate of 3%, how much money would be in the account in 2002?

14. Find the accumulated value of $100 at the end of 1 year at a) $j_1 = 12.55\%$; b) $j_2 = 12.18\%$; c) $j_4 = 12\%$; d) $j_{12} = 11.88\%$.

Part B

1. Melinda has a savings account that earns interest at 6% per annum. She opened her account with $1000 on December 31. How much interest will she earn during the first year if
   a) the interest is compounded daily;
   b) the interest is calculated daily and paid into the account on June 30 and December 31;
   c) the interest is calculated daily and paid into the account at the end of each month?

2. Prove the fundamental compound interest formula $S = P(1 + i)^n$ using mathematical induction.

3. Set up a table and plot the graph showing the growth of $100 at compound interest rates $j_{365} = 8\%$, 12%, 16% and time = 5, 10, 15, 20 and 25 years.

4. Find the compound interest earned on an investment of $10 000 for 10 years at a nominal rate of 12% compounded with frequencies $m = 1, 2, 4, 12, 52$ and 365.

| *Section 2.2* | **Equivalent Compound Interest Rates** |

The yearly nominal rate is meaningless until we specify the frequency of conversion m. The table below illustrates the effect of the frequency of conversion on the amount to which \$10 000 will accumulate in 10 years at a nominal rate of 8% compounded with frequencies $m = 1, 2, 4, 12$, and 365.

Frequency of Conversion	Rate per Period	Number of Periods	Amount
$m = 1$	$i = 8\%$	10	\$21 589.25
$m = 2$	$i = 4\%$	20	\$21 911.23
$m = 4$	$i = 2\%$	40	\$22 080.40
$m = 12$	$i = \frac{2}{3}\%$	120	\$22 196.40
$m = 365$	$i = \frac{8}{365}\%$	3650	\$22 253.41

At the same nominal rate the accumulated value depends on the frequency of conversion; it increases with the increased frequency of conversion.

For a given nominal rate j_m compounded m times per year, we define the corresponding **annual effective rate** to be that rate j which, if compounded annually, will produce the same amount of interest per year. To find the yearly effective rate j corresponding to a given nominal rate j_m we compare the accumulated values of \$1 at the end of 1 year. At the rate j, \$1 will accumulate at the end of 1 year to $1 + j$. At the rate $i = j_m/m$, \$1 will accumulate at the end of 1 year to $(1 + i)^m$.

Thus

$$1 + j = (1 + i)^m$$

$$\boxed{j = (1 + i)^m - 1} \tag{8}$$

EXAMPLE 1 Find the annual effective rate j corresponding to a) $j_2 = 10\%$; b) $j_{12} = 6\%$; c) $j_{365} = 13\frac{1}{4}\%$.

Solution a \$1 at rate j for 1 year will accumulate to $1 + j$
\$1 at rate $j_2 = 10\%$ for 1 year will accumulate to $(1.05)^2$
Comparing the accumulated values we obtain
$1 + j = (1.05)^2 = 1.1025$ and $j = 0.1025 = 10.25\%$.

Solution b We have $i = \frac{1}{2}\% = .005$ and applying equation **(8)**

$$j = (1.005)^{12} - 1 = .0616778 \doteq 6.17\%$$

Solution c Applying equation **(8)** we have

$$j = \left(1 + \tfrac{.1325}{365}\right)^{365} - 1 = .14165139 \doteq 14.17\%.$$

∎

In Chapter 1 we defined equivalent rates as the rates that have the same effect on money over the same period of time. (See p. 21.)

Applying the general definition specifically for compound interest rates we say:

> Two nominal compound interest rates are **equivalent** if they yield the same accumulated values at the end of one year, and hence at the end of any number of years.

EXAMPLE 2   Find the rate j_4 equivalent to a) $j_{12} = 12\%$; b) $j_2 = 10\%$.

Solution   We shall compare the accumulated values of $1 at the end of 1 year.

Solution a $1 at the rate j_4 will accumulate at the end of 1 year to $(1 + i)^4$. $1 at the rate $j_{12} = 12\%$ will accumulate at the end of 1 year to $(1.01)^{12}$. Thus

$$(1 + i)^4 = (1.01)^{12}$$
$$1 + i = (1.01)^3$$
$$i = (1.01)^3 - 1$$
$$i = 0.030301$$

and

$$j_4 = 4i = 0.121204 \doteq 12.12\%$$

Solution b $1 at the rate j_4 will accumulate at the end of 1 year to $(1 + i)^4$. $1 at the rate $j_2 = 10\%$ will accumulate at the end of 1 year to $(1.05)^2$. Thus

$$(1 + i)^4 = (1.05)^2$$
$$1 + i = (1.05)^{\frac{1}{2}}$$
$$i = (1.05)^{\frac{1}{2}} - 1$$
$$i = 0.02469508$$

and

$$j_4 = 4i = 0.0987803 \doteq 9.88\%$$

∎

EXAMPLE 3   What simple interest rate is equivalent to $j_2 = 9\%$ if money is invested for 3 years?

Solution   Let r be the unknown simple interest rate. $1 invested at rate r will accumulate at the end of 3 years to $1 + 3r$. $1 invested at $j_2 = 9\%$ will accumulate at the end of 3 years to $(1.045)^6$.
   Thus

$$1 + 3r = (1.045)^6$$
$$1 + 3r = 1.3022601$$
$$r = 0.10075338$$
$$r \doteq 10.08\%$$

∎

CALCULATION TIP: When calculating unknown yearly rates of interest, round off your answer to the nearest hundredth of a percent.

Exercise 2.2

Part A

1. Find the annual effective rate (two decimals) equivalent to the following rates:
   a) $j_2 = 7\%$; b) $j_4 = 16\%$; c) $j_4 = 8\%$; d) $j_{365} = 12\%$; e) $j_{12} = 18\%$.

2. Find the nominal rate (two decimals) equivalent to the given annual effective rate:
   a) $j = 6\%$ find j_2; b) $j = 9\%$ find j_4;
   c) $j = 10\%$ find j_{12}; d) $j = 17\%$ find j_{365};
   e) $j = 8\%$ find j_{52}.

3. Find the nominal rate (two decimals) equivalent to the given nominal rate:
   a) $j_2 = 8\%$ find j_4; b) $j_4 = 6\%$ find j_2;
   c) $j_{12} = 18\%$ find j_4; d) $j_6 = 10\%$ find j_{12};
   e) $j_4 = 8\%$ find j_2; f) $j_{52} = 11\%$ find j_2;
   g) $j_2 = 18\frac{1}{4}\%$ find j_{12}; h) $j_4 = 12.79\%$ find j_{365}.

4. What simple interest rate is equivalent to $j_{12} = 13\frac{1}{2}\%$ if money is invested for 2 years?

5. What simple interest rate is equivalent to $j_{365} = 12\%$ if money is invested for 3 years?

6. If the interest on the outstanding balance of a credit card account is charged at $1\frac{3}{4}\%$ per month, what is the annual effective rate of interest?

7. A trust company offers guaranteed investment certificates paying $j_2 = 8.9\%$ and $j_1 = 9\%$. Which option yields the higher annual effective rate of interest?

8. Which rate gives the best and the worst rate of return on your investment?
   a) $j_{12} = 15\%$, $j_2 = 15\frac{1}{2}\%$, $j_{365} = 14.9\%$
   b) $j_{12} = 6\%$, $j_2 = 6\frac{1}{2}\%$, $j_{365} = 5.9\%$

9. Bank A has an annual effective interest rate of 10%. Bank B has a nominal interest rate of $9\frac{3}{4}\%$. What is the minimum frequency of compounding for bank B in order that the rate at bank B be at least as attractive as that at bank A?

Part B

1. Find out what is the current best and worst interest rate available on 5-year guaranteed investment certificates by checking three different financial institutions. Calculate the difference in compound interest earned using the best and the worst rate available if you buy a $2000, 5-year certificate.

2. Find the annual effective rate of interest on the major credit cards and charge-accounts in major department stores.

3. For a given nominal rate $j_2 = 2i$, develop equations for equivalent nominal rates j_1, j_4, j_{12} and j_{365}.

4. For a given nominal rate $j_{12} = 12i$, develop equations for equivalent nominal rates $j_1, j_2, j_4, j_{52}, j_{365}$.

5. Twenty thousand dollars is invested for 5 years at a nominal rate of 6%. Find the accumulated value of the investment if the rate is compounded with frequencies $m = 1, 2, 4, 12$ and 365 using
   a) the fundamental compound interest formula;
   b) equivalent annual effective rates;
   c) equivalent nominal rates compounded monthly.
   Compare your answers in a), b) and c).

6. A bank pays 6% per annum on its savings accounts. At the end of every 3 years a 2% bonus is paid on the balance. Find the annual effective rate of interest, j_1, earned by an investor if the deposit is withdrawn:
   a) in 2 years,
   b) in 3 years,
   c) in 4 years.

7. A fund earns interest at the nominal rate of 8.04% compounded quarterly. At the end of each quarter, just after interest is credited, an expense charge equal to 0.50% of the fund is withdrawn. Find the annual effective yield realized by the fund.

8. An insurance company says you can pay for your life insurance by paying $100 at the beginning of each year or $51.50 at the beginning of each half-year. They say the rate of interest underlying this calculation is $j_2 = 3\%$. What is the true value of j_2?

9. In general, the *annual effective rate of interest* is the ratio of the amount of interest earned during the year to the amount of principal invested at the beginning of the year.
   a) Show that, at a simple interest rate r, the annual effective rate of interest for the nth year is $\dfrac{r}{1 + r(n - 1)}$, which is a decreasing function of n. Thus a constant rate of simple interest implies a decreasing annual effective rate of interest.
   b) Show that, at a compound interest rate i per year, the annual effective rate of interest for the nth year is i, which is independent of n. Thus a constant rate of compound interest implies a constant annual effective rate of interest.

| *Section 2.3* | **Discounted Value at Compound Interest** |

In business transactions it is frequently necessary to determine what principal P now will accumulate at a given interest rate to a specified amount S at a specified future date. From fundamental formula **(7)** we can obtain

$$P = \frac{S}{(1 + i)^n} = S(1 + i)^{-n} \tag{9}$$

P is called the **discounted value** of S, or the **present value** of S, or the **proceeds**. The process of finding P from S is called **discounting**. The difference $S - P$ is called **compound discount** on S. It is *compound discount at an interest rate* and it is the prevailing practice in compound discount problems. Compound discount at a discount rate is covered in Section 2.10.

> The factor $(1 + i)^{-n}$ in **(9)** is called the **discount factor** or the **discounted value of \$1**. To obtain a discounted value P of S for n periods at rate i, we multiply S by the corresponding discount factor $(1 + i)^{-n}$. Values of $(1 + i)^{-n}$ will be calculated by using the function y^x of your calculator.

EXAMPLE 1   Find the discounted value of $100 000 due in a) 10 years, b) 25 years, if money is worth $j_{12} = 6\%$.

Solution a   We have $S = 100\ 000$, $i = .005$, $n = 120$ and calculate

$$P = 100\ 000(1.005)^{-120} = \$54\ 963.27$$

Solution b   We have $S = 100\ 000$, $i = .005$, $n = 300$ and calculate

$$P = 100\ 000(1.005)^{-300} = \$22\ 396.57.$$

■

EXAMPLE 2   Let us suppose you can buy a lot for $84 000 cash or for payments of $50 000 now, $20 000 in 1 year and $20 000 in 2 years. If money is worth $j_{12} = 9\%$, which option is better for you?

Solution   We arrange the data on a time diagram below. Notice that time is in terms of interest conversion periods.

Option 1:   84 000

```
    0                      12                    24
                       (1 year)             (2 years)
```

Option 2:   50 000   20 000   20 000

Discounted value of option 1 is $84 000.
Discounted value of option 2 is $50\ 000 + 20\ 000\left(1 + \frac{.09}{12}\right)^{-12} + 20\ 000\left(1 + \frac{.09}{12}\right)^{-24}$

$$= 50\ 000 + 18\ 284.76 + 16\ 716.63$$
$$= \$85\ 001.39$$

You should take option 1 and save $85\ 001.39 - 84\ 000 = \$1\ 001.39$ at the present time.

A different rate of interest could lead to a different decision. If money were worth $j_{12} = 12\%$, the discounted value of option 2 would be

$$50\ 000 + 20\ 000(1.01)^{-12} + 20\ 000(1.01)^{-24}$$
$$= 50\ 000 + 17\ 748.98 + 15\ 751.32$$
$$= \$83\ 500.30$$

and you should take option 2 and save $84\ 000 - 83\ 500.30 = \$499.70$ at the present time.

■

EXAMPLE 3   A note for $2 000 dated September 1, 2001, is due with compound interest at $j_{12} = 12\%$, 3 years after the date. On December 1, 2002, the holder of the note has it discounted by a lender who charges $j_4 = 13\frac{1}{4}\%$. Find the proceeds and the compound discount.

Solution   We arrange the data on a time diagram below.

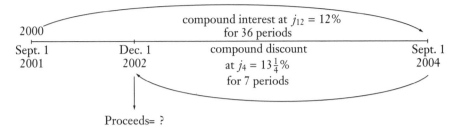

Proceeds= ?

The maturity value of the note is $S = 2000\left(1 + \frac{.12}{12}\right)^{36} = \$2861.54$.

The proceeds are $P = 2861.54(1 + \frac{.1325}{4})^{-7} = \$2277.87$.

The compound discount equals $2861.54 - 2277.87 = \$583.67$.

◼

CALCULATION TIP: Long-term promissory notes (with a term longer than one year) are usually subject to compound interest. There is no requirement to add three days of grace in determining the legal due date (maturity date) of a long-term promissory note.

Exercise 2.3

Part A

Find the discounted value in problems 1 to 5.

No.	Amount	Nominal Rate	Conversion	Time
1.	$ 100	6%	quarterly	3 years
2.	$ 50	$8\frac{1}{2}$%	monthly	2 years
3.	$2000	11.8%	yearly	10 years
4.	$ 500	10%	semi-annually	5 years
5.	$ 800	12%	daily	3 years

6. What amount of money invested today will grow to $1000 at the end of 5 years if $j_4 = 8\%$?

7. How much would have to be deposited today in an investment fund paying $j_{12} = 10.4\%$ to have $2000 in three years' time?

8. What is the discounted value of $2500 due in 10 years if $j_2 = 9.6\%$?

9. On her 20th birthday a woman receives $1000 as a result of a deposit her parents made on the day she was born. How large was that deposit if it earned interest at $j_2 = 6\%$?

10. An obligation of $2000 is due December 31, 2005. What is the value of this obligation on June 30, 2001, at $j_4 = 13\frac{1}{4}\%$?

11. A note dated October 1, 2001, calls for the payment of $800 in 7 years. On October 1, 2003, it is sold at a price that will yield the investor $j_4 = 12\%$. How much is paid for the note?

12. A note for $250 dated August 1, 2001, is due with compound interest at $j_{12} = 15\frac{1}{4}\%$ 4 years after date. On November 1, 2002, the holder of the note has it discounted by a lender who charges $j_4 = 13\frac{1}{2}\%$. What are the proceeds?

13. A note for $1000 dated January 1, 2001, is due with compound interest at 13% compounded semi-annually 5 years after date. On July 1, 2002, the holder of the note has it discounted by a lender who charges $14\frac{1}{2}\%$ compounded quarterly. Find the proceeds.

14. A man can buy a piece of land for $170 000 cash or payments of $120 000 down and $100 000 in 5 years. If he can earn $j_{365} = 10\%$, which plan is better?

15. Find the total value on July 1, 2001, of payments of $1000 on July 1, 1991, and $600 on July 1, 2008, if $j_2 = 9\%$.

16. If money is worth $j_1 = 10\%$, find the present value of a debt of $3000 with interest at $11\frac{1}{2}\%$ compounded semi-annually due in 5 years.

Part B

1. A note for $2500 dated January 1, 2002, is due with interest at $j_{12} = 12\%$ 40 months later. On May 1, 2002, the holder of the note has it discounted by Financial Consultants Inc. at $j_4 = 13\frac{1}{4}\%$. The same day the note is sold by Financial Consultants Inc. to a bank that discounts notes at $j_1 = 13\%$. What is the profit made by Financial Consultants Inc.?

2. Find the compound discount if $1000, due in 5 years with interest at $j_1 = 14\frac{1}{2}\%$, is discounted at a nominal rate of 10% compounded with frequencies $m = 1, 2, 4, 12, 52$ and 365.

3. The management of a company must decide between two proposals. The following information is available:

Proposal	Investment Now	Net Cash Inflow at the End of Year 1	Year 2	Year 3
A	80 000	95 400	39 000	12 000
B	100 000	35 000	58 000	80 000

Advise management regarding the proposal that should be selected, assuming that on projects of this type the company can earn $j_1 = 14\%$.

Section 2.4 — Accumulated and Discounted Value for a Fractional Period of Time

Formulas **(7)** and **(9)** were developed under the assumption that n is an integer. Theoretically, formulas **(7)** and **(9)** can be used when n is a fraction. When we calculate the accumulated or the discounted value using formulas **(7)** or **(9)** for a fractional part of an interest conversion period, we call it the **exact** or **theoretical method** of accumulating or discounting.

EXAMPLE 1 Find the accumulated and the discounted value of $1500 for 16 months at $j_4 = 12\%$, using the exact method.

Solution   To calculate the exact accumulated value S of $1500, we use formula **(7)**, substituting for $P = 1500$, $i = \frac{.12}{4} = 0.03$, $n = 5\frac{1}{3}$. We obtain
$S = 1500(1.03)^{5 \ 1/3} = \$1756.13$

To calculate the exact discounted value P of $1500, we use formula **(9)**, substituting for $S = 1500$, $i = 0.03$, $n = 5\frac{1}{3}$. We obtain
$P = 1500(1.03)^{-5 \ 1/3} = \$1281.23$

■

In practice compound interest for the full number of conversion periods and simple interest for the fractional part of a conversion period are often used. This method is called an **approximate** or **practical** method of accumulating or discounting and is illustrated in the following example. The use of simple interest for the fraction of a period is equivalent to linear interpolation as will be shown in problem 2 of Part B of Exercise 2.4. Linear interpolation is explained, in detail, in Appendix 3.

EXAMPLE 2   Find the accumulated and the discounted value of $1500 for 16 months at $j_4 = 12\%$, using the approximate method, and compare the results with those of Example 1.

Solution   To calculate the approximate accumulated value S of $1500 for 16 months at $j_4 = 12\%$ we first accumulate $1500 for 5 periods at $j_4 = 12\%$ ($i = 0.03$) and then accumulate this value for an additional 1 month at a simple interest rate of 12%. (See diagram.)

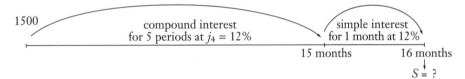

Numerically, we can combine the two accumulations and obtain

$$S = 1500(1.03)^5 \left[1 + (.12)\left(\tfrac{1}{12}\right)\right] = \$1756.30$$

Note that in the simple interest accumulation we multiply by an accumulation factor $(1 + rt)$ where r is a yearly rate j_4 and t is time in years.

To calculate the approximate discounted value P of $1500 for 16 months at $j_4 = 12\%$ we first discount $1500 for 6 periods at $j_4 = 12\%$ (in general we discount for the smallest number of whole periods containing the given time) and then accumulate this value for 2 months at a simple interest rate of 12%. (See diagram.)

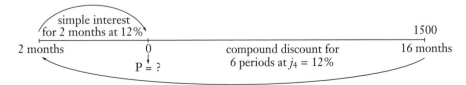

Numerically, we can combine the compound discount and simple interest calculations and obtain

$$P = 1500(1.03)^{-6}\left[1 + (.12)\left(\tfrac{2}{12}\right)\right] = \$1281.35$$

■

OBSERVATION: Comparing the results of Example 2 with those of Example 1 we can conclude that the approximate accumulated and discounted values are slightly greater than the exact accumulated and discounted values.

The proof that the approximate accumulated and discounted values are always greater than the exact accumulated and discounted values is left as an exercise. See problem 1 of Part B of Exercise 2.4.

CALCULATION TIP: Unless stated otherwise, it will be understood that the **practical** method is to be used throughout this textbook.

EXAMPLE 3 A note for $3000 without interest is due on August 18, 2003. On June 11, 2002, the holder of the note has it discounted by a lender who charges $j_{12} = 12\%$. What are the proceeds?

Solution The legal due date is August 18, 2003. We arrange the data on the diagram below.

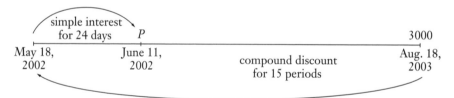

First we discount $3000 for 15 periods at rate $i = 1\%$ and then accumulate this value for 24 days at a simple interest rate of 12%. Numerically,

$$P = 3000(1.01)^{-15}\left[1 + (0.12)\left(\tfrac{24}{365}\right)\right] = \$2604.44$$

The proceeds are $2604.44.

■

Exercise 2.4

Part A

For each of problems 1 to 4 find the answer first using the exact method and then the practical method. Compare your answers in each case. In problems 5 to 9 use the practical method.

1. Find the accumulated value of $100 over 5 years 7 months if $j_2 = 13\tfrac{1}{2}\%$.

2. Find the accumulated value of $800 over 4 years 7 months if $j_4 = 8\%$.

3. Find the discounted value of $5000 due in 8 years 10 months if interest is at rate $j_2 = 12.73\%$.

4. Find the discounted value of $280 due in 3 years 7 months if interest is at rate $j_1 = 10\%$.

5. A note of $2000 face value is due without interest on October 20, 2006. On April 28, 2001, the holder of the note has it discounted at a bank that charges $j_4 = 12\%$. What are the proceeds?

6. On July 7, 2001, Mrs. Smith borrowed $1200 at $j_{12} = 12\%$. How much would she have to repay on September 18, 2004?

7. On April 7, 1997, a debt of $4000 was incurred at rate $j_2 = 10\%$. What amount will be required to settle the debt on September 19, 2002?

8. A noninterest-bearing note for $850 is due December 8, 2004. On August 7, 2001, the holder of the note has it discounted by a lender who charges $j_2 = 15\frac{1}{4}\%$. What are the proceeds?

9. A note for $1200 dated August 24, 2001, is due with interest at $j_{12} = 14\frac{3}{4}\%$ in 2 years. On June 18, 2002, the holder of the note has it discounted by a lender who charges $16\frac{1}{4}\%$ compounded quarterly. Find the proceeds and the compound discount.

Part B

1. a) Assuming that $0 < i < 1$, prove that
   i) $(1 + i)^t < 1 + it$ if $0 < t < 1$
   ii) $(1 + i)^t > 1 + it$ if $t > 1$
   [Hint: Use the binomial theorem.]
   b) Give a geometric illustration of the relationships in a) by graphing $(1 + i)^t$ and $1 + it$.
   c) Show that the approximate accumulated and discounted values are greater than the exact accumulated and discounted values when fractional parts of interest periods are involved.

2. A common method for finding an unknown value between two known values is called linear interpolation. It is based on the following formula:

$$f(n + k) = (1 - k)f(n) + kf(n + 1) 0 < k < 1$$

e.g.,
$$f\left(2\tfrac{2}{3}\right) = \tfrac{1}{3}f(2) + \tfrac{2}{3}f(3)$$

Applying this to the compound interest formula where $f(n) = (1 + i)^n$ we get:

$$(1 + i)^{n+k} = (1 - k)(1 + i)^n + k(1 + i)^{n+1} 0 < k < 1$$

Prove that using linear interpolation is equivalent to assuming simple interest in the final fractional period, i.e., prove

$$(1 - k)(1 + i)^n + k(1 + i)^{n+1} = (1 + i)^n(1 + ki)$$

Students not familiar with linear interpolation should read Appendix 3.

3. A promissory note for $2000 dated April 5, 2002, is due on October 4, 2006, with interest at $j_1 = 12\%$. On June 7, 2003, the holder of the note has it discounted at a bank that charges $j_4 = 14\%$. Find the proceeds and the compound discount.

Section 2.5   **Finding the Rate and the Time**

Finding the Rate: When S, P and n are given, we can substitute the given values into the fundamental compound interest formula $S = P(1 + i)^n$ and solve it for the unknown interest rate i.

$$S = P(1 + i)^n$$
$$(1 + i)^n = \frac{S}{P}$$
$$1 + i = \left(\frac{S}{P}\right)^{1/n}$$

$$\boxed{i = \left(\frac{S}{P}\right)^{1/n} - 1}$$

CALCULATION TIP: Using the power key of our calculator, we calculate the exact value of i. The nominal annual rate of interest is determined by multiplying the conversion rate i by the number of conversion periods per year, and then rounded off to the nearest hundredth of a percent.

EXAMPLE 1   At what nominal rate j_{12} will money triple itself in 12 years?

Solution   We can use any sum of money as the principal. Let $P = x$, then $S = 3x$, and $n = 144$.

Substituting in $S = P(1 + i)^n$ we obtain an equation for the unknown interest rate i per month

$$3x = x(1 + i)^{144}$$
$$(1 + i)^{144} = 3$$

Solving the exponential equation $(1 + i)^{144} = 3$ directly for i and then using a pocket calculator we have

$$(1 + i)^{144} = 3$$
$$1 + i = 3^{1/144}$$
$$i = 3^{1/144} - 1$$
$$i = .007658429$$
$$j_{12} = 12i = .091901147$$
$$j_{12} \doteq 9.19\%$$

∎

Finding the Time: When S, P and i are given, we can substitute the given values into the fundamental compound amount formula $S = P(1 + i)^n$ and solve it for the unknown n using one of the following methods:

a) Logarithms may be used to solve the exponential equation $P(1 + i)^n = S$ for the unknown exponent n. If the theoretical method of accumulation is used, i.e., compound interest is allowed for the fractional part of the conversion period, a logarithmic solution gives the correct value of n. An explanation of common logarithms together with applications to compound interest problems will be found in Appendix 1. In this textbook we assume that students have pocket calculators with a built-in common logarithmic function $\log x$ and its inverse function 10^x. This eliminates the use of logarithmic tables.

Solving the formula $S = P(1 + i)^n$ for n we obtain

$$(1 + i)^n = \frac{S}{P}$$

$$n \log(1 + i) = \log\left(\frac{S}{P}\right)$$

$$\boxed{n = \frac{\log\left(\frac{S}{P}\right)}{\log(1 + i)}}$$

b) Method of interpolation. Linear interpolation is explained in Appendix 3. If the practical method of accumulation is used, i.e., simple interest is allowed for the fractional part of the conversion period, interpolation gives the correct value of n.

The actual accuracy of the solution depends on the number of decimal places in the accumulation factors used in the interpolation. In our examples we shall round off the factors to 4 decimal places, since additional places do not significantly increase the accuracy.

EXAMPLE 2   How long will it take \$500 to accumulate to \$850 at $j_{12} = 12\%$? Assume that a) the theoretical method of accumulation is in effect; b) the practical method of accumulation is in effect.

Solution   Let n represent the number of months, then we have

$$500(1.01)^n = 850$$
$$(1.01)^n = \frac{850}{500}$$
$$(1.01)^n = 1.7$$

a) If compound interest is allowed for the fractional part of the interest conversion period, we can solve the equation $(1.01)^n = 1.7$ by logarithms. We have

$$n \log 1.01 = \log 1.7$$
$$n = \frac{\log 1.7}{\log 1.01}$$
$$n = 53.3277 \text{ months}$$
$$n \doteq 4 \text{ years 5 months 10 days}$$

b) If simple interest is allowed for the fractional part of the interest conversion period, then we find n by interpolation. Arranging the data in an interpolation table we have

$$\begin{array}{c|c} & (1.01)^n & n \\ \hline .0169 \left\{ .0055 \left\{ \begin{array}{c} 1.6945 \\ 1.7000 \\ \\ 1.7114 \end{array} \right. \right. & \begin{array}{c} 53 \\ n \\ \\ 54 \end{array} \left. \right\} d \left. \right\} 1 \end{array}$$

$$\frac{d}{1} = \frac{.0055}{.0169}$$
$$d = .33$$
and $n = 53.33$ months
$$n = 4 \text{ years } 5 \text{ months } 10 \text{ days}$$

Alternative Solution   The accumulated value of $500 for 53 periods at $j_{12} = 12\%$ is $500(1.01)^{53} = \$847.23$. Now we calculate how long it will take $847.23 to accumulate $2.77 simple interest at rate 12%.

$$t = \frac{I}{Pr} = \frac{2.77}{847.23 \times .12} = .027245652 \text{ years} \doteq 10 \text{ days}$$

Thus the time is 4 years 5 months and 10 days.

In this example both methods, theoretical and practical, give the same answer, when the time is expressed in days. The actual difference in time between the two solutions is less than $\frac{3}{100}$ of a day.

If no interest is allowed for part of the interest period it would take 54 months or $4\frac{1}{2}$ years to accumulate at least $850.

∎

CALCULATION TIP: When calculating unknown n from the fundamental compound interest formula **(7)**, use logarithms (that is, the theoretical method) in all exercises. The difference between the answers by the theoretical and the practical method is negligible.

Exercise 2.5

Part A

In problems 1 to 4 find the nominal rate of interest.

No.	Principal	Amount	Time	Conversion
1.	$2000	$3000.00	3 years 9 months	quarterly
2.	$ 100	$ 150.00	4 years 7 months	monthly
3.	$ 200	$ 600.00	15 years	annually
4.	$1000	$1581.72	3 years 6 months	semi-annually

In problems 5 to 8 find the time.

No.	Principal	Amount	Interest Rate	Conversion
5.	$2000	$2800	10%	quarterly
6.	$ 100	$ 130	9%	semi-annually
7.	$ 500	$ 800	12%	monthly
8.	$1800	$2200	8%	quarterly

9. An investment fund advertises that it will guarantee to double your money in 10 years. What rate of interest j_1 is implied?

10. If an investment grows 50% in 4 years, what rate of interest j_4 is being earned?

11. From 1997 to 2002, the earnings per share of common stock of a company increased from \$4.71 to \$9.38. What was the compounded annual rate of increase?

12. At what rate j_{365} will an investment of \$4000 grow to \$6000 in 3 years?

13. How long will it take to double your deposit in a savings account that accumulates at
    a) $j_1 = 4.56\%$?
    b) $j_{365} = 7\%$?

14. How long will it take for \$800 to grow to \$1500 in a fund earning interest at rate 9.8% compounded semi-annually?

15. How long will it take to increase your investment by 50% at rate $14\frac{1}{4}\%$ compounded daily?

Part B

1. At a given rate of interest, j_2, money will double in value in 8 years. If you invest \$1000 at this rate of interest, how much money will you have
   a) in 5 years?
   b) in 10 years?

2. If money doubles at a certain rate of interest compounded daily in 6 years, how long will it take for the same amount of money to triple in value?

3. Draw a graph showing the time needed to double your money at rate j_1 for the rates 2%, 4%, 6%,...20%.

4. Determine how long \$1 must be left to accumulate at $j_{12} = 18\%$ for it to amount to twice the accumulated value of another \$1 deposited at the same time at $j_2 = 10\%$.

5. Money doubles in t years at rate of interest j_1. At what rate of interest j_1^* will money double in $\frac{t}{2}$ years?

6. You deposit \$800 in an account paying $j_2 = 9\%$, and \$600 in a second account paying $j_2 = 7\%$. When will the first account have twice the accumulated value of the second account?

7. Account A starts now with \$100 and pays $j_1 = 4\%$. After 2 years an additional \$25 is deposited in account A. Account B is opened 1 year from now with a deposit of \$95 and pays $j_1 = 8\%$. When (in years and days) will account B have $1\frac{1}{2}$ times the accumulated value in Account A if simple interest is allowed for part of a year?

8. In how many years and days should a single payment of \$2000 be made in order to be equivalent to payments of \$500 now and \$800 in 3 years if interest is $j_1 = 8\%$ and simple interest is allowed for part of a year?

Section 2.6   Equations of Value

In Chapter 1, Section 1.3 we dealt with equations of value at a simple interest rate. We suggest that the student read over Section 1.3 before studying this section, since most of the principles and procedures from Section 1.3 will apply here as well.

In general, we compare dated values by the following **definition of equivalence**:

> $X due on a given date is equivalent at a given compound interest rate i to Y due n periods later, if
> $$Y = X(1 + i)^n \quad \text{or} \quad X = Y(1 + i)^{-n}$$

The following diagram illustrates dated values equivalent to a given dated value X.

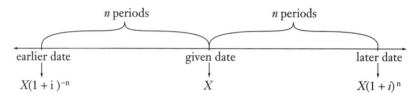

Based on the time diagram above we can state the following simple rules:

> 1. When we move money forward, we accumulate, i.e., we multiply the sum by an accumulation factor $(1 + i)^n$.
> 2. When we move backward, we discount, i.e., we multiply the sum by a discount factor $(1 + i)^{-n}$.

The following property of equivalent dated values, called the property of transitivity, holds at compound interest:

At a given compound interest rate, if X is equivalent to Y, and Y is equivalent to Z, then X is equivalent to Z.

To prove this property we arrange our data on a time diagram.

If X is equivalent to Y, then $Y = X(1 + i)^{n_2 - n_1}$

If Y is equivalent to Z, then $Z = Y(1 + i)^{n_3 - n_2}$

Eliminating Y from the second equation we obtain

$$Z = X(1 + i)^{n_2 - n_1}(1 + i)^{n_3 - n_2} = X(1 + i)^{n_3 - n_1}$$

Thus Z is equivalent to X.

As a result, the solutions to the problems by equations of value at compound interest do not depend on the selection of the focal date.

> **OBSERVATION:** In mathematics, an equivalence relation must satisfy the property of transitivity. Thus, strictly speaking, the equivalence of dated values at a simple interest rate is not an equivalence relation. The equivalence of dated values at a compound interest rate is an equivalence relation.

EXAMPLE 1 An obligation of $500 falls due at the end of 3 years. Find an equivalent debt at the end of a) 3 months; b) 3 years 9 months, at $j_4 = 12\%$.

Solution We arrange the data on a time diagram below.

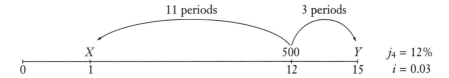

By definition of equivalence
$$X = 500(1.03)^{-11} = \$361.21$$
$$Y = 500(1.03)^3 = \$546.36$$

Note that X and Y are equivalent by verifying
$$Y = X(1.03)^{14} \text{ or } 546.36 = 361.21(1.03)^{14}$$

∎

The sum of a set of dated values, due on different dates, has no meaning. We have to replace all the dated values by equivalent dated values, due on the same date. The sum of the equivalent values is called the **dated value of the set**.

At compound interest the following property is true: *The various dated values of the same set are equivalent.* The proof is left as an exercise. See problem 1 of Part B of Exercise 2.6.

EXAMPLE 2 A person owes $200 due in 6 months and $300 due in 15 months. What single payment a) now, b) in 12 months, will liquidate these obligations if money is worth $j_{12} = 15\%$?

Solution We arrange the data on the diagram below. Let X be the single payment now and Y be the single payment in 12 months.

We calculate the equivalent dated values X and Y

$$X = 200(1.0125)^{-6} + 300(1.0125)^{-15} = 185.63 + 249.00 = \$434.63$$
$$Y = 200(1.0125)^{6} + 300(1.0125)^{-3} = 215.48 + 289.03 = \$504.51$$

We can verify the property of equivalence of X and Y by showing that

$$Y = X(1.0125)^{12} = 434.63(1.0125)^{12} = \$504.50$$
$$\text{or } X = Y(1.0125)^{-12} = 504.51(1.0125)^{-12} = \$434.64$$

The 1-cent error is due to rounding off in calculating X and Y.

∎

As stated in Section 1.3, one of the most important problems in the mathematics of finance is the replacing of a given set of payments by an equivalent set. Two sets of payments are equivalent at a given compound interest rate if the dated values of the sets, on any common date, are equal. An equation stating that the dated values, on a common date, of two sets of payments are equal is called an **equation of value** or an **equation of equivalence**. The date used is called the **focal date** or the **comparison date** or the **valuation date**.

The procedure described in Section 1.3 for solving problems in the mathematics of finance by equations of value applies the same way when compound interest is used. The answers, however, will not depend on the location of the focal date.

The following examples illustrate the use of equations of value in the mathematics of finance.

EXAMPLE 3   A debt of $1000 with interest at $j_4 = 10\%$ will be repaid by a payment of $200 at the end of 3 months and 3 equal payments at the ends of 6, 9, and 12 months. What will these payments be?

Solution   We arrange all the dated values on a time diagram.

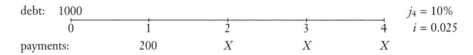

debt: 1000 $j_4 = 10\%$

0   1   2   3   4 $i = 0.025$

payments: 200 X X X

Any date can be selected as a focal date. We show the calculation using the end of 12 months and the present time.

Equation of value at the end of 12 months:

$$\text{dated value of the payments} = \text{dated value of the debts}$$
$$200(1.025)^3 + X(1.025)^2 + X(1.025)^1 + X = 1000(1.025)^4$$
$$215.38 + 1.050625\,X + 1.025\,X + X = 1103.81$$
$$3.075625\,X = 888.43$$
$$X = \$288.86$$

Equation of value at the present time:

$$200(1.025)^{-1} + X(1.025)^{-2} + X(1.025)^{-3} + X(1.025)^{-4} = 1000.00$$
$$195.12 + 0.951814396\,X + 0.928599411\,X + 0.905950645\,X = 1000.00$$
$$2.786364452\,X = 804.88$$
$$X = \$288.86$$

■

CALCULATION TIP: Choose a convenient focal date, one that simplifies your calculations, when using equations of value at compound interest.

EXAMPLE 4   A man leaves an estate of $50 000 that is invested at $j_{12} = 9\%$. At the time of his death, he has two children aged 13 and 18. Each child is to receive an equal amount from the estate when they reach age 21. How much does each child get?

Solution   The older child will get X in 3 years; the younger child will get X in 8 years. We arrange the dated values on a time diagram.

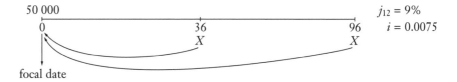

Equation of value at present:

$$X(1.0075)^{-36} + X(1.0075)^{-96} = 50\ 000$$
$$0.764148961\,X + 0.488061711\,X = 50\ 000$$
$$1.252210671\,X = 50\ 000$$
$$X = \$39\ 929.38$$

Each child will receive $39 929.38.

The following calculation checks the correctness of the answer:

$$\text{Amount in fund at the end of 3 years} = 50\ 000(1.0075)^{36} = \$65\ 432.27$$
$$\text{Payment to the older child} = \underline{\$39\ 929.38}$$
$$\text{Balance in the fund} = \$25\ 502.89$$
$$\text{Amount in fund after 5 more years} = 25\ 502.88(1.0075)^{60} = \$39\ 929.39$$

The 1-cent difference is due to rounding.

■

Exercise 2.6

Part A

1. If money is worth $j_4 = 6\%$, find the sum of money due at the end of 15 years equivalent to $1000 due at the end of 6 years.

2. What sum of money, due at the end of 5 years, is equivalent to $1800 due at the end of 12 years if money is worth $j_2 = 11\frac{3}{4}\%$?

3. An obligation of $2500 falls due at the end of 7 years. Find an equivalent debt at the end of a) 3 years and b) 10 years, if $j_{12} = 10\%$.

4. One thousand dollars is due at the end of 2 years and $1500 at the end of 4 years. If money is worth $j_4 = 8\%$, find an equivalent single amount at the end of 3 years.

5. Eight hundred dollars is due at the end of 4 years and $700 at the end of 8 years. If money is worth $j_{12} = 12\%$, find an equivalent single amount at a) the end of 2 years; b) the end of 6 years; c) the end of 10 years. Show your answers are equivalent.

6. A debt of $2000 is due at the end of 8 years. If $1000 is paid at the end of 3 years, what single payment at the end of 7 years would liquidate the debt if money is worth $j_2 = 12\%$?

7. A person borrows $4000 at $j_4 = 12\%$. He promises to pay $1000 at the end of one year, $2000 at the end of 2 years and the balance at the end of 3 years. What will the final payment be?

8. A consumer buys goods worth $1500. She pays $300 down and will pay $500 at the end of 6 months. If the store charges $j_{12} = 18\%$ on the unpaid balance, what final payment will be necessary at the end of one year?

9. A debt of $1000 is due at the end of 4 years. If money is worth $j_{12} = 8\%$, and $375 is paid at the end of 1 year, what equal payments at the end of 2 and 3 years respectively would liquidate the debt?

10. On September 1, 2000, Paul borrowed $3000, agreeing to pay interest at 12% compounded quarterly. He paid $900 on March 1, 2001, and $1200 on December 1, 2001.
    a) What equal payments on June 1, 2002, and December 1, 2002, will be needed to settle the debt?
    b) If Paul paid $900 on March 1, 2001, $1200 on December 1, 2001, and $900 on March 1, 2002, what would be his outstanding balance on September 1, 2002?

11. A woman's bank account showed the following deposits and withdrawals:

	Deposits	Withdrawals
January 1, 2001	$200	
July 1, 2001	$150	
January 1, 2002		$250
July 1, 2002	$100	

If the account earns $j_2 = 6\%$, find the balance in the account on January 1, 2003.

12. Instead of paying $400 at the end of 5 years and $300 at the end of 10 years, a man agrees to pay $X at the end of 3 years and $2X at the end of 6 years. Find X if $j_1 = 10\%$.

13. A man stipulates in his will that $50 000 from his estate is to be placed in a fund from which his three children are to each receive an equal amount when they reach age 21. When the man dies, the children are ages 19, 15 and 13. If this fund earns interest at $j_2 = 12\%$, how much does each receive?

14. To pay off a loan of $4000 at $j_{12} = 15\%$, Ms. Fil agrees to make three payments in three, seven and twelve months respectively. The second and third payment are to be double the first. What is the size of the first payment?

Part B

1. a) Prove the following property in the case of a set of two dated values: The various dated values of the same set are equivalent at compound interest.
   b) Show, algebraically, why this is not true for simple interest.

2. If money is worth $j_1 = 8\%$, what single sum of money payable at the end of 2 years will equitably replace $1000 due today plus a $2000 debt due at the end of 4 years with interest at $12\frac{1}{2}\%$ per annum compounded semi-annually?

3. On January 1, 2001, Mr. Planz borrowed $5000 to be repaid in a lump-sum payment with interest at $j_4 = 9\%$ on January 1, 2007. It is now January 1, 2003. Mr. Planz would like to pay $500 today and complete the liquidation with equal payments on January 1, 2005, and January 1, 2007. If money is now worth $j_4 = 8\%$, what will these payments be?

4. You are given two loans, with each loan to be repaid by a single payment in the future. Each payment will include both principal and interest. The first loan is repaid by a $3000 payment at the end of 4 years. The interest is accrued at 10% per annum compounded semi-annually. The second loan is repaid by a $4000 payment at the end of 5 years. The interest is accrued at 8% per annum compounded semi-annually. These two loans are to be consolidated. The consolidated loan is to be repaid by two equal instalments of X, with interest at 12% per annum compounded semi-annually. The first payment is due immediately and the second payment is due one year from now. Calculate X.

5. You are given the following data on three series of payments:

	Payment at End of Year			Accumulated Value at End of Year 18
	6	12	18	
Series A	240	200	300	X
Series B	0	360	700	X + 100
Series C	Y	600	0	X

Assume interest is compounded annually. Calculate Y.

Section 2.7   Compound Interest at Changing Interest Rates

The previous sections have assumed that the rate of compound interest relevant to any particular problem remains unchanged throughout the term of the problem. However, this need not be the case and, in practice, interest rates vary with considerable frequency. For instance, banks, trust companies and credit unions vary their deposit rates with changes in market conditions.

This complication of changing interest rates is not really a difficulty as the problems may be solved by completing the appropriate compound interest calculations in stages. As shown in the following examples, intermediate values are found at each date that has an interest rate change. No new compound interest techniques are required and these problems may be considered to be two or more compound interest problems, simply expressed in one question.

EXAMPLE 1   How much will \$1000 accumulate to in 8 years if it earns $j_1 = 12\%$ for 6 years and $j_1 = 9\%$ for 2 years?

Solution   The time diagram below sets out the 2 stages in the calculation and the 2 interest rates.

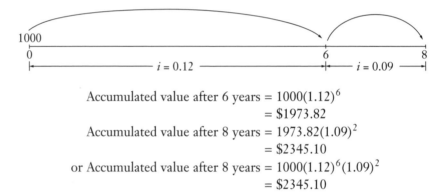

$$\text{Accumulated value after 6 years} = 1000(1.12)^6$$
$$= \$1973.82$$
$$\text{Accumulated value after 8 years} = 1973.82(1.09)^2$$
$$= \$2345.10$$
$$\text{or Accumulated value after 8 years} = 1000(1.12)^6(1.09)^2$$
$$= \$2345.10$$

■

OBSERVATION: When compound interest rates vary, an average interest rate must *never* be calculated. Each compound interest rate must be allowed to have its full effect.

EXAMPLE 2   Find the present value of \$10 000 due in 10 years' time, if $j_4 = 8\%$ for the first 3 years, $j_2 = 10\%$ for the next 5 years and $j_1 = 9\%$ for the last 2 years.

Solution   This problem should be tackled in three stages, moving from the value in 10 years' time towards the present value, stopping at each date that has an interest rate change. The following time diagram illustrates the problem:

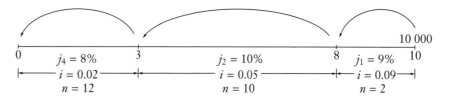

The three stages in this solution are:

$$\textit{Step 1} \quad \text{Value in 8 years' time} = 10\ 000(1.09)^{-2}$$
$$= \$8416.80$$

$$\textit{Step 2} \quad \text{Value in 3 years' time} = 8416.80(1.05)^{-10}$$
$$= \$5167.19$$

$$\textit{Step 3} \quad \text{Value now} = 5167.19(1.02)^{-12}$$
$$= \$4074.29$$

This solution may also be expressed as:

$$\text{Present value} = 10\ 000(1.02)^{-12}(1.05)^{-10}(1.09)^{-2}$$
$$= \$4074.29$$

■

Equations of value may also involve more than one interest rate. The above methods are appropriate for these problems such that when there is a change of interest rate, intermediate values should be calculated.

EXAMPLE 3   A student owes \$200 due in 6 months and \$300 in 15 months. What single payment now will repay these debts if the interest rate is $j_4 = 12\%$ for 9 months and $j_4 = 8\%$ thereafter?

Solution   We arrange the data as set out in the following time diagram, where X is the single payment and time is expressed in quarters.

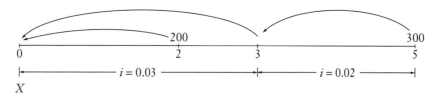

$$\text{The present value of the \$200 debt} = 200(1.03)^{-2}$$
$$= \$188.52$$

$$\text{The value at time 3 of the \$300 debt} = 300(1.02)^{-2}$$
$$= \$288.35$$

$$\text{The present value of the \$300 debt} = 288.35(1.03)^{-3}$$
$$= \$263.88$$

$$\text{Hence the present value of both debts} = 188.52 + 263.88$$
$$X = \$452.40$$

This solution may also be expressed as

$$X = 200(1.03)^{-2} + 300(1.02)^{-2}(1.03)^{-3}$$
$$= 188.52 + 263.88$$
$$= \$452.40$$

∎

Exercise 2.7

Part A

1. How much will $2000 accumulate to in 12 years' time if the interest rate is $j_1 = 11\%$ for the first 6 years and $j_1 = 9\%$ for the next 6 years?

2. What is the present value of $1000 in 6 years if $j_1 = 0.08$ for 2 years and $j_1 = 0.07$ thereafter?

3. Carol deposited $500 into her credit union account on January 1, 1998. What will be in the account on January 1, 2003, if $j_2 = 5\%$ for 1998, $j_2 = 6\%$ for 1999 and 2000 and $j_2 = 4.5\%$ for 2001 and 2002?

4. Two thousand dollars are invested for 10 years at the following interest rates:
   $j_2 = 10\%$ for years 1, 2 and 3;
   $j_4 = 8\%$ for years 4, 5, 6 and 7; and
   $j_{12} = 12\%$ for years 8, 9 and 10.
   Find the accumulated value and the compound interest earned.

5. What sum of money due on July 1, 2001, is equivalent to $2000 due on January 1, 1995, if $j_2 = 10\%$ for 1995 and 1996, and $j_2 = 9\%$ thereafter?

6. A debt of $5000 is due at the end of 5 years. It is proposed that $X be paid now with another $X paid in 10 years' time to liquidate the debt. Calculate the value of X if the effective interest rate is 12% for the first 6 years and 8% for the next 4 years.

7. A company wishes to replace the following three debts:
   $20 000 due on July 1, 1999
   $30 000 due on January 1, 2002 and
   $35 000 due on July 1, 2005,
   with a single debt of $Y payable on January 1, 2002. Calculate the value of Y if $j_2 = 12\%$ prior to January 1, 2002, and $j_2 = 10\%$ after January 1, 2002.

8. A young couple owns a block of land worth $29 000. They are offered a 20% deposit and 2 equal payments of $15 000 each at the end of years 2 and 4. If money is worth $j_2 = 8\%$ for the first 2 years and $j_4 = 12\%$ for the next 2 years, should they accept the offer?

9. A person can buy a lot for $13 000 cash outright or $6000 down, $6000 in 2 years and $6000 in 5 years. Which option is better if money can be invested at
   a) $j_{12} = 18\%$
   b) $j_4 = 12\%$ for the first 3 years and $j_4 = 14\%$ for the next 2 years?

10. Find the annual effective rate of interest that is equivalent to $j_4 = 6\%$ for 2 years followed by $j_{12} = 8\%$ for 4 years.

Part B

1. Show algebraically that

$$(1 + i)^n \times (1 + j)^n \neq \left(1 + \frac{i + j}{2}\right)^{2n}$$

and hence conclude that compound interest rates should not be averaged.

2. Gus invests $500 for 4 years. The nominal interest rate remains 8% each year although in the first year it is convertible half-yearly, in the second year convertible quarterly, in the third year convertible monthly and in the fourth year convertible daily. What is the accumulated value? How much greater is this value than the corresponding value assuming that the first rate had remained unchanged for the 4 years?

3. Calculate the payment due on July 1, 2002, that is equivalent to $1000 due on January 1, 1999, plus $2000 due on March 1, 2004, if $j_4 = 8\%$ prior to July 1, 2002, and $j_{12} = 12\%$ thereafter.

4. Find the total present value of $20 000 due in 8 years and $30 000 due in 15 years if

$j_1 = 12\%$ for years 1, 2 and 3;

$j_2 = 10\%$ for years 4, 5, 6, 7, 8 and 9;

$j_4 = 8\%$ for years 10, 11 and 12; and

$j_{12} = 9\%$ for years 13, 14 and 15.

5. $1000 was deposited on January 1, 1997, and $2000 was deposited in an account on July 1, 1999. Interest was paid on the account at $j_4 = 7\%$ from January 1, 1997, to October 1, 1999, and at $j_2 = 5\%$ from October 1, 1999, until April 1, 2001. Find the amount in the account on April 1, 2001, and find the equivalent nominal interest rate compounded monthly actually earned on the investment over the period.

6. A sum of money is left invested for 3 years. In the first year it earns interest at $j_{12} = 15\%$. In the second year, the rate of interest earned is $j_4 = 10\%$ and in the third year the rate of interest changes to $j_{365} = 12\%$. Find the level rate of interest, j_1, that would give the same accumulated value at the end of three years.

7. What compound interest rate j_4 is equivalent over a 3-year period to a simple interest rate of 6% the first year, followed by a simple discount rate of 8% for the next 2 years?

Section 2.8

Other Applications of Compound Interest Theory, Inflation and the "Real" Rate of Interest

We know that the more money you invest at some given interest rate, i, the more dollars of interest you will earn. Further, once you earn a dollar's worth of interest, it becomes a part of the invested money and earns interest itself. The latter characteristic is referred to as multiplicative or geometric growth and is what differentiates compound interest from simple interest.

Any time we have geometric growth we can use the theory of compound interest.

EXAMPLE 1   A tree, measured in 1998, contains an estimated 150 cubic metres of wood. If the tree grows at a rate of 3% per annum, how much wood would it produce in 2008?

Solution   This is just geometric growth, so we can use compound interest theory. Thus, in 2008

$$\text{Amount of wood} = 150(1.03)^{10}$$
$$\doteq 202 \text{ cubic metres}$$

∎

EXAMPLE 2   The population of Canada in July 1987 was 26.5 million people. In July 1997 it was 30.3 million people.

a) What was the annual growth rate from July 1987 to July 1997?

b) At this rate of growth, when will the population reach 40 million people?

Solution a   This is the same as finding the unknown rate of interest in a compound interest question. Hence,

$$26.5(1 + i)^{10} = 30.3$$
$$(1 + i)^{10} = \tfrac{30.3}{26.5}$$
$$1 + i = \left(\tfrac{30.3}{26.5}\right)^{1/10}$$
$$1 + i = 1.013490484$$
$$i \doteq 1.35\%$$

Solution b   This is the same as finding unknown time in a compound interest question. Therefore,

$$30.3(1.0135)^t = 40$$
$$(1.0135)^t = \tfrac{40}{30.3}$$
$$t \log (1.0135) = \log \tfrac{40}{30.3}$$
$$t \doteq 20.71 \text{ years}$$
$$\doteq 20 \text{ years } 9 \text{ months}$$

Therefore our population will reach 40 million sometime in March 2018 assuming the same rate of growth.

∎

Inflation and the "Real" Rate of Interest

One very valuable use of compound interest theory is the analysis of rates of inflation. A widely used measure of inflation is the annual change in the Consumer Price Index. It measures the annual effective rate of change in the cost of a specified "basket" of consumer items.*

* The following Web site describes the CPI and also includes an inflation calculator.
http://www.bank-banque-canada.ca/english/inflation_calc.htm

EXAMPLE 3   In June 1992 the Canadian Consumer Price Index was set at 100. In June 1999 the index was 110.5. That means that if goods cost an average of $100 in June 1992, they cost $110.50 in June 1999.

a) Over that 7-year period, what was the average annual compound percentage rate of change?

b) If that rate of inflation were to continue, how long would it take before the purchasing power of a June 1992 Canadian dollar was only 80¢?

Solution a   Find i such that

$$100(1 + i)^7 = 110.5$$
$$(1 + i)^7 = 1.105$$
$$i = (1.105)^{1/7} - 1$$
$$i \doteq 1.44\%$$

Solution b   Find t such that

$$(1.0144)^t = \frac{1}{.80}$$
$$t \log 1.0144 = -\log .8$$
$$t \doteq 15.61 \text{ years}$$

■

The last example illustrated the effect of the rate of inflation as measured by the Consumer Price Index (CPI). Inflation rates vary from country to country and from time to time and can be relatively difficult to predict very far into the future. It is also possible to experience a period of time where prices drop and the CPI decreases. This is called deflation.

There is often a relationship between interest rates and rates of inflation. Investors want a "real" rate of return on their investment over and above the rate of inflation.

If i is the annual rate of interest being paid in the marketplace and r is the annual rate of inflation, then $1 invested at the beginning of the year will grow to $(1 + i)$ at the end of the year. However, its purchasing power is only equal to $\left(\frac{1+i}{1+r}\right)$. Hence, the real rate of return is

$$\boxed{i_{real} = \frac{1+i}{1+r} - 1 = \frac{i-r}{1+r}.}$$

For low inflation rates r, $1 + r$ is close to 1 and i_{real} is close to $i - r$. To make a meaningful comparison of interest and inflation, both rates should refer to the same one-year period.

EXAMPLE 4   Jack invested $1000 for one year at $j_1 = 8\%$. The annual inflation rate for that year was 2%.

a) What was the real annual rate of return on Jack's investment?

b) What was Jack's annual real after-tax rate of return, if he paid tax at a 40% marginal rate?

c) Repeat part b) for a 50% marginal tax rate.

Solution a

$$i_{real} = \frac{.08 - .02}{1 + .02} = .058823529 \doteq 5.88\%$$

Solution b   Jack's annual real after-tax rate of return was

$$i_{real\ after-tax} = \frac{.08(.6) - .02}{1 + .02} = .02745098 \doteq 2.75\%$$

Solution c   Jack's annual real after-tax rate of return was

$$i_{real\ after-tax} = \frac{.08(.5) - .02}{1 + .02} = .019607843 \doteq 1.96\%.$$

∎

EXAMPLE 5

a) Suppose that the forecast for next year's annual inflation rate is $r = .05$, and for the annual interest rate it is $i = .04$. What will be the corresponding real rate of interest for the next year?

b) Using the values of r and i in part a), suppose you borrow $10 000 for a year at $i = .04$ and buy 5000 units of a certain item that has a current cost of $2 per unit. If the price of this item is tied to a rate of inflation $r = .05$ and you sell the items one year from now at the inflated price, what will be your net gain on this transaction?

Solution a

$$i_{real} = \frac{.04 - .05}{1.05} \doteq -.00952381 \doteq -.95\%$$

Solution b

$$\text{Net gain} = \text{Increase in value of 5000 units} - \text{Cost of borrowing}$$
$$= 5000(2)(.05) - 10\ 000(.04) = 500 - 400 = 100$$

∎

OBSERVATION: Example 5 illustrates that during times when inflation rates exceed interest rates there is an incentive to borrow at the negative real rate of interest. Similarly, it can be shown that at low real rates of interest there is an incentive to consume rather than to save.

Exercise 2.8

Part A

1. A city increased in population 4% a year during the period 1985 to 1995. If the population was 40 000 in 1985, what is the estimated population in 2005, assuming the rate of growth remains the same?

2. The population of Happy Town on December 31, 1997, was 15 000. The town is growing at a rate of 2% per annum. What would be the increase in population in the calendar year 2005?

3. At what annual growth rate will the population of a city double in 11 years?

4. If the cost of living rises 8% a year, how long will it take for the purchasing power of $1 to fall to 60¢?

5. The cost of living rises 2.1% a year for 5 years. Over that period of time, what would be the increase in value of a $160 000 house due to inflation only?

6. A university graduate starts his new job on his 22nd birthday at an annual salary of $28 000. If his salary goes up 5% a year (on his birthday), how much will he be making when he retires one day before his 65th birthday?

7. Calculate the real annual rate of return for the following pairs of annual interest rates i and annual inflation rates r:
a) $i = 6\%$ $r = 2\%$
b) $i = 8\%$ $r = 4\%$
c) $i = 10\%$ $r = 6\%$

Part B

1. a) The number of fruit-flies in a certain lab increases at the compound rate of 4% every 40 minutes. If there are 100 000 flies at 1 p.m. today, what will be the increase in the number of flies between 7 a.m. and 11 a.m. tomorrow?
b) At what time will there be 200 000 flies in the lab?

2. The population of a county was 200 000 in 1980 and 250 000 in 1990. Estimate the change in population of the county between 2000 and 2005.

3. You will need $X U.S. one year from now and can invest funds in a U.S. dollar account for the next year at $j_1 = 3\%$. Alternatively, you can invest in a Canadian dollar account for the next year at $j_1 = 6\%$. If the current exchange rate* is $0.82 U.S. = $1 Cdn., what is the implied exchange rate one year from now? Assume that both alternatives require the same amount of currency today.

4. Let i be the annual effective interest rate and r be the annual effective inflation rate. Show that the present value of $(1 + r)^n$ due in n years at an annual effective rate i is equal to the present value of 1 due in n years at an annual effective rate $\frac{i-r}{1+r}$.

Section 2.9 **Continuous Compounding**

In any text on calculus one will find the equation

$$\lim_{m \to \infty} \left(1 + \frac{x}{m}\right)^m = e^x$$

where the number $e \doteq 2.718$ has an infinite expansion and is the base of the natural logarithms.

This equation will be useful in compound interest problems where a nominal rate of interest is compounded continuously. Continuous compounding is an important topic for actuaries to develop theoretical models in actuarial science.

* For a current exchange rate, see *www.quicken.ca* or *www.canoe.ca/money*.

We have already dealt with problems where the nominal rate j has been compounded as often as daily. For example, consider the nominal rate of interest $j_m = 12\%$ compounded at different frequencies and compare the accumulated value of $1 over a 1-year period (so-called annual accumulation factors). The results can be summarized in the following table.

$$j_m = 12\%$$

m	Annual Accumulation Factor
1	$(1.12)^1 = 1.12$
2	$(1 + \frac{.12}{2})^2 = 1.1236$
4	$(1 + \frac{.12}{4})^4 = 1.12550881$
12	$(1 + \frac{.12}{12})^{12} = 1.12682503$
52	$(1 + \frac{.12}{52})^{52} = 1.127340987$
365	$(1 + \frac{.12}{365})^{365} = 1.127474614$

From the above table we can see that as the frequency m of compounding increases, the annual accumulation factor also increases and approaches an upper bound as m is increased without limit, i.e., $m \to \infty$. To determine this upper bound that represents the annual accumulation factor at nominal rate 12% compounded continuously (i.e., $j_\infty = 12\%$), we wish to calculate

$$\lim_{m \to \infty} \left(1 + \frac{.12}{m}\right)^m$$

Using the equation

$$\lim_{m \to \infty} \left(1 + \frac{x}{m}\right)^m = e^x$$

we obtain

$$\lim_{m \to \infty} \left(1 + \frac{.12}{m}\right)^m = e^{.12} = 1.127496852$$

CALCULATION TIP: To calculate e^x on a pocket calculator without the function e^x you may use inverse $\ln x$. Consult your owner's manual for details.

EXAMPLE 1   Find the rate j_{12} equivalent to $j_\infty = 15\%$.

Solution   We shall compare the annual accumulation factors at these rates.

$$(1 + i)^{12} = e^{.15}$$
$$i = e^{.15/12} - 1$$
$$i = .012578452$$
$$j_{12} = .150941418$$
$$j_{12} \doteq 15.09\%$$

∎

The accumulated value S of principal P at rate j_m for t years is given by the fundamental compound interest formula **(7)**

$$S = P(1 + i)^n = P\left(1 + \frac{j_m}{m}\right)^{mt} = P\left[\left(1 + \frac{j_m}{m}\right)^m\right]^t$$

If interest is compounded continuously

$$S = \lim_{m \to \infty} P\left[\left(1 + \frac{j_m}{m}\right)^m\right]^t = Pe^{j_\infty t}$$

Similarly we can develop the formula for the discounted value P, given S, j_∞, and t.

$$P = Se^{-j_\infty t}$$

The following examples illustrate how the formula $S = P\,e^{j_\infty t}$ can be used to find the accumulated value S, the discounted value P, the rate j_∞, the effective rate j and time in years t.

EXAMPLE 2   Find the accumulated and the discounted value of $5000 over 15 months at a nominal rate 8% compounded continuously.

Solution   We have $j_\infty = .08$, $t = \frac{15}{12} = 1.25$ and calculate the accumulated value S of $5000

$$S = 5000\,e^{\,(.08)(1.25)} = \$5525.85$$

and then the discounted value P of $5000

$$P = 5000\,e^{\,-(.08)(1.25)} = \$4524.19$$

∎

EXAMPLE 3   A mutual fund deposit of $1000 increased in value by $560 over 30 months. Find

  a) the continuous rate of increase;

  b) the annual effective rate of increase.

Solution a   We have $P = 1000$, $S = 1560$, $t = \frac{30}{12} = 2.5$ and solve the equation

$$1000\,e^{\,j_\infty(2.5)} = 1560$$
$$e^{\,2.5j_\infty} = 1.560$$
$$2.5j_\infty = \ln 1.560$$
$$j_\infty = \frac{\ln 1.560}{2.5}$$
$$j_\infty = .177874329$$
$$j_\infty \doteq 17.79\%$$

Solution b   We want to find the equivalent annual effective rate j for a given rate $j_\infty = .177874329$ by comparing the accumulated value of $1 at the end of 1 year

$1 at j will accumulate to $1 + j$
$1 at $j_\infty = .177874329$ will accumulate to $e^{.177874329}$

Thus

$$1 + j = e^{.177874329}$$
$$j = e^{.177874329} - 1$$
$$j = .194675175$$
$$j \doteq 19.47\%$$

We can also find j by solving the equation

$$1000(1 + j)^{2.5} = 1560$$
$$(1 + j)^{2.5} = 1.560$$
$$1 + j = (1.560)^{1/2.5}$$
$$j = (1.560)^{1/2.5} - 1$$
$$j = .194675175$$
$$j \doteq 19.47\%$$

∎

EXAMPLE 4   How long will it take to triple your investment at 15% compounded continuously?

Solution   We have $P = x$, $S = 3x$, $j_\infty = .15$ and solve the equation below for time t in years.

$$x\, e^{.15t} = 3x$$
$$e^{.15t} = 3$$
$$.15t = \ln 3$$
$$t = \frac{\ln 3}{.15}$$
$$t = 7.324081924 \text{ years}$$
$$t \doteq 7 \text{ years } 118 \text{ days}$$

∎

Exercise 2.9

Part A

1. Fifteen hundred dollars is invested for 18 months at a nominal rate of 13%. Find the accumulated value if interest is compounded a) annually; b) monthly; c) continuously.

2. A debt of $8000 is due in 5 years. Find the discounted value at a nominal rate of 8% compounded a) quarterly; b) daily; c) continuously.

3. At what nominal rate compounded continuously will your investment increase 50% in value in 3 years? Find also the equivalent annual effective rate.

4. By what date will $800 deposited on February 4, 2000, be worth at least $1200
   a) at 6% compounded daily;
   b) at 6% compounded continuously?

5. If money doubles at a certain rate of interest compounded continuously in 5 years, how long will it take for the same amount of money to triple in value?

6. Shirley must borrow $1000 for 2 years. She is offered the money at
   a) 13% compounded continuously;
   b) $13\frac{1}{4}$% compounded semi-annually;
   c) $14\frac{1}{2}$% simple interest.
   Which offer should she accept?

Part B

Actuaries use δ for $j_\infty = i^{(\infty)}$ and call it the "force of interest."

1. What simple interest rate is equivalent to the force of interest $\delta = 7\%$ if money is invested for 5 years?

2. If the population of the world doubles every 25 years, how long does it take to increase by 50%? Assume population growth takes place continuously at a uniform rate.

3. The force of interest is $j_\infty = 10\%$. At what time should a single payment of $2500 be made so as to be equivalent to payments of $1000 in 1.25 years and $1500 in 6.5 years?

4. How long will it take $250 to accumulate to $400, if the force of interest is $\delta = 7\%$ for the first 2 years and $\delta = 8\%$ thereafter?

5. Jill borrowed $1000 and wants to pay off the debt by payments of $400 at the end of 3 months and 2 equal payments at the end of 6 months and 12 months. What will these payments be, if the force of interest is $\delta = 10\%$?

6. Find a simple discount rate equivalent to 8% compounded continuously if money is invested for 4 years.

Section 2.10 | Compound Discount at a Discount Rate and Equivalent Discount Rates

In Chapter 1 we introduced simple discount at a discount rate. Similarly, we can introduce compound discount at a discount rate. Let $d^{(m)}$ be the nominal rate of discount compounded m times per year. Then the discount rate per conversion period is $\frac{d^{(m)}}{m}$, and the discounted value P of a future amount S due in n periods is

$$P = S\left(1 - \frac{d^{(m)}}{m}\right)^n$$

From the above equation we may determine the accumulated value S of a principal P for n periods at $d^{(m)}$

$$S = P\left(1 - \frac{d^{(m)}}{m}\right)^{-n}$$

Compound discount at a discount rate is an important topic in developing theoretical models in actuarial science.

EXAMPLE 1 Find the discounted value of $1000 due in 2 years at

a) $d^{(12)} = 12\%$;

b) $d^{(365)} = 7\%$.

Solution a We have $S = 1000$, $n = 24$, discount rate per month $= \frac{.12}{12} = .01$ and

$$P = 1000(1 - .01)^{24} = \$785.68$$

Solution b We have $S = 1000$, $n = 2 \times 365 = 730$, discount rate per day $= \frac{.07}{365}$ and

$$P = 1000\left(1 - \tfrac{.07}{365}\right)^{730} = \$869.35$$

∎

EXAMPLE 2 Find the accumulated value of $500 at the end of 3 years

a) at 8% simple discount;

b) at $d^{(2)} = 8\%$.

Solution a We have $P = 500$, $t = 3$, $d = .08$ and using equation **(6)** from Chapter 1

$$S = P(1 - dt)^{-1} = 500[1 - (.08)(3)]^{-1} = \$657.89$$

Solution b We have $P = 500$, $n = 3 \times 2 = 6$, discount rate per half-year $= \frac{.08}{2} = .04$ and

$$S = P\left(1 - \frac{d^{(m)}}{m}\right)^{-n} = 500\left(1 - \frac{.08}{2}\right)^{-6} = \$638.77$$

∎

For a given nominal rate of discount $d^{(m)}$ compounded m times per year, we calculate the equivalent **annual effective discount rate** d by comparing the discounted values of $1 due in 1 year

$$1 - d = \left(1 - \frac{d^{(m)}}{m}\right)^{m}$$

$$\boxed{d = 1 - \left(1 - \frac{d^{(m)}}{m}\right)^{m}}$$

We can also calculate d by comparing the accumulated values of $1 at the end of 1 year.

EXAMPLE 3 Find the annual effective discount rate d corresponding to $d^{(12)} = 9\%$.

Solution

$$1 - d = \left(1 - \tfrac{.09}{12}\right)^{12}$$

$$d = 1 - \left(1 - \tfrac{.09}{12}\right)^{12}$$

$$d = 0.086378765$$

$$d \doteq 8.64\%$$

∎

> Two nominal compound rates of discount are equivalent if they yield the same discounted (or accumulated) values over the same period of time.

EXAMPLE 4 Find $d^{(12)}$ equivalent to a) $d^{(2)} = 6\%$; b) $d^{(365)} = 10\%$.

Solution a Comparing discounted values of $1 due in 1 year we obtain

$$\left(1 - \frac{d^{(12)}}{12}\right)^{12} = \left(1 - \tfrac{.06}{2}\right)^{2}$$

$$1 - \frac{d^{(12)}}{12} = \left(1 - \tfrac{.06}{2}\right)^{1/6}$$

$$d^{(12)} = 12\left[1 - \left(1 - \tfrac{.06}{2}\right)^{1/6}\right]$$

$$d^{(12)} = 0.060764049$$

$$d^{(12)} \doteq 6.08\%$$

Solution b Comparing the accumulated values of $1 for 1 year we obtain

$$\left(1 - \frac{d^{(12)}}{12}\right)^{-12} = \left(1 - \tfrac{.1}{365}\right)^{-365}$$

$$1 - \frac{d^{(12)}}{12} = \left(1 - \tfrac{.1}{365}\right)^{365/12}$$

$$d^{(12)} = 12\left[1 - \left(1 - \tfrac{.1}{365}\right)^{365/12}\right]$$

$$d^{(12)} = 0.099598075$$

$$d^{(12)} \doteq 9.96\%$$

∎

EXAMPLE 5 Find $d^{(4)}$ equivalent to an annual effective compound interest rate of 9%.

Solution Comparing discounted values of $1 due in 1 year we obtain

$$\left(1 - \frac{d^{(4)}}{4}\right)^4 = (1.09)^{-1}$$

$$1 - \frac{d^{(4)}}{4} = (1.09)^{-1/4}$$

$$d^{(4)} = 4\left[1 - (1.09)^{-1/4}\right]$$

$$d^{(4)} = 0.085256003$$

$$d^{(4)} \doteq 8.53\%$$

∎

Treasury Bills (or T-bills)

Treasury bills are popular short-term and low-risk securities issued by the Federal Government of Canada every Thursday with maturities 13 or 26 weeks later (91 days or 182 days later). There is an active secondary market in T-bills. Sales of T-bills are reported in the financial press in terms of a nominal interest rate (to the nearest .01%) convertible every 91 (or 182) days, and a quoted price per 100 of face value (to the nearest $.001).

EXAMPLE 6 A 91-day T-bill with a face value of $100 000 is purchased for $97 250. Find the 91-day effective interest and discount rates, the nominal interest rate convertible every 91 days, and the quoted price per 100 of face value.

Solution Interest earned for the 91-day period is $100 000 − 97 250 = 2750. The 91-day effective interest rate is $\frac{2750}{97\ 250} = 0.028277635$.
The 91-day effective discount rate is $\frac{2750}{100\ 000} = 0.0275$.
The nominal interest rate convertible every 91 days is

$$\left(\tfrac{365}{91}\right)\left(\tfrac{2750}{97\ 250}\right) = .113421283 \text{ or } 11.34\%.$$

The quoted price is 97.250.

∎

EXAMPLE 7 Show that the 91-day effective interest and discount rates of Example 6 are equivalent.

Solution We compare the accumulated values at the end of a 91-day period:
At the 91-day effective interest rate: $1 + \frac{2750}{97\ 250} = 1.028277635$.

At the 91-day effective discount rate: $\left(1 - \frac{2750}{100\ 000}\right)^{-1} = 1.028277635$.

∎

Exercise 2.10

Part A

1. Find the discounted value of $2000 due in 3 years at
   a) $d^{(2)} = 8\%$
   b) $d^{(12)} = 10.3\%$

2. Find the accumulated value of $3000 for 2 years at
   a) $d^{(52)} = 10\%$
   b) $d^{(365)} = 10\%$

3. Find the annual effective discount rate corresponding to
   a) $d^{(4)} = 8.2\%$
   b) $d^{(365)} = 6\%$

4. Find $d^{(4)}$ equivalent to
   a) $d^{(12)} = 9.8\%$
   b) $d^{(52)} = 8.5\%$
   c) $j_4 = 8\%$
   d) $j_\infty = 7.3\%$

5. Calculate the value at the end of 4 years of an $18 000 car, using an annual effective rate of depreciation of 30%. (Hint: This is the same as an annual effective rate of compound discount of 30%.)

6. A 91-day T-bill has a quoted price of 97.792. What is the corresponding nominal interest rate convertible every 91 days?

7. A 182-day T-bill has a nominal interest rate convertible every 182 days quoted at 9.23%. What is the corresponding quoted price?

8. A noninterest-bearing note of amount X is due in 3 months. A finance company calculates the value of the note today to be $3825. Find X under each of the following interest calculation methods:
   a) compound interest at an annual effective interest rate of 9%.
   b) compound discount at an annual effective discount rate of 9%.
   c) a simple interest rate of 9%.
   d) a simple discount rate of 9%.

9. At what
   a) annual effective rate of discount;
   b) nominal discount rate compounded monthly, will money double itself in 10 years?

Part B

1. Let $i = \dfrac{j_m}{m} = \dfrac{i^{(m)}}{m}$ be an effective interest rate per period

   $d = \dfrac{d^{(m)}}{m}$ be an effective discount rate per period

   $v = (1 + i)^{-1}$

   Show that:
   a) $v = 1 - d$
   b) $d = iv$
   c) $\left(1 + \dfrac{i^{(m)}}{m}\right)^m = \left(1 - \dfrac{d^{(m)}}{m}\right)^{-m}$

2. Find $i^{(4)}$ equivalent to $d^{(12)} = 6\%$ for 2 years followed by $d^{(365)} = 5\%$ for 2 years.

3. What is the difference between the annual effective rate of interest and the annual effective rate of discount corresponding to a nominal rate of interest of 12% per annum compounded monthly?

4. Show that
   a) $\lim\limits_{m \to \infty} \left(1 - \dfrac{d^{(m)}}{m}\right)^{-m} = e^{d^{(\infty)}}$
   b) $d^{(\infty)} = i^{(\infty)} = \delta$

5. Show that equivalent nominal interest and discount rates $i^{(m)}$ and $d^{(m)}$ satisfy the relationships

a) $d^{(m)} = \dfrac{i^{(m)}}{1 + \frac{i^{(m)}}{m}}$ and   b) $i^{(m)} = \dfrac{d^{(m)}}{1 - \frac{d^{(m)}}{m}}$

| *Section 2.11* | **Summary and Review Exercises** |

- Accumulated value of X at the end of n periods at the interest rate $i = \frac{j_m}{m}$ per interest period is given by $X(1 + i)^n$ where $(1 + i)^n$ is called an accumulation factor at compound interest.

- Discounted value of X due at the end of n periods at the interest rate $i = \frac{j_m}{m}$ per interest period is given by $X(1 + i)^{-n}$ where $(1 + i)^{-n}$ is called a discount factor at compound interest.

- Exact (theoretical) method of accumulating or discounting uses compound interest for the fractional part of an interest period.

  Approximate (practical) method of accumulating or discounting uses simple interest for the fractional part of an interest period.

- Equivalence of dated values at compound interest rate $i = \frac{j_m}{m}$:
  X is due on a given date is equivalent to Y due n periods later,
  if $Y = X(1 + i)^n$ or $X = Y(1 + i)^{-n}$.

Note: The above definition of equivalence of dated values at compound interest rate i satisfies the property of transitivity. At compound interest rate i, any choice of a comparison date to set up an equation of value will result in the same answer.

- Accumulated value of X at the end of t years at force of interest $\delta \equiv j_\infty$ is given by $Xe^{\delta t}$ where $e^{\delta t}$ is called an accumulation factor at continuous compounding.

- Discounted value of X due in t years at force of interest $\delta \equiv j_\infty$ is given by $Xe^{-\delta t}$ where $e^{-\delta t}$ is called a discount factor at continuous compounding.

- Discounted value of X due at the end of n periods at the discount rate $\frac{d^{(m)}}{m}$ per period is given by $X\left(1 - \frac{d^{(m)}}{m}\right)^n$ where $\left(1 - \frac{d^{(m)}}{m}\right)^n$ is called a discount factor at compound discount.

- Accumulated value of X due at the end of n periods at the discount rate $\frac{d^{(m)}}{m}$ per period is given by $X\left(1 - \frac{d^{(m)}}{m}\right)^{-n}$ where $\left(1 - \frac{d^{(m)}}{m}\right)^{-n}$ is called an accumulation factor at compound discount.

Review Exercises 2.11

1. Find the total value on June 1, 2002, of $1000 due on December 1, 1997, and $800 due on December 1, 2007, at $j_2 = 11.38\%$.

2. A person deposits $1500 into a mutual fund. If the fund earns 9.8% a year compounded daily for 10 years, what will be the accumulated value of the initial deposit?

3. On the birth of their first grandchild, the Guimonds bought a $100 savings bond that paid interest at $j_1 = 8\%$. How much money did their grandchild receive upon cashing this bond on her 20th birthday?

4. Find a) the accumulated value; b) the discounted value of $2000 for 14 months at $j_2 = 8\%$ using both the theoretical and practical method.

5. The XYZ company has had an increase in sales of 4% per annum. If sales in 2001 are $680 000, what would be the estimated sales for 2006?

6. Find the amount of interest earned between 5 and 10 years after the date of an investment of $100 if interest is paid semi-annually at $j_2 = 7\%$.

7. Ahmed buys goods worth $1500. He wants to pay $500 at the end of 3 months, $600 at the end of 6 months, and $300 at the end of 9 months. If the store charges $j_{12} = 21\%$ on the unpaid balance, what downpayment will be necessary?

8. A trust company offers guaranteed investment certificates paying $j_2 = 6\frac{3}{4}\%$, $j_4 = 6\frac{1}{4}\%$, or $j_{12} = 6\frac{1}{8}\%$. Rate the options from best to worst.

9. A note with face value $3000 is due without interest on November 21, 2005. On April 3, 2001 the holder of the note has it discounted at a bank charging $j_4 = 6\%$. What are the proceeds?

10. How long will it take $1000 to accumulate to $2500 at $j_{365} = 6\%$?

11. An investment fund advertises that it will triple your money in 10 years. What rate of interest, j_4, is implied?

12. A loan of $10 000, taken on January 1, 2000, is to be repaid on January 1, 2006. The debtor would like to pay $2000 on January 1, 2003, and make equal payments on January 1, 2005, and January 1, 2006. What will the size of these payments be if interest is assumed to be at $j_{12} = 15\frac{1}{4}\%$?

13. By what date will $1000 deposited on November 20, 2001, at $j_{365} = 12\frac{1}{2}\%$ be worth at least $1250?

14. To pay off a loan of $5000 at $j_{12} = 9\%$, Mrs. Leung agrees to make three payments in two, five and ten months respectively. The second payment is to be double the first, and the third payment is to be triple the first. What is the size of the first payment?

15. At what nominal rate compounded a) monthly; b) daily; c) continuously, will money triple in value in 10 years?

16. Paul has deposited $1000 in a savings account paying interest at $j_1 = 4.5\%$ and now finds that his deposits have accumulated to $1246.18. If he had been able to invest the $1000 over the same period in a guaranteed investment certificate paying interest at $j_1 = 6\%$, to what sum would his $1000 now have accumulated?

17. A piece of land can be purchased by paying $50 000 cash or $20 000 down and equal payments of $20 000 at the end of two and four years respectively. To pay cash, the buyer would have to withdraw money from an investment earning interest at rate $j_2 = 8\%$. Which option is better and by how much?

18. A note for $2000 dated October 6, 2001 is due with compound interest at $j_{12} = 8\%$, two years after the date. On January 16, 2003, the holder of the note has it discounted by a lender who charges $j_4 = 9\%$. Find the proceeds and the compound discount.

19. Jackie invested $500 in an investment fund. Find the accumulated value of her investment at the end of 6 years, if the interest rate was $j_2 = 12.3\%$ for the first two years, $j_{12} = 11.2\%$ for the next three years and $j_{365} = 10\%$ for the last year.

20. Calculate the present value of $2000 due in $5\frac{1}{2}$ years if $j_4 = 8\%$ for the first 2 years and $j_2 = 10\%$ thereafter.

21. Calculate the accumulated value of $1000 at the end of 5 years using:
    a) annual effective interest rate of 6%;
    b) nominal interest rate of 6% compounded monthly;
    c) force of interest 6% per annum;
    d) nominal discount rate of 6% compounded monthly;
    e) annual effective discount rate of 6%.

22. Julie bought $2000 in Canada Savings Bonds that paid interest at $j_1 = 5\%$ with interest accruing at a simple interest rate for each month before November 1 (i.e., for any partial year). How much will she receive for the bonds if they are cashed in 5 years and 3 months after the date of issue?

23. $5000 is invested at $j_{12} = 11.8\%$ on May 10, 2001. What is the accumulated value on April 5, 2003, if the contract states that a) simple interest is calculated for a fraction of a month and b) compound interest is calculated for a fraction of a month?

24. An obligation of $2000 is due on February 4, 2003. Find the value of this obligation on December 13, 2000, at 10% compounded quarterly, if simple interest is used for part of an interest conversion period.

25. You borrow $4000 now and agree to pay X in 3 months, $2X$ in 7 months, and $2X$ in 12 months. Find X, if interest is at 15% compounded a) monthly; b) continuously.

26. How long will it take to double your money at a nominal interest rate of 10% compounded a) daily; b) continuously?

Chapter Three

Simple Annuities

Definitions

An **annuity** is a sequence of periodic payments, usually equal, made at equal intervals of time. Premiums on insurance, mortgage payments, interest payments on bonds, payments of rent, payments on instalment purchases, deposits to and withdrawals from mutual fund accounts, and dividends are just a few examples of annuities.

The time between successive payments of an annuity is called the **payment interval**. The time from the beginning of the first payment interval to the end of the last payment interval is called the **term** of an annuity. When the term of an annuity is fixed, i.e., the dates of the first and the last payments are fixed, the annuity is called an **annuity certain**. When the term of the annuity depends on some uncertain event, the annuity is called a **contingent annuity**. Bond interest payments form an annuity certain; life-insurance premiums form a contingent annuity (they cease with the death of the insured). Unless otherwise specified, the word annuity will refer to an annuity certain.

When the payments are made at the ends of the payment intervals, the annuity is called an **ordinary annuity** (or **immediate annuity**). When the payments are made at the beginning of the payment intervals, the annuity is called an **annuity due**. A **deferred annuity** is an annuity whose first payment is due at some later date.

When the payment interval and interest conversion period coincide, the annuity is called a **simple annuity**, otherwise, it is a **general annuity**.

We define the **accumulated value** of an annuity as the equivalent dated value of the set of payments due at the end of the term. Similarly, the **discounted value** of an annuity is defined as the equivalent dated value of the set of payments due at the beginning of the term.

We shall use the following notation:

R = the periodic payment of the annuity.

n = the number of interest conversion periods during the term of an annuity (in the case of a simple annuity, n equals the total number of payments).

i = interest rate per conversion period (assume $i > 0$).
S = the accumulated value, or the amount of an annuity.
A = the discounted value, or the present value of an annuity.

Section 3.2

Accumulated Value of an Ordinary Simple Annuity

The accumulated value S of an ordinary simple annuity is defined as the equivalent dated value of the set of payments due at the end of the term, i.e., *on the date of the last payment*. Below we display an ordinary simple annuity on a time diagram with the interest period as the unit of measure.

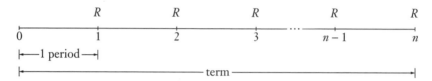

We can calculate the accumulated value S by repeated application of the compound interest formula as shown in the following example.

EXAMPLE 1 Find the accumulated value of an ordinary simple annuity consisting of 4 quarterly payments of $250 each if money is worth 12% per annum compounded quarterly.

Solution We arrange the data on a time diagram below.

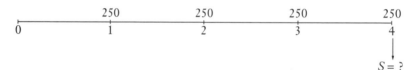

To obtain S, we write an equation of value using the end of the term as the focal date. This gives

$$S = 250 + 250(1.03)^1 + 250(1.03)^2 + 250(1.03)^3$$
$$= 250 + 257.50 + 265.23 + 273.18 = \$1045.91$$

∎

Now we develop a formula for the accumulated value S of an ordinary simple annuity using the sum of a geometric progression. Geometric progressions are dealt with in Appendix 2.

Let us consider an ordinary simple annuity of n payments of $1 each as shown on a time diagram below.

Let us denote the accumulated value of this annuity $s_{\overline{n}|i}$ (read "s angle n at i"). To obtain $s_{\overline{n}|i}$ we write an equation of value at the end of the term, accumulating each \$1 payment to the date of the last payment.

$$s_{\overline{n}|i} = 1 + 1(1 + i)^1 + 1(1 + i)^2 + 1(1 + i)^3 + \dots + 1(1 + i)^{n-1}$$

The expression on the right side of the equal sign in the above equation is a geometric progression of n terms whose first term is $t_1 = 1$ and whose common ratio is $r = (1 + i) > 1$. Then, applying the formula for the sum of a geometric progression, we obtain

$$s_{\overline{n}|i} = t_1 \frac{r^n - 1}{r - 1} = 1 \frac{(1 + i)^n - 1}{(1 + i) - 1} = \frac{(1 + i)^n - 1}{i}$$

The factor $s_{\overline{n}|i} = \frac{(1+i)^n - 1}{i}$ is called an **accumulation factor for n payments**, or the **accumulated value of \$1 per period**.

To obtain the accumulated value S of an ordinary simple annuity of n payments of \$R each, we simply multiply R by $s_{\overline{n}|i}$. Thus the basic formula for the accumulated value S of an ordinary simple annuity is

$$\boxed{S = Rs_{\overline{n}|i} = R\frac{(1 + i)^n - 1}{i}} \qquad \text{(10)}$$

Using **(10)** in Example 1 above, we calculate

$$S = 250\, s_{\overline{4}|.03} = 250\frac{(1.03)^4 - 1}{.03} = \$1045.91$$

CALCULATION TIP: In this textbook we calculate the factor $s_{\overline{n}|i} = \frac{(1+i)^n - 1}{i}$ directly on a calculator using all the digits of the display of the calculator to achieve the highest possible accuracy in our results.

EXAMPLE 2 A couple deposits \$500 every 3 months into a savings account that pays interest at $j_4 = 4\%$. They made the first deposit on March 1, 2002. How much money will they have in the account just after they make their deposit on September 1, 2006?

Solution We arrange the data on a time diagram below, noting that the ordinary annuity starts one interest period before the 1st payment, i.e., on December 1, 2001.

	500	500	500	500	500
0	1	2	3	18	19
Dec. 1	March 1	June 1	Sept. 1	June 1	Sept. 1
2001	2002	2002	2002	2006	2006

The time elapsed from December 1, 2001, to September 1, 2006, is exactly 4 years 9 months, or 19 quarters of a year. Thus we calculate the accumulated value S of an ordinary simple annuity of 19 payments of $500 each at $j_4 = 4\%$ or $i = 0.01$

$$S = 500 \, s_{\overline{19}|.01} = 500\frac{(1.01)^{19} - 1}{.01} = \$10\,405.45$$

∎

EXAMPLE 3 Mrs. Simpson has deposited $1000 at the end of each year into her Registered Retirement Savings Plan for the last 10 years. Her investments earned $j_1 = 9\%$ for the first 4 years and $j_1 = 9\frac{1}{2}\%$ for the last 6 years. What is the value of her RRSP 5 years after the last deposit, assuming that her RRSP earns $j_1 = 9\frac{1}{2}\%$ for the 5-year period after the last deposit?

Solution We arrange the data on a time diagram below.

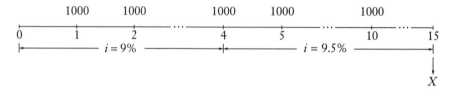

The equation of value at the end of 15 years gives the accumulated value X of all deposits.

$$X = 1000s_{\overline{4}|.09}(1.095)^{11} + 1000s_{\overline{6}|.095}(1.095)^5$$
$$= 1000\frac{(1.09)^4 - 1}{.09}(1.095)^{11} + 1000\frac{(1.095)^6 - 1}{.095}(1.095)^5$$
$$= 12\,409.91 + 11\,993.90 = \$24\,403.81$$

The value of the RRSP 5 years after the last deposit is $24 403.81.

∎

When equation **(10)** is solved for R, we obtain

$$\boxed{R = \frac{S}{s_{\overline{n}|i}} = \frac{S}{\frac{(1+i)^n - 1}{i}}}$$

as the periodic payment of an ordinary simple annuity whose accumulated value S is given.

EXAMPLE 4 A man wants to accumulate a $200 000 retirement fund. He made the first deposit on March 1, 1994, and his plan calls for the last deposit to be made on September 1, 2015. Find the size of each deposit if
a) he makes the deposits semi-annually in a fund that pays $12\frac{1}{2}\%$ per annum compounded semi-annually.
b) he makes the deposits monthly in a fund that pays $12\frac{1}{2}\%$ per annum compounded monthly.

Solution a Arranging the data on a time diagram below we have $S = 200\ 000$, $i = \frac{.125}{2} = .0625$, $n = 44$.

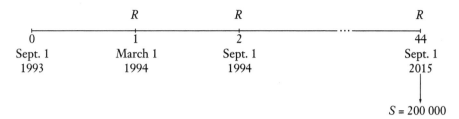

We calculate $R = \dfrac{200\ 000}{s_{\overline{44}|.0625}} = \dfrac{200\ 000}{\frac{(1.0625)^{44}-1}{.0625}} = \$932.58$

Semi-annual deposits of \$932.58 will accumulate at $j_2 = 12\frac{1}{2}\%$ to \$200 000 by September 1, 2015.

Solution b Arranging the data on a time diagram below we have $S = 200\ 000$, $i = \frac{.125}{12}$, $n = 259$.

	R	R		R
0	1	2	...	259
Feb. 1	March 1	April 1		Sept. 1
1994	1994	1994		2015

$S = 200\ 000$

We calculate $R = \dfrac{200\ 000}{s_{\overline{259}|i}} = \dfrac{200\ 000}{\frac{(1+i)^{259}-1}{i}} = \$152.70$

Monthly deposits of \$152.70 will accumulate at $j_{12} = 12\frac{1}{2}\%$ to \$200 000 by September 1, 2015.

■

Exercise 3.2

Part A

1. Find the accumulated value of an ordinary simple annuity of \$2000 per year for 5 years if money is worth a) $j_1 = 9\%$, b) $j_1 = 12\frac{1}{2}\%$.

2. Find the accumulated value of an annuity of \$500 at the end of each month for 4 years at 9% per annum payable monthly.

3. Lauren deposits \$100 every month in a savings account that pays interest at $j_{12} = 4.5\%$. If she makes her first deposit on July 1, 2003, how much will she have in her account just after she makes her deposit on January 1, 2006?

4. One hundred dollars at the end of each year for 5 years is equivalent to what single payment at the end of 5 years if interest is at a) $j_1 = 10\%$ and at b) $j_1 = 6.3\%$?

5. Jason is repaying a debt with payments of $120 a month. If he misses his payments for June, July, August and September, what payment will be required in October to put him back on schedule if interest is at 12% per annum convertible monthly?

6. Find the accumulated value of an annuity of $50 a month for 25 years if interest is a) 8% per annum compounded monthly, b) $j_{12} = 11\%$.

7. Find the accumulated value of annual deposits of $1000 each immediately after the 10th deposit if the deposits earned 5% per annum in the first 5 years and 6% per annum in the last 5 years.

8. Michael deposits $1000 at the end of each year for 5 years and then $2000 at the end of each year for 8 years. Find the accumulated value of these deposits at the end of 13 years if interest is $j_1 = 7\%$.

9. Ashley deposits $500 into an investment fund each January 1 starting in 1999 and continuing to 2008 inclusive. If the fund has an average annual growth rate of $j_1 = 10\%$, how much will be in her account on January 1, 2013?

10. Mr. Juneau has deposited $800 at the end of each year into a RRSP investment fund for the last 10 years. His investments earned $j_1 = 10\%$ for the first 7 years and $j_1 = 9\%$ for the last 3 years. How much money does he have in his account 10 years after his last deposit if rates of return have remained level at $j_1 = 9\%$?

11. Rebecca has deposited $80 at the end of each month for 7 years. For the first 5 years the deposits earned 6% compounded monthly. After 5 years they earned 4.5% compounded monthly. Find the value of the annuity after a) 7 years, b) 10 years.

12. What quarterly deposits should be made into a savings account paying $j_4 = 4\%$ to accumulate $10 000 at the end of 10 years?

13. It is estimated that a machine will need replacing 10 years from now at a cost of $80 000. How much must be set aside each year to provide that money if the company's savings earn interest at an 8% annual effective rate?

14. Jack has made semi-annual deposits of $500 for 5 years into an investment fund growing at $j_2 = 13\frac{1}{4}\%$. What semi-annual deposits for the next 2 years will bring the fund up to $10 000?

15. Marie deposited $150 at the end of each month into an account earning $j_{12} = 8\%$. She made these deposits for 14 years except that in the fifth year she was unable to make any deposits. Find the value of the account two years after the last deposit.

Part B

1. Jane opened an investment account with a deposit of $1000 on January 1, 1997. She then made monthly deposits of $200 for 10 years (first deposit February 1, 1997). She then made monthly withdrawals of $300 for 5 years (first withdrawal February 1, 2007). Find the balance in this account just after the last $300 withdrawal (i.e., January 1, 2012) if $j_{12} = 6\%$.

2. Frank has deposited $1000 at the end of each year into his Registered Retirement Savings Plan for the last 10 years. His deposits earned $j_1 = 9\%$ for the first 3 years, $j_1 = 11\%$ for the next 4 years and $j_1 = 8\%$ for the last 3 years. What is the value of his plan after his last deposit?

3. Prove
   a) $(1 + i)s_{\overline{m}|i} = s_{\overline{n+1}|i} - 1$
   b) $s_{\overline{m+n}|i} = s_{\overline{m}|i} + (1 + i)^m s_{\overline{m}|i} = (1 + i)^n s_{\overline{m}|i} + s_{\overline{m}|i}$

   Illustrate both a) and b) using a time diagram.

4. If $s_{\overline{m}|i} = 10$ and $i = 10\%$, find $s_{\overline{n+2}|i}$ and $s_{\overline{2n}|i}$.

5. Beginning June 30, 2000, and continuing every three months until December 31, 2004, Albert deposits $300 into a mutual fund account. Starting September 30, 2005, he makes quarterly withdrawals of $500. What is Albert's balance after the withdrawal on June 30, 2007, if growth of the mutual fund is at $j_4 = 8\%$ until March 31, 2003 and $j_4 = 6\%$ afterward?

6. It is desired to check a column of values of $s_{\overline{m}|i}$ from $n = 20$ through $n = 40$ by verifying their sum by means of an independent formula. Derive an expression for the sum of these values.

7. Show that $s_{\overline{m}|i} = n + \frac{n(n-1)}{1 \cdot 2}i + \frac{n(n-1)(n-2)}{1 \cdot 2 \cdot 3}i^2 + \dots$

8. a) Show that $(1 + i)^n = 1 + i\, s_{\overline{m}|i}$.
   b) Verbally interpret this formula.

9. Find an expression for an accumulation factor for n equal payments of $1 assuming simple interest at rate i per payment period.

10. Prove that $\dfrac{s_{\overline{2n}|}}{s_{\overline{m}|}} + \dfrac{s_{\overline{m}|}}{s_{\overline{2n}|}} - \dfrac{s_{\overline{3n}|}}{s_{\overline{2n}|}} = 1$

11. a) Barbara wants to accumulate $10 000 by the end of 10 years. She starts making quarterly deposits in her investment account, which pays $j_4 = 8\%$. Find the size of these deposits.
   b) After 4 years, the rate of return changes to $j_4 = 6\%$. Find the size of the quarterly deposits now required if the $10 000 goal is to be met.

12. George wants to accumulate $7000 in a fund at the end of 10 years. He deposits $300 at the end of each year for the first 5 years and then $(300 + x)$ at the end of each year for the next 5 years. Find x if $j_1 = 13\frac{1}{4}\%$.

13. You want to accumulate $100 000 at the end of 20 years. You deposit $1000 at the end of each year for the first 10 years and $(1000 + x)$ at the end of each year for the second 10 years. Rate of return is $j_1 = 10\frac{1}{4}\%$.
   a) Find x.
   b) If the last 4 payments of $1000 (at the end of years 7 through 10) were missed, what would be the value of x?

14. Beginning on June 1, 2000, and continuing until December 1, 2005, a company will need $250 000 semi-annually to retire a series of bonds. What equal semi-annual deposits in a fund paying $j_2 = 10\%$ beginning on June 1, 1995, and continuing until December 1, 2005, are necessary to retire the bonds as they fall due?

Section 3.3 — Discounted Value of an Ordinary Simple Annuity

The discounted value A of an ordinary simple annuity is defined as the equivalent dated value of the set of payments due at the beginning of the term, i.e., *1 period before the first payment*. Below we display an ordinary simple annuity on a time diagram.

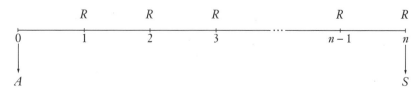

It is possible to derive the discounted value A in at least two different ways. Note first that A is the sum of a series of single discounted payments of \$$R$ each. That is,

$$A = R(1 + i)^{-1} + R(1 + i)^{-2} + \cdots + R(1 + i)^{-n}$$

The value of A can be found as the sum of a geometric progression of n terms whose first term is $t_1 = R(1 + i)^{-1}$ and whose common ratio is $r = (1 + i)^{-1}$. We obtain

$$A = t_1 \frac{1 - r^n}{1 - r} = R(1 + i)^{-1} \frac{1 - (1 + i)^{-n}}{1 - (1 + i)^{-1}}$$

If we then multiply by $\frac{1+i}{1+i}$, we get

$$A = R \frac{1 - (1 + i)^{-n}}{(1 + i) - 1} = R \frac{1 - (1 + i)^{-n}}{i}$$

Alternatively, we note that A and S are both dated values of the same set of payments and thus they can be made equivalent to each other using

$$A = S(1 + i)^{-n}$$

Substituting for S from equation **(10)** we have

$$A = Rs_{\overline{n}|i}(1 + i)^{-n} = R \frac{(1 + i)^n - 1}{i}(1 + i)^{-n} = R \frac{1 - (1 + i)^{-n}}{i}$$

If we set

$$a_{\overline{n}|i} = \frac{1 - (1 + i)^{-n}}{i}$$

we obtain

$$\boxed{A = Ra_{\overline{n}|i} = R \frac{1 - (1 + i)^{-n}}{i}} \tag{11}$$

The factor $a_{\overline{n}|i} = \frac{1-(1+i)^{-n}}{i}$ (read "a angle n at i") is called a **discount factor for n payments**, or **the discounted value of \$1 per period.**

CALCULATION TIP: We calculate the factor $a_{\overline{n}|i} = \frac{1-(1+i)^{-n}}{i}$ directly on a calculator and use all the digits of the display of the calculator to achieve the highest possible accuracy in our results.

EXAMPLE 1   How much money is needed now to provide \$500 at the end of each month (first payment 1 month from now) for 15 years if money earns interest at
   a) $j_{12} = 7\%$ for the next 15 years;
   b) $j_{12} = 6\%$ for the first 8 years and $j_{12} = 9\%$ for the next 7 years?

Solution a   We have $R = 500$, $i = \frac{.07}{12}$, $n = 15 \times 12 = 180$ and using **(11)** we calculate

$$A = 500 \, a_{\overline{180}|i} = 500\frac{1 - (1 + i)^{-180}}{i} = \$55\ 627.98$$

Note that the face value of 180 payments of $500 each is $90 000 whereas only $55 627.98 is required at present to furnish these payments.

Solution b   We arrange the data on a time diagram below.

The equation of value at present gives the discounted value X of all payments

$$X = 500a_{\overline{96}|.005} + 500a_{\overline{84}|.0075}(1.005)^{-96}$$

$$= 500\frac{1 - (1.005)^{-96}}{.005} + 500\frac{1 - (1.0075)^{-84}}{.0075}(1.005)^{-96}$$

$$= 38\ 047.61 + 19\ 252.93 = \$57\ 300.54 \ .$$

∎

EXAMPLE 2   Mr. Li signed a contract that calls for a down payment of $1500 and for the payment of $200 a month for 10 years. The interest rate is 12% per annum compounded monthly.

  a) What is the cash value of the contract?
  b) If Mr. Li missed the first 8 payments, what must he pay at the time the ninth payment is due to bring himself up to date?
  c) If Mr. Li missed the first 8 payments, what must he pay at the time the ninth payment is due to discharge his indebtedness completely?
  d) If, at the beginning of the 5th year (just after the 48th payment is made), the contract is sold to a buyer at a price that will yield $j_{12} = 15\%$, what does the buyer pay?

Solution a   Let C denote the cash value of the contract. Then C is $1500 plus the discounted value of 120 monthly payments of $200 each

We calculate

$$C = 1500 + 200a_{\overline{120}|.01} = 1500 + 200\frac{1 - (1.01)^{-120}}{.01}$$
$$= 1500 + 13\ 940.10 = \$15\ 440.10$$

Solution b   Let X denote the required payment. Mr. Li must pay the accumulated value of the first 9 payments at the time the 9th payment is due.

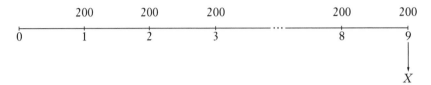

We calculate

$$X = 200\ s_{\overline{9}|.01} = 200\frac{(1.01)^9 - 1}{.01} = \$1873.71$$

Solution c   Let Y denote the required payment. Mr. Li must pay the accumulated value of the first 9 payments plus the discounted value of the last $120 - 9 = 111$ payments at the time the 9th payment is due, in order to discharge his indebtedness completely.

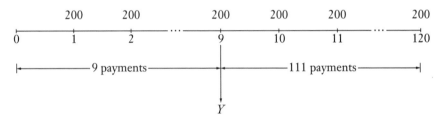

$$Y = 200s_{\overline{9}|.01} + 200a_{\overline{111}|.01}$$
$$= 200\frac{(1.01)^9 - 1}{.01} + 200\frac{1 - (1.01)^{-111}}{.01}$$
$$= 1873.71 + 13\ 372.38 = \$15\ 246.09$$

Alternatively, we can calculate Y by finding the discounted value of all 120 payments and then accumulating it to the end of the 9th period, i.e.,

$$Y = 200a_{\overline{120}|.01}(1.01)^9 = 200\frac{1 - (1.01)^{-120}}{.01}(1.01)^9 = \$15\ 246.09$$

Solution d   Let Z be the price of the contract the buyer must pay. Then Z is the discounted value of the remaining $120 - 48 = 72$ payments at $j_{12} = 15\%$. We calculate

$$Z = 200a_{\overline{72}|.0125} = 200\frac{1 - (1.0125)^{-72}}{.0125} = \$9458.49$$ ∎

The concept of the discounted value has important applications in **investment decision making** and will be discussed further in Chapter 7. In making investment decisions such as which asset to acquire or whether to buy or lease, we should compare the **net present value** (or net discounted value) of the cash flows of the different alternatives. This principle of comparison of the net present value of cash flows applies whether the asset is an investment in a machine, real estate, an entire business or an investment in bonds.

EXAMPLE 3   A company is considering the possibility of acquiring new computer equipment for $400 000 cash. The salvage value is estimated to be $50 000 at the end of the 6-year life of the equipment. Maintenance costs will be $4000 per month, payable at the end of each month. The company could lease the equipment for $12 000 per month, payable at the end of each month. Under the 6-year lease agreement the lessor would pay the maintenance costs. If the company can earn $j_{12} = 12\%$ on its capital, advise the company whether to buy or to lease.

Solution   We consider two alternatives: BUY or LEASE, and calculate the net present value (NPV) of the cash flows for each alternative.

NPV = Present Value of Cash Inflows − Present Value of Cash Outflows

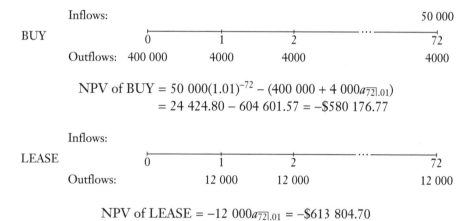

$$\text{NPV of BUY} = 50\ 000(1.01)^{-72} - (400\ 000 + 4\ 000 a_{\overline{72}|.01})$$
$$= 24\ 424.80 - 604\ 601.57 = -\$580\ 176.77$$

$$\text{NPV of LEASE} = -12\ 000 a_{\overline{72}|.01} = -\$613\ 804.70$$

The negative values of NPV represent costs to the company and since NPV of the costs to buy are less than NPV of the costs to lease, the company should buy the equipment.

∎

When equation **(11)** is solved for R, we obtain

$$R = \frac{A}{a_{\overline{n}|i}} = \frac{A}{\dfrac{1 - (1 + i)^{-n}}{i}}$$

as the periodic payment of an ordinary simple annuity whose discounted value A is given.

EXAMPLE 4   A car selling for $22 500 may be purchased by paying $2500 down and the balance in equal monthly payments for 5 years. Find these monthly payments at a) $j_{12} = 12\%$; b) $j_{12} = 4.9\%$

Solution a   We have $A = 20\ 000$, $i = \frac{.12}{12} = .01$, $n = 60$ and calculate

$$R = \frac{20\ 000}{a_{\overline{60}|.01}} = \frac{20\ 000}{\frac{1-(1.01)^{-60}}{.01}} = \$444.89$$

Solution b   Using $A = 20\ 000$, $i = \frac{.049}{12}$, $n = 60$

$$R = \frac{20\ 000}{a_{\overline{60}|i}} = \frac{20\ 000}{\frac{1-(1+i)^{-60}}{i}} = \$376.51$$

∎

EXAMPLE 5   With the death of the insured on September 1, 2002, a life insurance policy pays out $80 000 as a death benefit. The beneficiary is to receive monthly payments, with the first payment on October 1, 2002. Find the size of the monthly payments and the date of the concluding payment, if interest is earned at $j_{12} = 8\%$ and the beneficiary is to receive 120 payments.

Solution   We have $A = 80\ 000$, $i = \frac{.08}{12}$, $n = 120$ and calculate

$$R = \frac{80\ 000}{a_{\overline{120}|i}} = \frac{80\ 000}{\frac{1-(1+i)^{-120}}{i}} = \$970.62$$

The date of the last payment is September 1, 2012.

∎

Exercise 3.3

Part A

1. Find the discounted value of an ordinary simple annuity of $1000 per year for 5 years if money is worth a) $j_1 = 8\%$, b) $j_1 = 16\%$, c) $j_1 = 12.79\%$.
2. Find the discounted value of an annuity of $380 at the end of each month for 3 years at a) $j_{12} = 8\%$, b) $j_{12} = 12\%$, c) $j_{12} = 10.38\%$.
3. Mr. Goldberg wants to save enough money to send his two children to university. Since they are three years apart in age, he wants to have a sum of money that will provide $6000 a year for six years. Find the single sum required one year before the first withdrawal if interest is at $j_1 = 8\%$.
4. Diana has an insurance policy whose cash value at age 65 will provide payments of $1500 a year for 15 years, first payment at age 66. If the insurance company pays $j_1 = 5\%$ on its funds, what is the cash value at age 65?
5. Ahmed buys a used car by paying $500 down plus $180 a month for 3 years. What was the price of the car if the interest rate on the loan is $j_{12} = 18\%$?
6. An annuity pays R per month starting February 1, 2002, and ending January 1, 2005 (inclusive). If the value of this annuity on January 1, 2005, is $8000 and $j_{12} = 11\%$, what is its value on January 1, 2002?

7. Maria receives an inheritance of $700 each half year for 15 years, the first payment being made in 6 months. If money is worth $j_2 = 10\%$, what is the cash value of this inheritance?

8. Find the discounted value of annual payments of $1000 each over 10 years if interest is $j_1 = 11\%$ for the first 4 years and $j_1 = 13\%$ for the last 6 years.

9. An annuity pays $2000 at the end of each year for 5 years and then $1000 at the end of each year for the next 8 years. Find the discounted value of these payments if $j_1 = 10\%$.

10. An annuity pays 10 annual payments of $2000 each starting January 1, 2006. Find the discounted value of these payments on January 1, 2003, if $j_1 = 7\%$.

11. A contract calls for payments of $250 a month for 10 years. At the beginning of the 5th year (just after the 48th payment is made) the contract is sold to a buyer at a price that will yield $j_{12} = 14\%$. What does the buyer pay?

12. A used car is purchased for $2000 down and $200 a month for 6 years. Interest is at $j_{12} = 10\%$.
    a) Determine the price of the car.
    b) If the first 4 payments are missed, what payment at the time of the 5th payment will update the payments?
    c) Assuming no payments are missed, what single payment at the end of 2 years will completely pay off the debt?
    d) After 27 payments have been made, the contract is sold to a buyer who wishes to yield $j_{12} = 12\%$. Determine the sale price.

13. An insurance policy is worth $10 000 at age 65. What monthly annuity will this fund provide for 15 years if the insurance company pays interest at $j_{12} = 6\%$?

14. On the birth of their first child a couple puts $2000 in an investment account earning $j_1 = 11\%$. This fund is used to pay university fees and allows for three withdrawals corresponding to the 18th through 20th birthdays. What size are these withdrawals?

15. A television set worth $780 may be purchased by paying $80 down and the balance in monthly instalments for 2 years. Find the size of the monthly instalments if $j_{12} = 15\%$.

16. A family needs to borrow $5000 for some home renovations. The loan is to be repaid with monthly payments over 5 years. If they go to a finance company, the interest rate will be $j_{12} = 21\%$; if they use their credit card, the interest rate will be $j_{12} = 18\%$; and if they go to the bank, the rate will be $j_{12} = 15\%$. Find the respective monthly payments and the total of all payments for each loan.

17. At age 65 Mrs. Bergeron takes her life savings of $120 000 and buys a 15-year annuity certain with monthly payments. Find the size of these payments a) at 6% compounded monthly; b) at 9% compounded monthly.

18. A person buys a boat with a cash price of $4500. She pays $500 down and the balance is financed at $j_{12} = 14.79\%$. If she is to make 24 equal monthly payments, what will be the size of each payment?

Part B

1. Prove
    a) $(1 + i)a_{\overline{m}|i} = a_{\overline{n-1}|i} + 1$
    b) $\frac{1}{s_{\overline{m}|i}} + i = \frac{1}{a_{\overline{m}|i}}$
    c) $a_{\overline{m+n}|i} = a_{\overline{m}|i} + (1 + i)^{-m}a_{\overline{n}|i} = (1 + i)^{-n}a_{\overline{m}|i} + a_{\overline{n}|i}$

2. If $a_{\overline{m}|i} = 10$ and $i = .08$, find $s_{\overline{m}|i}$ and $a_{\overline{2n}|i}$.

3. Explain logically why $1 = ia_{\overline{m}|i} + (1 + i)^{-n}$ and illustrate on a time diagram.

4. If $a_{\overline{2n}|i} = 1.6a_{\overline{m}|i}$ and $i = 10\%$, find $s_{\overline{2n}|i}$.

5. If $a_{\overline{m}|i} = 6$ when $i = \frac{1}{9}$, find $a_{\overline{n+2}|i}$.

6. Find an expression for a discounted factor for n equal payments of \$1 assuming simple interest at rate i per payment period.

7. If money doubles itself in n years at rate i, show that $a_{\overline{m}|i}$, $a_{\overline{2n}|i}$, and $s_{\overline{m}|i}$ are in arithmetic progression.

8. Prove that $(1 - ia_{\overline{m}|i})$, 1, $(is_{\overline{m}|i} + 1)$ are in geometric progression.

9. If $X(s_{\overline{2n}|i} + a_{\overline{2n}|i}) = (s_{\overline{3n}|i} + a_{\overline{m}|i})$, what is the value of X?

10. Mrs. Cheung signed a contract that calls for payments of \$150 a month for 5 years. The interest rate is $j_{12} = 15\%$.
    a) What is the cash value of the contract?
    b) If Mrs. Cheung missed making the first 6 payments, what must she pay at the time of the 7th payment to be fully up to date?
    c) If Mrs. Cheung missed the first 6 payments, what must she pay at the time of the 7th payment to fully discharge all indebtedness? (Can you find the answer to c) directly from the answer to a)?)
    d) If, at the beginning of the 3rd year (after 24 payments have been made) the contract is sold to a buyer at a price that will yield $j_{12} = 18\%$, what does the buyer pay? Is this value larger or smaller than the discounted value of Mrs. Cheung's indebtedness at that time?

11. An annuity pays \$200 at the end of each month for 2 years, then \$300 at the end of each month for the next year and then \$400 at the end of each month for the next 2 years. Find the discounted value of these payments at $j_{12} = 10\%$.

12. An oil company requires an arctic drilling machine and is deciding whether to purchase it for \$2 000 000 cash or to lease it for \$240 000, payable at the end of each half-year. The salvage value is \$100 000 at the end of the machine's 6-year life. Maintenance costs are \$10 000 each 6 months, but payable by the lessor, if the machine is leased. If the company can earn 10% on its capital, compounded semi-annually, advise the company whether to lease or buy.

13. The Ace Manufacturing Company is considering the purchase of two machines. Machine A costs \$200 000 and Machine B costs \$400 000. The machines are projected to have a life of five years and to yield the following revenues:

	Cash Revenue	
End of Year	Machine A	Machine B
1	none	\$90 000
2	\$100 000	90 000
3	100 000	90 000
4	100 000	90 000
5	100 000	300 000

The market rate of interest is 14% per annum compounded yearly. Which machine should the company purchase?

14. A man aged 30 wishes to accumulate a fund for retirement by depositing \$1000 at the end of each year for the next 35 years. Starting on his 66th birthday he will make 15 annual withdrawals of equal amount. Find the amount of each withdrawal if $j_1 = 10\%$ throughout.

15. To prepare for early retirement, a self-employed consultant makes deposits of $5500 into her Registered Retirement Savings Plan each year for 20 years, starting on her 31st birthday. When she is 51 she wishes to draw out 30 equal annual payments. What is the size of each withdrawal if interest is at $j_1 = 12\%$ until the 10th deposit and at $j_1 = 11\%$ for the remaining time?

16. Mr. Horvath has been accumulating a retirement fund at an annual effective rate of 9% that will provide him with an income of $12 000 per year for 20 years, the first payment on his 65th birthday. If he now wishes to reduce the number of retirement payments to 15, what should he receive annually?

17. Today Wilma purchased a cottage and a boat worth a combined present value of $65 000. To purchase the cottage, Wilma agreed to pay $10 000 down and $350 at the end of each month for 20 years. To buy the boat, Wilma was also required to pay $1000 down and $R at the end of each month for 4 years. Find R if $j_{12} = 12\%$.

Section 3.4	# Other Simple Annuities

In this section we introduce other simple annuities that can be treated as minor modifications of ordinary annuities.

Annuities Due

An **annuity due** is an annuity whose periodic payments are due at the beginning of each payment interval. The term of an annuity due starts at the time of the first payment and ends on the payment period after the date of the last payment. The diagram below shows the simple case (payment intervals and interest periods coincide) of an annuity due of n payments.

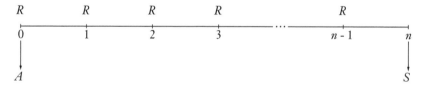

It is easy to recognize an annuity due as a "slipped" ordinary annuity. Thus we can simply write the formulas for the accumulated value S and discounted value A of an annuity due, by adjusting equations **(10)** and **(11)** from Sections 3.1 and 3.2.

Since the accumulated value of an annuity was defined as the equivalent dated value of the payments at the end of the term, it means that the accumulated value S of an annuity due is an equivalent value *due one period after the last payment.*

The accumulated value of the payments at the end of the $(n - 1)$th period is $Rs_{\overline{n}|i}$. We then accumulate $Rs_{\overline{n}|i}$ for 1 interest period, to obtain

$$\boxed{S = Rs_{\overline{n}|i}(1 + i)}$$

To find A, we recall that the discounted value of an annuity was defined as the equivalent dated value of the payments at the beginning of the term.

The discounted value of the payments 1 period before the 1st payment is $Ra_{\overline{n}|i}$. We then accumulate $Ra_{\overline{n}|i}$ for 1 interest period to obtain

$$\boxed{A = Ra_{\overline{n}|i}(1 + i)}$$

EXAMPLE 1   Mary Jones deposits $100 at the beginning of each month for 3 years into an account paying $j_{12} = 4.5\%$. How much is in her account at the end of 3 years?

Solution   We arrange the data on a time diagram below.

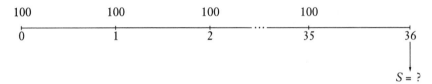

We have $R = 100$, $i = \frac{.045}{12} = .00375$, $n = 36$ and calculate

$$S = 100s_{\overline{36}|.00375}(1.00375) = \$3861.03$$

■

EXAMPLE 2   The monthly rent for a townhouse is $820 payable at the beginning of each month. If money is worth $j_{12} = 9\%$,
  a) what is the equivalent yearly rentable payable in advance,
  b) what is the cash equivalent of 5 years of rent?

Solution a   We calculate the discounted value A of an annuity due of 12 payments of $820 each at $j_{12} = 9\%$

$$A = 820\,a_{\overline{12}|.0075}(1.0075) = \$9446.95$$

Solution b   We calculate the discounted value A of an annuity due of 60 payments of $820 each at $j_{12} = 9\%$

$$A = 820\,a_{\overline{60}|.0075}(1.0075) = \$39\,798.43$$

■

EXAMPLE 3   A debt of $10 000 with interest at $j_4 = 11\%$ is to be paid off by 8 equal quarterly payments, the first due today. Find the quarterly payment.

Solution   Arranging the data on a time diagram below:

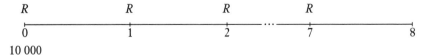

We have $A = 10\ 000$, $i = .0275$, $n = 8$ and calculate R from the equation

$$Ra_{\overline{8}|.0275}(1.0275) = 10\ 000$$
$$R = \frac{10\ 000}{a_{\overline{8}|.0275}(1.0275)}$$
$$R = \$1371.85$$

∎

Deferred Annuities

A **deferred annuity** is an annuity whose payment is due some time later than the end of the first interest period. It is customary to analyze all deferred annuities as ordinary deferred annuities. Thus, an ordinary deferred annuity is an ordinary annuity whose term is deferred for (let's say) k periods. The time diagram below shows the simple case of an ordinary deferred annuity.

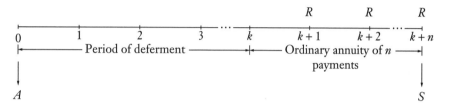

OBSERVATION: Note that in the above diagram the period of deferment is k periods and the first payment of the ordinary annuity is $k + 1$. This is because the term of an ordinary annuity starts one period before its first payment. Thus, when the time of the first payment is given, it is necessary to determine the period of deferment by moving back one interest period.

To find the discounted value A of an ordinary deferred annuity we find the discounted value of n payments one period before the first payment and discount this sum for k periods. We obtain

$$\boxed{A = Ra_{\overline{n}|i}(1 + i)^{-k}}$$

If you now return to Exercise 3.3, Part A, you will see that we have already handled questions of this nature (questions 8, 9 and 10, for example).

EXAMPLE 4 What sum of money should be set aside on a child's birth to provide 8 semi-annual payments of $1500 to cover the expenses for university education if the first payment is to be made on the child's 19th birthday? The fund will earn interest at $j_2 = 12\%$.

Solution We arrange the data on a time diagram below.

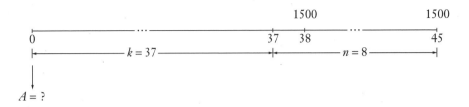

$A = ?$

We have $R = 1500$, $i = .06$, $n = 8$, $k = 37$ and calculate

$$A = 1500 \, a_{\overline{8}|.06}(1.06)^{-37} = \$1078.58$$

∎

EXAMPLE 5 Jaclyn wins $100\ 000 in a provincial lottery. She takes only $20\ 000 in cash and invests the balance at $j_{12} = 8\%$ with the understanding that she will receive 180 equal monthly payments with the first one to be made in 4 years. Find the size of the payments.

Solution

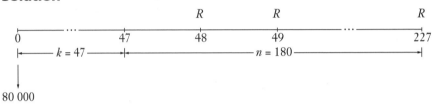

80 000

We have $A = 80\ 000$, $i = \frac{.08}{12}$, $n = 180$, $k = 47$ and calculate R from the equation

$$80\ 000 = Ra_{\overline{180}|i}(1 + i)^{-47}$$

$$R = \frac{80\ 000(1 + i)^{47}}{a_{\overline{180}|i}} = \$1044.76$$

∎

CALCULATION TIP: In general, problems involving any simple annuities can be efficiently solved by applying equations **(10)** and **(11)** for ordinary annuities to find equivalent lump sums and then moving to a requested point in time.

EXAMPLE 6 A couple deposits $200 a month in a savings account paying interest at $j_{12} = 4\frac{1}{2}\%$. The first deposit is made on February 1, 2000 and the last deposit on July 1, 2006.
  a) How much money is in the account on
    i) January 1, 2004 (after the payment is made)
    ii) January 1, 2005 (before the payment is made)
    iii) January 1, 2008?
  b) If they want to draw down their account with equal monthly withdrawals from February 1, 2008 to February 1, 2009, how much will they get each month?

Solution a We arrange the data on a time diagram below.

i) We have $R = 200$, $i = \frac{.045}{12}$ and calculate the accumulated value of 48 payments of an ordinary annuity.

$$S_1 = 200 \, s_{\overline{48}|i} = \$10\ 496.77$$

ii) We calculate the accumulated value of 60 payments of an ordinary annuity and subtract the last payment.

$$S_2 = 200 \, s_{\overline{60}|i} - 200 = \$13\ 229.11$$

(Or, we calculate the accumulated value of 59 payments of an annuity due

$$S_2 = 200 \, s_{\overline{59}|i}(1 + i) = \$13\ 229.11.)$$

iii) We calculate the accumulated value of 78 payments moved forward 18 periods.

$$S_3 = 200 \, s_{\overline{78}|i}(1 + i)^{18} = \$19\ 342.37$$

Solution b Using result iii) from solution a) the accumulated value of their deposits on January 1, 2008, becomes the discounted value of their 13 future withdrawals. We calculate the monthly withdrawal

$$R = \frac{19\ 342.37}{a_{\overline{13}|i}} = \$1527.22$$

■

Exercise 3.4

Part A

1. Find the discounted value and the accumulated value of $500 payable semi-annually at the beginning of each half-year over 10 years if interest is 8% per annum payable semi-annually.

2. A couple wants to accumulate $10 000 by December 31, 2010. They make 10 annual deposits starting January 1, 2001. If interest is at $j_1 = 12\%$, what annual deposits are needed?

3. The premium on a life insurance policy can be paid either yearly or monthly in advance. If the annual premium is $120, what monthly premium would be equivalent at $j_{12} = 5\%$?

4. A life insurance policy allows the option of paying your premium yearly or monthly in advance. If the monthly premium quoted is $45, what annual premium would be equivalent at $j_{12} = 6\%$?

5. An insurance policy provides a death benefit of $100 000 or payments at the beginning of each month for 10 years. What size would these monthly payments be if $j_{12} = 5.5\%$?

6. A used car sells for $9550. Brent wishes to pay for it in 18 monthly instalments, the first due on the day of purchase. If 18% compounded monthly is charged, find the size of the monthly payment.

7. A realtor rents office space for $5800 every three months, payable in advance. He immediately invests half of each payment in a fund paying 13% compounded quarterly. How much is in the fund at the end of 5 years?

8. A refrigerator is bought for $60 down and $60 a month for 15 months. If interest is charged at $j_{12} = 18\frac{1}{2}\%$, what is the cash price of the refrigerator?

9. Find the discounted value of an ordinary annuity deferred 5 years paying $1000 a year for 10 years if interest is at $j_1 = 8\%$.

10. Find the discounted value of an ordinary annuity deferred 3 years six months that pays $500 semi-annually for 7 years if interest is 7% per annum payable semi-annually.

11. What sum of money must be set aside at a child's birth to provide for 6 semi-annual payments of $1500 to cover the expenses for a university education if the first payment is to be made on the child's 19th birthday and interest is at $j_2 = 8\%$?

12. On Mr. Pimentel's 55th birthday, the Pimentels decide to sell their house and move into an apartment. They realize $150 000 on the sale of the house and invest this money in a fund paying $j_1 = 9\%$. On Mr. Pimentel's 65th birthday they make their first of 15 annual withdrawals that will exhaust the fund. What is the dollar size of each withdrawal?

13. Mrs. Howlett changes employers at age 46. She is given $8500 as her vested benefits in the company's pension plan. She invests this money in an RRSP (Registered Retirement Savings Plan) paying $j_1 = 8\%$ and leaves it there until her ultimate retirement at age 60. She plans on 25 annual withdrawals from this fund, the first withdrawal on her 61st birthday. Find the size of these withdrawals.

14. Find the value on January 1, 2005, of quarterly payments of $100 each over 10 years if the first payment is on January 1, 2007, and interest is 7% per annum compounded quarterly.

15. Find the value on July 1, 1999, of semi-annual payments of $500 each over 6 years if the first payment is on January 1, 2003, and interest is at $11\frac{1}{4}\%$ per annum payable semi-annually.

16. The XYZ Furniture Store sells a chesterfield for $950. It can be purchased for $50 down and no payments for 3 months. At the end of the third month you make your first payment and continue until a total of 18 payments are made. Find the size of each payment if interest is at $j_{12} = 18\%$.

17. An 8-year-old child wins $1 000 000 from a lottery. The law requires that this money be set aside in a trust fund until the child reaches 18. The child's parents decide that the money should be paid out in 20 equal annual payments with the first payment at age 18. Find these payments if the trust fund pays interest at $j_1 = 10\%$.

Part B

1. A man aged 40 deposits $5000 at the beginning of each year for 25 years into an RRSP paying interest at $j_1 = 9\%$. Starting on his 65th birthday he makes 15 annual withdrawals from the fund at the beginning of each year. During this period (i.e., from his 65th birthday on) the fund pays interest at $j_1 = 12\%$. Find the amount of each withdrawal starting at age 65.

2. Jacques signed a contract that calls for payments of $500 at the beginning of each 6 months for 10 years. If money is worth $j_2 = 13\%$, find the value of the remaining payments
a) just after he makes the 4th payment;
b) just before he makes the 6th payment.
If after making the first 3 payments he failed to make the next 3 payments,
c) what would he have to pay when the next payment is due to bring himself back on schedule?

3. Using a geometric progression, derive the formula for the accumulated value S of an annuity due
$$S = Rs_{\overline{m}|i}(1 + i)$$
Show that it is equivalent to
$$S = R(s_{\overline{n+1}|} - 1)$$

4. Using a geometric progression derive the formula for the discounted value A of an annuity due
$$A = Ra_{\overline{m}|i}(1 + i)$$
Show that it is equivalent to
$$A = R(a_{\overline{n-1}|i} + 1)$$

5. A mutual fund promises a rate of growth of 10% a year on funds left with it. How much would an investor who makes deposits of $1000 at the beginning of each year have on deposit by the time the first deposit has grown by 159%? You may assume that the time is approximately an integral number of years and that the investor is about to make, but has not made, an annual deposit at that time.

6. Show that the discounted value A of an ordinary deferred annuity is equivalent to
$$A = R(a_{\overline{k+n}|i} - a_{\overline{k}|i})$$

7. Actuaries use the following notation:
$\ddot{a}_{\overline{m}|i} \equiv$ the discounted value of a $1 annuity due
$\ddot{s}_{\overline{m}|i} \equiv$ the accumulated value of a $1 annuity due
Show that
a) $\ddot{a}_{\overline{m}|i} = (1 + i)a_{\overline{m}|i} = 1 + a_{\overline{n-1}|i}$
b) $\ddot{s}_{\overline{m}|i} = (1 + i)s_{\overline{m}|i} = s_{\overline{n+1}|i} - 1$
c) $s_{\overline{n+1}|i} - \ddot{a}_{\overline{n+1}|i} = \ddot{s}_{\overline{m}|i} - a_{\overline{m}|i}$

8. Starting on his 45th birthday, a man deposits $1000 a year in an investment account that pays interest at $j_1 = 13\%$. He makes his last deposit on his 64th birthday. On his 65th birthday he transfers his total savings to a special retirement fund that pays $j_1 = 14\frac{1}{2}\%$. From this fund he will receive level payments of $X at the beginning of each year for 15 years (first payment on his 65th birthday). Find X.

9. Given the following diagram

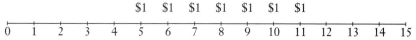

find simplified expressions for a single sum equivalent to the seven payments shown at times 1, 5, 8, 12 and 15, assuming rate i per period.

10. Provide symbolic answers (using $a_{\overline{n}|i}$, $s_{\overline{n}|i}$, i) in simplified form for the value of the following payments at the time indicated, assuming rate i per payment period.

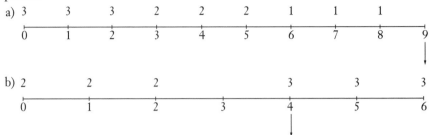

a) 3   3   3   2   2   2   1   1   1

b) 2   2   2   3   3   3

11. Deposits are $100 per month for 3 years, nothing for 2 years and then $200 per month for 3 years. Interest rates start at $j_{12} = 8\%$ and fall to $j_{12} = 6\%$ on the date of the first $200 deposit. Find the accumulated value at the time of the final $200 deposit.

12. How much must you deposit at the end of each month for 25 years to fund an annuity of $2500 per month for 20 years? The first annuity payment is made 5 years after the last deposit, and interest changes from $j_{12} = 8\%$ to $j_{12} = 7\%$ at the time of the first annuity payment.

13. The present value of an annuity of $1000 payable for n years commencing one year from now is $6053. The annual effective rate of interest is 12.5%. Find the present value of an annuity of $1000 commencing one year from now and payable for $n + 2$ years.

14. A person deposits $100 at the beginning of each year for 20 years. Simple interest at an annual rate of i is credited to each deposit from the date of deposit to the end of the 20-year period. The total amount thus accumulated is $2840. If, instead, compound interest had been credited at an annual effective rate of i, what would the accumulated value of these deposits have been at the end of 20 years?

Section 3.5 **Finding the Term of an Annuity**

In some problems the accumulated value S or the discounted value A, the periodic payment R, and the rate i are specified. This leaves the number of payments n to be determined. Formulas **(10)** and **(11)** may be solved for n by the use of logarithms. Normally, when given a value of S, or A or R and a rate i, you will not find an integer time period n for the annuity. Algebraically this means that there is usually no integer n such that $S = Rs_{\overline{n}|i}$ or $A = Ra_{\overline{n}|i}$. It is necessary to make the concluding payment different from R in order to have equivalence. One of the following procedures is followed in practice.

Procedure 1: The last regular payment is increased by a sum that will make the payments equivalent to the accumulated value S or the discounted value A.

Procedure 2: A smaller concluding payment is made one period after the last full payment. Sometimes, when a certain sum of money is to be accumulated, a smaller concluding payment will not be required because the interest after the last full payment will equal or exceed the balance needed (Example 1).

CALCULATION TIP: Unless specified otherwise, we shall use Procedure 2 throughout this book. Procedure 2 is more often used in practice.

EXAMPLE 1   A couple wants to accumulate $10 000 by making payments of $800 at the end of each half year into an investment account that earns interest at $j_2 = 9\%$. Find the number of full payments required and the size of the concluding payment using both Procedures 1 and 2.

Solution   We have $S = 10\ 000$, $R = 800$, $i = 4\frac{1}{2}\%$, and we want to calculate n. Substituting in equation **(10)** we obtain

$$800 s_{\overline{n}|.045} = 10\ 000$$
$$s_{\overline{n}|.045} = 12.5$$
$$\frac{(1.045)^n - 1}{.045} = 12.5$$
$$(1.045)^n - 1 = (12.5)(.045)$$
$$(1.045)^n = 1.5625$$
$$n \log 1.045 = \log 1.5625$$
$$n = 10.13899776$$

Thus there will be 10 full deposits.

Procedure 1   Let X be the sum that will be added to the last regular payment to make the payments equivalent to the accumulated value $S = 10\ 000$. We arrange the data on a time diagram below.

Using 10 as a focal date, we obtain the equation of value for unknown X

$$800 s_{\overline{10}|.045} + X = 10\ 000$$
$$X = 10\ 000 - 800 s_{\overline{10}|.045}$$
$$X = 10\ 000 - 9830.57$$
$$X = \$169.43$$

Thus the 10th deposit will be $969.43

Procedure 2   Let Y be the size of a smaller concluding payment made a half year after the last full deposit. We arrange the data on a time diagram below.

Using 11 as the focal date we obtain the equation of value for unknown Y

$$800 s_{\overline{10}|.045}(1.045) + Y = 10\ 000$$
$$10\ 272.94 + Y = 10\ 000$$
$$Y = -\$272.94$$

The negative value of Y indicates that there is no concluding payment required; the interest after the last full payment will exceed the required balance by $272.94.

Check: Carrying the accumulated value of 10 payments forward for one-half year will result in the accumulated value
$$800s_{\overline{10}|.045}(1.045) = \$10\ 272.94$$

∎

EXAMPLE 2   A man dies and leaves his wife an estate of $50 000. The money is invested at $j_{12} = 12\%$. How many monthly payments of $750 would the widow receive and what would be the size of the concluding payment?

Solution   We have $A = 50\ 000$, $R = 750$, $i = 1\%$. Substituting in equation **(11)** we obtain

$$750a_{\overline{n}|.01} = 50\ 000$$
$$a_{\overline{n}|.01} = \frac{50\ 000}{750}$$
$$\frac{1 - (1.01)^{-n}}{.01} = \frac{200}{3}$$
$$1 - (1.01)^{-n} = \frac{2}{3}$$
$$(1.01)^{-n} = \frac{1}{3}$$
$$-n\log 1.01 = \log \frac{1}{3}$$
$$n = 110.409624$$

Thus the widow will receive 110 full payments of $750 and 1 smaller concluding payment, say X, 1 month after the last full payment. We arrange the data on a time diagram below.

Using 111 as the focal date, we obtain the equation of value for unknown X

$$750s_{\overline{110}|.01}(1.01) + X = 50\ 000(1.01)^{111}$$
$$150\ 575.63 + X = 150\ 883.76$$
$$X = \$308.13$$

Using 0 as the focal date, we obtain the equation of value

$$750a_{\overline{110}|.01} + X(1.01)^{-111} = 50\ 000$$
$$49\ 897.89 + X(1.01)^{-111} = 50\ 000$$
$$X(1.01)^{-111} = 102.11$$
$$X = 102.11(1.01)^{111}$$
$$X = \$308.13$$

∎

Exercise 3.5

Part A

1. A debt of $4000 bears interest at $j_2 = 12\%$. It is to be repaid by semi-annual payments of $400. Find the number of full payments needed and the final smaller payment after the last full payment.

2. Melissa takes her inheritance of $25 000 and invests it at $j_{12} = 9\%$. How many monthly payments of $250 can she expect to receive and what will be the size of the concluding payment? Use both Procedure 1 and Procedure 2.

3. A couple wants to accumulate $10 000. If they deposit $250 at the end of each quarter year in an account paying $j_4 = 6\%$, how many deposits must they make and what will be the size of the final deposit? Use both Procedure 1 and Procedure 2.

4. A firm buys a machine for $30 000. They pay $5000 down and $5000 at the end of each year. If interest is at $j_1 = 10\%$, how many full payments must they make and what will be the size of the concluding smaller payment?

5. A fund of $20 000 is to be accumulated by n annual payments of $2500 plus a final smaller payment made one year after the last regular payment. If interest is at $j_1 = 8\%$, find n and the final irregular payment.

6. A loan of $10 000 is to be repaid by monthly payments of $400, the first payment due in one year's time. If $j_{12} = 12\%$, find the number of regular monthly payments needed and the size of the final smaller payment.

7. A fund of $8000 is to be accumulated by semi-annual payments of $2000. If $j_2 = 12\%$, find the number of full deposits required and the final smaller payment.

8. On July 1, 2003, Shannon has $10 000 in an account paying interest at $j_4 = 12\frac{1}{2}\%$. She plans to withdraw $500 every three months with the first withdrawal on October 1, 2003. How many full withdrawals can she make and what will be the size and the date of the concluding withdrawal?

9. A parcel of land, valued at $350 000, is sold for $150 000 down. The buyer agrees to pay the balance with interest at $j_{12} = 12\%$ by paying $5000 monthly as long as necessary, the first payment due 2 years from now.
   a) Find the number of full payments needed and the size of the concluding payment one month after the last $5000 payment.
   b) Find the monthly payment needed to pay off the balance by 36 equal payments, if the first payment is 1 year from now and interest is at $j_{12} = 12\%$.

10. From July 1, 1996, to January 1, 2001, a couple made semi-annual deposits of $500 into an investment account paying $j_2 = 11\%$. Starting July 1, 2005, they start making semi-annual withdrawals of $800.
    a) How many withdrawals can they make and what is the size and date of the last withdrawal?
    b) If the couple decided to exhaust their savings with equal semi-annual withdrawals from July 1, 2005, to July 1, 2015, inclusive, how much will they get each half-year?

11. In October 1997, an industrialist gave your school $30 000 to be used for future scholarships of $4000 at each fall convocation starting the year the industrialist dies. If the industrialist died May 2000 and the money earns interest at $j_1 = 8\%$, for how many years will full scholarships be awarded?

12. A debt of $18 000 is to be repaid in annual instalments of $3000 with the first instalment due at the end of the second year and a final instalment of less than $3000. Interest is 14% per year compounded annually. Find the final payment.

Part B

1. Antonio is accumulating a $10 000 fund by depositing $100 each month, starting September 1, 1997. If the interest rate on the fund is $j_{12} = 12\%$ until May 1, 2000 and then it drops to $j_{12} = 10\frac{1}{2}\%$, find the time and amount of the reduced final deposit.

2. A widow, as beneficiary of a $100 000 insurance policy, will receive $20 000 immediately and $1800 every three months thereafter. The company pays interest at $j_4 = 6\%$; after 3 years, the rate is increased to $j_4 = 7\%$.
   a) How many full payments of $1800 will she receive?
   b) What additional sum paid with the last full payment will exhaust her benefits?
   c) What payment 3 months after the last full payment will exhaust her benefits?

3. On his 25th birthday, Yves deposited $2000 in a fund paying $j_1 = 10\%$ and continued to make such deposits each year, the last on his 49th birthday. Beginning on his 50th birthday, Yves plans to make equal annual withdrawals of $20 000.
   a) How many such withdrawals can be made?
   b) What additional sum paid with the last withdrawal will exhaust the fund?
   c) What sum paid one year after the last full withdrawal will exhaust the fund?

4. A couple bought land worth $300 000. They paid $50 000 down and signed a contract agreeing to repay the balance with interest at $j_1 = 12\%$ by annual payments of $50 000 for as long as necessary and a smaller concluding payment one year later. The contract was sold just after the 4th annual payment to an investor who wants to realize a yield of $j_1 = 13\%$. Find the selling price.

5. A loan of $20 000 is to be repaid by annual payments of $4000 per annum for the first 5 years and payments of $4500 per year thereafter for as long as necessary. Find the total number of payments and the amount of the smaller final payment made one year after the last regular payment. Assume an annual effective rate of 18%.

6. Solve the equation $S = Rs_{\overline{n}|i}$ for n.

7. Solve the equation $A = Ra_{\overline{n}|i}$ for n.

8. A friend agrees to lend you $2000 on September 1 each year for 4 years to help with education costs. One year after the last payment you are expected to start annual repayments of $800 for as long as necessary. Interest on the loan is $j_1 = 6\frac{1}{2}\%$. Find the number of repayments needed, and the amount of the final payment.

9. A car loan of $10 000 at $j_{12} = 12\%$ is being paid off by n payments. The first $(n - 1)$ payments are $263.34 per month. The final monthly payment is $263.24. Find n.

Section 3.6 | Finding the Interest Rate

A very practical application of equations **(10)** and **(11)** is finding the interest rate. In many business transactions the true interest rate is concealed in one way or another. In order to compare different propositions (options, investments), it is necessary to determine the true interest rate of each proposition and make the decision based on true interest rates.

When R, n and either S or A are given, the interest rate i may be determined approximately by linear interpolation. For most practical purposes, linear interpolation gives sufficient accuracy and will be used throughout this textbook.

> **CALCULATION TIP:** To obtain a starting value to solve the equation $s_{\overline{n}|i} = k$ by linear interpolation, we may use the formula $i = \dfrac{(\frac{k}{n})^2 - 1}{k}$. To obtain a starting value to solve the equation $a_{\overline{n}|i} = k$ by linear interpolation, we may use the formula $i = \dfrac{1 - (\frac{k}{n})^2}{k}$.

> **OBSERVATION:** The interest rates i and $j_m = mi$ may be computed directly using either a financial calculator or a spreadsheet.

EXAMPLE 1   Find the interest rate j_4 at which deposits of $250 at the end of every 3 months will accumulate to $5000 in 4 years.

Solution   We have $S = 5000$, $R = 250$, $n = 16$ and using **(10)** we obtain

$$250 s_{\overline{16}|i} = 5000$$
$$s_{\overline{16}|i} = 20$$

We want to find the rate $j_4 = 4i$ such that $s_{\overline{16}|i} = \frac{(1+i)^{16}-1}{i} = 20$. A starting value to solve $s_{\overline{16}|i} = 20$ is $i = \frac{(\frac{20}{16})^2 - 1}{20} = .028125$ or $j_4 = 4i = 11.25\%$.

By successive trials we find two factors $s_{\overline{16}|i}$, one greater than 20 and one less than 20. The corresponding rates $j_4 = 4i$ will provide an upper and lower bound on the unknown rate j_4, which is then approximated by a linear interpolation.

> **CALCULATION TIP:** The values of factors $s_{\overline{n}|i}$ and $a_{\overline{n}|i}$ will be rounded off to 4 decimal places, since additional places do not really increase the accuracy. For fixed n, factors $s_{\overline{n}|i}$ increase when i increases, whereas factors $a_{\overline{n}|i}$ decrease when i increases. Closer bounds on the nominal rate j_m generally provide better approximations of the unknown rate j_m by linear interpolation. In this textbook we will interpolate between two nominal rates that are 1% apart.

For $j_4 = 11\%$, we calculate $s_{\overline{16}|i} = 19.7640$
For $j_4 = 12\%$, we calculate $s_{\overline{16}|i} = 20.1569$
Now we have two rates, $j_4 = 11\%$ and $j_4 = 12\%$, 1% apart, that provide upper and lower bounds for interpolation.

Arranging our data in an interpolation table we have:

	$s_{\overline{16}|i}$	j_4
	19.7640	11%
	20.0000	j_4
	20.1569	12%

$$.2360 \left\{ \begin{array}{l} \\ \\ \end{array} \right.$$

$$.3929 \left\{ \begin{array}{l} \\ \\ \\ \end{array} \right.$$

$$\left. \begin{array}{l} \\ \end{array} \right\} d \quad \left. \begin{array}{l} \\ \\ \end{array} \right\} 1\%$$

$$\frac{d}{1\%} = \frac{.2360}{.3929}$$
$$d = .60\%$$
$$\text{and } j_4 = 11.60\%$$

We may check the accuracy of this answer by substituting $R = 250$, $n = 16$ and $i = \frac{.1160}{4}$ into **(10)** and calculate the accumulated value

$$S = 250\frac{(1 + i)^{16} - 1}{i} = \$4999.65$$

Linear interpolation between two nominal rates, 1% apart, gave us a very good approximation of the unknown rate j_4.

Note: Using a BA-35 financial calculator, we enter $N = 16$, $PMT = -250$, $FV = 5000$ and compute $j_4 = 4i = 11.60\%$.

Using Excel's Financial function Wizard = RATE (16, −250, 0, 5000)*4 we calculate $j_4 = 11.60\%$.

■

EXAMPLE 2 A used car sells for $6000 cash or $1000 down and $900 a month for 6 months. Find the interest rate j_{12} if the purchaser buys the car on the instalment plan.

Solution For any instalment plan, the following equation of value must hold to have the cash option equivalent to the instalment option.

cash price = down payment + discounted value of instalments

We have

$$6000 = 1000 + 900a_{\overline{6}|i}$$

$$a_{\overline{6}|i} = \frac{5000}{900}$$

$$a_{\overline{6}|i} = 5.5556$$

We want to find the rate $j_{12} = 12i$ such that $a_{\overline{6}|i} = \frac{1-(1+i)^{-6}}{i} = 5.5556$.

A starting value to solve the equation $a_{\overline{6}|i} = 5.5556$ is $i = \frac{1-(\frac{5.5556}{6})^2}{5.5556} = .025676338$ or $j_{12} = 12i = 30.81\%$.

By successive trials we find two factors $a_{\overline{6}|i}$, one greater than 5.5556 and one less than 5.5556, such that the corresponding rates j_{12} are 1% apart.

For $j_{12} = 26\%$, we calculate $a_{\overline{6}|i} = 5.5701$
For $j_{12} = 27\%$, we calculate $a_{\overline{6}|i} = 5.5545$
Arranging our data in an interpolation table we have:

	$a_{\overline{6}|i}$	j_{12}
	5.5701	26%
.0145	5.5556	j_{12}
.0156	5.5545	27%

$$\frac{d}{1\%} = \frac{.0145}{.0156}$$
$$d = .93\%$$
and $j_{12} = 26.93\%$

Checking the accuracy of our answer we calculate the discounted value of the instalment plan at $j_{12} = 26.93\%$

$$1000 + 900a_{\overline{6}|i} = \$6000$$

Note: Using a BA-35 financial calculator, we enter $N = 6$, $PMT = 900$, $PV = 5000$ and compute $j_{12} = 12i = 26.93\%$.

Using Excel's Financial function Wizard = RATE (6, −900, 5000)*12 gives $j_{12} = 26.93\%$.

Exercise 3.6

Part A

1. Find the interest rate j_2 at which semi-annual deposits of $500 will accumulate to $6000 in 5 years.

2. What rate of interest j_1 must be earned for deposits of $500 at the end of each year to accumulate to $12 000 in 10 years?

3. An insurance company will pay $80 000 to a beneficiary or monthly payments of $1000 for 10 years. What rate j_{12} is the insurance company using?

4. A television set sells for $700. Sales tax of 7% is added to that. The T.V. may be purchased for $100 down and monthly payments of $60 for one year. What is the interest rate j_{12}? What is the annual effective interest rate?

5. You borrow $1600 from a licensed small loan company and agree to pay $160 a month for 12 months. What nominal rate j_{12} is the company charging?

6. A store offers to sell a watch for $55 cash or $5 a month for 12 months. What nominal rate j_{12} is the store actually charging on the instalment plan, if the first payment is made immediately?

7. On February 1, 1985, Andreas made the first of a sequence of regular annual deposits of $1000 into an investment account. The last deposit was made February 1, 2001. If the account earned $j_1 = 10\frac{1}{2}\%$, the balance after the last deposit would have been $42 472.13, while it would have been $44 500.84 at $j_1 = 11\%$. In fact, the balance in the account immediately after the last deposit was $43 500. What annual effective rate of interest did the account earn?

Part B

The following problems are examples of situations that could arise if there were no government legislation concerning disclosure of interest rates. In Canada, most provinces have "truth in lending" laws setting down regulations on the disclosure of the rate of interest involved in financial transactions. It is very important to check out all loan clauses fully.

1. The "Fly By Night" Used Car Lot uses the following to illustrate their 12% finance plan on a car paid for over 3 years.

   Cost of car   12 000.00
   12% finance charge   4 320.00 (12% of 12 000 × 3 years)
   Total cost   16 320.00

   Monthly payment $= \frac{16\ 320}{36} = \$453.33$

   What is the true interest rate j_{12} being charged?

2. A dealer sells an article for $6000. He will allow a customer to buy it by paying $2400 down and the balance by paying $300 a month for a year. If you pay cash for the item he will give you a 10% discount. Find the interest rate j_{12} paid by the purchaser who uses the instalment plan described above.

3. Goods worth $1000 are purchased using the following carrying-charge plan: A down payment of $100 is required after which 18% of the unpaid balance is added on and the amount is then divided into 12 equal monthly instalments. What rate of interest j_{12} does the plan include?

4. A finance company charges 15% "interest in advance" and allows the client to repay the loan in 12 equal monthly payments. The monthly payment is calculated as one-twelfth of the total of principal and interest (15% of principal). Find the nominal rate compounded monthly and the annual effective rate charged.

5. To buy a car costing $13 600 you can pay $1600 down and the balance in 36 monthly payments of $450 each. You can also borrow the money from a loan company and repay $13 600 by making quarterly payments of $1060 over 5 years, first payment in 3 months. Compare the annual effective rates of interest charged and determine which option is better.

6. A T.V. rental company uses the following illustration to prove that renting a T.V. at $25 a month is cheaper than buying.

   Cost of T.V. $600
   Sales Tax   42
   Total $642

   Therefore monthly payments over 3 years at 21% are $\frac{642 + (.21)(642)(3)}{36} = \$29.07$.

   Redo this illustration properly at $j_{12} = 21\%$ and comment.

7. You are offered a loan of $10 000 with no payments for 6 months, then $600 per month for 1 year and $500 per month for the following year. What annual effective rate of interest does this loan charge?

8. On July 1, 2001, $8500 is deposited in Fund X. On July 1, 2001, the first of 10 consecutive semi-annual payments of $1000 each is deposited in fund Y. Both funds earn interest at a nominal rate j_2. The balances in the two funds are equal on July 1, 2006. Find j_2.

Section 3.7   Summary and Review Exercises

● Accumulated value S of n payments of R per interest period, at rate i per period, *at the time of the last payment*, is given by $S = Rs_{\overline{n}|i}$ where

$$s_{\overline{n}|i} = \frac{(1 + i)^n - 1}{i}$$

is called an accumulation factor for n payments.

● Discounted value A of n payments of R per interest period, at rate i per period, *1 period before the first payment*, is given by $A = Ra_{\overline{n}|i}$ where

$$a_{\overline{n}|i} = \frac{1 - (1 + i)^{-n}}{i}$$

is called a discount factor for n payments.

● When calculating factors $s_{\overline{n}|i}$ and $a_{\overline{n}|i}$, do not round off the value of i. Use all digits provided by your calculator and, when necessary, store the value of i in the memory of your calculator. Note that for $i > 0$, $s_{\overline{n}|i} > n$ and $a_{\overline{n}|i} < n$.

● For a simple annuity due (payments are due at the beginning of each of n interest periods), the accumulated value S (*one payment period after the date of the last payment*) is given by

$$S = Rs_{\overline{m}|i}(1 + i)$$

and the discounted value A (*at the time of the first payment*) is given by

$$A = Ra_{\overline{m}|i}(1 + i)$$

● For an ordinary simple annuity of n payments deferred k interest periods, the discounted value, A is given by

$$A = Ra_{\overline{m}|i}(1 + i)^{-k}$$

Review Exercises 3.7

1. Find the accumulated and the discounted value of an annuity of $500 at the end of each month at $j_{12} = 9\%$ for a) 10 years, b) 20 years.

2. At age 65 Mrs. Papadopoulos takes her life savings of $100 000 and buys a 20-year annuity certain with quarterly payments. Find the size of these payments
   a) at 5% compounded quarterly,
   b) at 7% compounded quarterly.

3. An annuity provides $60 at the end of each month for 3 years and $80 at the end of each month for the following 2 years. If $j_{12} = 5\%$, find the present value of the annuity.

4. To finance the purchase of a car, Wendy agrees to pay $400 at the end of each month for 4 years. The interest rate is $j_{12} = 15\%$.
   a) If Wendy misses the first 4 payments, what must she pay at the time of the 5th payment to be fully up to date?
   b) Assuming Wendy has missed no payments, what single payment at the end of 2 years would completely discharge her of her indebtedness?

5. A couple needs a loan of $10 000 to buy a boat. One lender will charge $j_{12} = 18\%$ while a second lender charges $j_{12} = 16\%$. What will be the monthly savings in interest using the lower rate if the monthly payments are to run for 5 years?

6. Find the accumulated and the discounted value of semi-annual payments of $500 at the end of every half year over 10 years if interest is $j_2 = 10\%$ for the first four years and $j_2 = 12\%$ for the last six years.

7. Instead of paying $900 rent at the beginning of each month for the next 10 years, a couple decides to buy a townhouse. What is the cash equivalent of the 10 years of rent at $j_{12} = 9\%$?

8. A company sets aside $15 000 at the beginning of each year to accumulate a fund for future expansion. What is the amount in the fund at the end of 5 years if the fund earns $j_1 = 11\%$?

9. According to Mr. Peterson's will, the $100 000 life insurance benefit is invested at $j_1 = 13\%$ and from this fund his widow will receive $15 000 each year, the first payment immediately, so long as she lives. On the payment date following the death of his wife, the balance of the fund is to be donated to a local charity. If his wife died 4 years, 3 months later, how much did the charity receive?

10. Five years from now a company will need $150 000 to replace worn-out equipment. Starting now, what monthly deposits must be made in a fund paying $j_{12} = 9\%$ for 5 years to accumulate this sum?

11. Doreen bought a car on September 1, 2002, by paying $4000 down and agreeing to make 36 monthly payments of $350, the first due on December 1, 2002. If interest is at 18% compounded monthly, find the equivalent cash price.

12. An office space is renting for $30 000 a year payable in advance. Find the equivalent monthly rental payable in advance if money is worth 6% compounded monthly.

13. A farmer borrowed $80 000 to buy some farm equipment. He plans to pay off the loan with interest at $j_1 = 13\frac{3}{4}\%$ in 8 equal annual payments, the first to be made 5 years from now. Find the annual payment.

14. Charlie wants to accumulate $100 000 by making monthly deposits of $1000 into a fund that accumulates interest at $j_{12} = 12\%$. Find the number of full deposits required and the size of the concluding deposit using both procedures 1 and 2.

15. Lise borrows $10 000 at $j_4 = 18\%$. How many $800 quarterly payments will she pay and what will be the size of the partial concluding payment, 3 months after the last $800 payment?

16. On June 1, 2003, Ms. Kaminski purchased furniture for $2200. She paid $400 down and agreed to pay the balance by monthly payments of $100 plus a smaller final payment, the first payment due on July 1, 2003. If interest is at $j_{12} = 18\%$, when is the final payment made and what is the amount of the final payment?

17. A used car sells for $5000 cash or $1000 down and $800 a month for 6 months. Find the interest rate j_{12} if the purchaser buys the car on the instalment plan.

18. Find j_{12} at which deposits of $200 at the end of each month will accumulate to $10 000 in 3 years.

19. The XYZ Finance Company charges 10% "interest in advance" and allows the client to repay the loan in 12 equal monthly payments. Thus for a loan of $6000, they would charge $600 interest and have 12 monthly repayments of $550 each. What is the corresponding rate of interest convertible monthly?

20. A refrigerator is listed at $650. If a customer pays $200 down, the balance plus a carrying charge of $50 can be paid in 12 equal monthly payments. If the customer pays cash, he can get a discount of 15% off the list price. What is the nominal rate converted monthly if the refrigerator is bought on time?

21. Paul wants to accumulate $5000 by depositing $300 every 3 months into an account paying 8% per annum converted quarterly. He makes the first deposit on July 1, 2003. How many full deposits should he make and what will be the size and the date of the concluding deposit?

22. How much a month for 5 years at $j_{12} = 6\%$ would you have to save in order to receive $800 a month for 3 years afterward?

23. A bank account paying $j_{12} = 5\frac{1}{2}\%$ contains $5680 on March 1, 2003. Beginning April 1, 2003, the first of a sequence of monthly withdrawals of $400 is made. What is the date of the last $400 withdrawal? By what date (first of the month) will the balance again exceed $400?

24. Jones agrees to pay Smith $800 at the end of each quarter for 5 years, but is unable to do so until the end of the 15th month when he wins $100 000 in a provincial lottery. Assuming money is worth $j_4 = 6\%$, what single payment at the end of fifteen months liquidates his debt?

25. A deposit of $1000 is made to open an account on March 1, 2000. Monthly deposits of $300 are then made for 5 years, starting April 1, 2000. Starting April 1, 2005, the first of a sequence of 20 monthly withdrawals of $1000 is made. Find the balance in the account on December 1, 2007, assuming $j_{12} = 12\%$.

26. Fred needs a loan of $15 000 to buy a new car. The dealer offers him the loan at $j_{12} = 16\%$. Fred can get the loan at his bank at $j_{12} = 15\%$.
    a) What will be the monthly savings using the lower rate at his bank if the monthly payments are to run for 3 years?
    b) What will be the total interest on the loan at his bank?
    c) Assume that Fred takes the loan of $15 000 at his bank and decides to pay $600 at the end of each month. How many months will it take him to repay the loan, assuming a smaller final payment 1 month after the last $600 payment? Find the concluding payment.

27. On November 10, 2000, I.M. Broke obtained a bank loan of $4000 at 10% compounded monthly. I.M. Broke will repay the loan by making monthly payments of $250 beginning on December 10, 2000. Determine the number of full payments, the date and the size of the smaller concluding payment required.

28. The Sound Warehouse advertises a "no interest for 1 year" option. For example, you can buy a $1200 stereo by paying $100 at the end of each month for 1 year. However, if you pay in full at the time of the purchase, you get a 10% cash discount. What rate of interest j_{12} is the Sound Warehouse actually charging?

29. a) Elsa wants to accumulate $20 000 by the end of 6 years by making quarterly deposits in a fund that pays interest at 10% compounded quarterly. Find the size of these deposits.
    b) After 4 years the fund rate changes to $j_4 = 8\%$. Find the size of the quarterly deposits now required if the $20 000 goal is to be met.
    c) What is the total interest earned on the fund over 6 years?

30. Today, Walt purchased a cottage and a yacht worth a combined present value of $600 000. To purchase the cottage, Walt agreed to pay $100 000 down and $4000 at the end of each month for 15 years. To buy the yacht, he was also required to pay $20 000 down and $R at the end of each month for 5 years. Find R if $j_{12} = 12\%$.

31. Erika deposited $100 monthly in a fund earning $j_{12} = 12\%$. The first deposit was made on June 1, 1990, and the last deposit on November 1, 2000.
    a) Find the value of the fund on
       i) September 1, 1995 (after the payment is made);
       ii) December 1, 2002.
    b) From May 1, 2005, she plans to draw down the fund with monthly withdrawals of $1000. Find the date and the size of the smaller concluding withdrawal one month after the last $1000 withdrawal.

32. Home appliances are on sale with the following payment terms: either pay cash and receive a 5% discount, or pay monthly for 15 months with no money down. To calculate the monthly payments, add a $50 finance charge per $1000 purchase price to the purchase price, then divide this amount into 15 equal payments. What annual effective rate of interest is being charged for buying on time?

33. Peter wants to invest in a mutual fund. His goal is to accumulate enough money over the next 20 years so that he can withdraw $30 000 annually for 25 years afterward. From a website **www.globefund.com** he found out that the Trimark Fund earned $j_1 = 15.84\%$ over the past 15 years (as of December 31, 1999). What annual deposits for 20 years would provide him with an annuity of $30 000 per year, if he invests in the Trimark Fund, assuming that the first withdrawal will be 1 year after the last deposit and the fund will keep earning $j_1 = 15.84\%$ over the next 45 years?

CASE STUDY I

Car loans

The following article appeared in the *Kitchener-Waterloo Record* on Saturday, October 5, 1985.

This material has been copied under licence from CANCOPY. Resale or further copying of this material is strictly prohibited.

Car loan can make more cents than using cash

By S. J. DIAMOND
Los Angeles Times

At first, psychology Prof. Geoffrey Keppel of Berkeley, Calif., rejected out of hand his Toyota dealer's offer of financing: He had saved the money for a new Corolla and, like many people, didn't want a loan because "that's the way I'd been brought up."

The dealer even told him he could earn more on the same $8,300 in an eight per cent savings certificate than he'd pay out on a 14.2 per cent car loan, "but you just don't believe it," he says, "It's counter-intuitive: Anyone can see 14 is bigger than eight."

He's not unusual. Many people who spend their adult lives avoiding debt have heard from (usually richer) friends that it's generally better to spend borrowed money than savings, but they can never clearly see why, particularly when loan rates are higher than investment yields.

What's more, Keppel says, "when you've just spent several hours nickel-and-diming with an auto dealer, you're not inclined to jump when they say they have another good deal for you."

That night, however, Keppel awoke and went to his computer to "work it out for myself, month by month." Calculating different investment yields and weighing them against the total interest that he'd pay on the 14.2 per cent loan, he saw that he'd break even with only a 7 per cent investment, and, if he could earn 10 per cent, he'd make almost $1,500.

A couple of points must be interjected here. First of all, this seems a rather exclusive quandary, the concern only of affluent people who have the option of paying outright for their car.

But it really isn't: According to J.D. Power & Associates, an automotive market research firm, one-third of the people who buy cars do pay cash, and some of the two-thirds who don't probably could.

Second, it's a quandary only today's affluent can easily solve: Yesterday's consumers didn't have the calculators and personal computers for such analysis, and it's laborious by hand.

They might otherwise have seen the advantage in borrowing without taking anyone's word for it. Keppel, for example, calculated that 48 months of interest on a 14.2 per cent loan of $8,239.05 would be $2,607.82, while the same principal invested at 8 per cent, compounded monthly, would earn interest of $3,095.06—a profit of $487.44.

Tracing both transactions month by month, he could also see that the reason it worked to his advantage was that "the 14 per cent is applied to a declining balance, and the eight per cent is on an increasing balance."

Copyright, 1985, Los Angeles Times. Reprinted with permission.

Note: The price of $8239.05 for a new Toyota Corolla is the 1985 price in U.S. funds.

Required:
a) Verify the figures in the second last paragraph of the article.
b) Is the dealer correct in recommending the car loan as a better option than the cash purchase? Justify your answer.

Chapter Four

General and Other Annuities

General Annuities

So far, we have assumed that periodic payments have been made at the same dates as the interest is compounded. This is not always the case. In this chapter we will consider annuities for which payments are made more or less frequently than interest is compounded. Such a series of payments is called a **general annuity.**

One way to solve general annuity problems is to **replace the given interest rate** by an equivalent rate for which the interest conversion period is the same as the payment period. (A review of Section 2.2 will reacquaint you with this process.) In effect, the general annuity problems are transformed into simple annuity problems and the methods outlined in Chapter 3 can be directly used, and thus no new theory is required.

The second, and historical, approach used in solving general annuity problems is to **replace the given payments** by equivalent payments made on the stated interest conversion dates.

OBSERVATION: In this textbook we follow the "change of rate" approach, which is the preferred method of solution when calculators are used. The "change of payment" approach is illustrated in problems 2 through 5 of Part B of Exercise 4.1.

In order to solve a general annuity problem, we use the following two-step procedure:

> **1.** Convert the general annuity into an equivalent ordinary simple annuity by replacing the given interest rate (see Section 2.2).
> **2.** Solve a simple annuity problem using the methods outlined in Chapter 3.

EXAMPLE 1   At the start of each month for 6 years, Jackie deposits $100 into an account earning 9% compounded semi-annually. Find the value of the account at the end of 6 years.

Solution   First, find the rate i per month equivalent to 4.5% per half-year, such that

$$(1 + i)^{12} = (1.045)^2$$
$$i = (1.045)^{1/6} - 1$$
$$i = .007363123$$

Second, calculate the accumulated value S of a simple annuity due with $R = 100, n = 72, i = .007363123$.

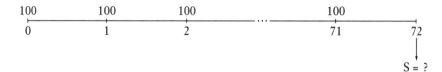

$$S = 100s_{\overline{72}|i}(1 + i) = \$9520.49$$

CALCULATION TIP: Do not round off the value of i but use all digits provided by your calculator. Store the value of i in the memory of your calculator.

∎

EXAMPLE 2   A contract calls for payments of $100 at the end of every month for 5 years and an additional payment of $2000 at the end of 5 years. What is the present worth of the contract at a) 15% compounded quarterly; b) 8% compounded continuously?

Solution a   First, find the rate i per month equivalent to $\frac{.15}{4} = .0375$ per quarter-year, such that

$$(1 + i)^{12} = (1.0375)^4$$
$$1 + i = (1.0375)^{1/3}$$
$$i = (1.0375)^{1/3} - 1$$
$$i = .012346926$$

Second, find the discounted value A of an ordinary simple annuity and of the additional payment of $2000.

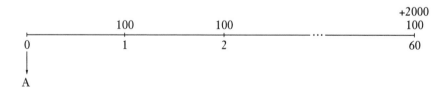

$$A = 100a_{\overline{60}|i} + 2000(1 + i)^{-60} = 4220.55 + 957.78 = \$5178.33$$

The present value of the contract is $5178.33.

Solution b   First, find the rate i per month equivalent to $j_\infty = 8\%$, such that

$$(1 + i)^{12} = e^{.08}$$
$$i = e^{.08/12} - 1$$
$$i = .006688938$$

Second, calculate the discounted value A of an ordinary simple annuity and of the additional payment of $2000.

$$A = 100a_{\overline{60}|i} + 2000(1 + i)^{-60} = 4928.73 + 1340.64 = \$6269.37$$

The present value of the contract is $6269.37.

■

EXAMPLE 3   What equal monthly payments for 10 years will pay off a loan of $20 000 with interest at 10% compounded daily, if the first payment is made 6 months from now?

Solution   First find the rate i per month equivalent to $\frac{1}{365}$ per day, such that

$$(1 + i)^{12} = (1 + \tfrac{1}{365})^{365}$$
$$i = (1 + \tfrac{1}{365})^{365/12} - 1$$
$$i = .008367001$$

The monthly payment R of a simple deferred annuity with $A = 20\ 000$, $n = 120$, $k = 5$, $i = .008367001$ is obtained from the equation $Ra_{\overline{120}|i}(1 + i)^{-5} = 20\ 000$ and

$$R = \frac{20\ 000}{a_{\overline{120}|i}(1 + i)^{-5}} = \$276.01$$

■

EXAMPLE 4   A couple would like to accumulate $20 000 in 3 years as a down payment on a house, by making deposits at the end of each week in an account paying interest at $j_{12} = 6\%$. Find the size of the weekly deposit, assuming that compound interest is given for part of a conversion period.

Solution   First, find the rate i per week equivalent to $\frac{.06}{12} = .005$ per month, such that

$$(1 + i)^{52} = (1.005)^{12}$$
$$1 + i = (1.005)^{12/52}$$
$$i = (1.005)^{12/52} - 1$$
$$i = .001151634$$

Second, find the weekly deposit R of an ordinary simple annuity with $S = 20\,000$, $n = 3 \times 52 = 156$, and $i = .001151634$.

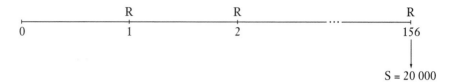

$$R = \frac{20\,000}{s_{\overline{156}|i}} = \$117.11$$

∎

> **CALCULATION TIP:**   When finding the unknown interest rate in a general annuity problem, we first find the nominal rate whose frequency of compounding matches the frequency of payments and then change it to the desired equivalent nominal rate.

EXAMPLE 5   A loan of \$2000 is repaid with 8 quarterly payments of \$300. What nominal rate compounded monthly is being charged?

Solution   First, we find the nominal rate j_4 by solving

$$300a_{\overline{8}|i} = 2000$$
$$a_{\overline{8}|i} = 6.6667$$

A starting value to solve $a_{\overline{8}|i} = 6.6667$ is

$$i = \frac{1 - \left(\frac{6.6667}{8}\right)^2}{6.6667} = .045832063$$

or $j_4 = 4i = 18.33\%$. Using linear interpolation (as in Chapter 3, Section 3.6) we have

	$a_{\overline{8}|i}$	j_4	
	6.7327	16%	
.0689 { .0660 {	6.6667	j_4	} d } 1%
	6.6638	17%	

$$\frac{d}{1\%} = \frac{.0660}{.0689}$$
$$d = .96\%$$
and $j_4 = 16.96\%$

Check: $300\,a_{\overline{8}|.1696/4} = \$1999.96$

Second, we find the nominal rate j_{12} equivalent to $j_4 = 16.96\%$.

$$(1 + i)^{12} = (1 + \tfrac{.1696}{4})^4$$
$$1 + i = (1 + \tfrac{.1696}{4})^{1/3}$$
$$i = (1 + \tfrac{.1696}{4})^{1/3} - 1$$
$$\text{and } j_{12} = 12i = 12[(1 + \tfrac{.1696}{4})^{1/3} - 1] \doteq 16.73\%$$

■

In cases in which payments are made more frequently than interest is compounded, financial institutions use different regulations for calculating interest for parts of a conversion period. The two most common procedures are as follows:

1. *Simple interest* is given for part of a conversion period. In this case the payments must be accumulated to the end of the conversion period by using simple interest.

2. *Compound interest* is given for part of a conversion period. (This situation is, by far, the most common in Canada.)

EXAMPLE 6   Payments of $200 are made at the end of each month in an account paying $j_4 = 8\%$. How much money will be accumulated in the account at the end of 5 years if a) simple interest is paid for part of a period? b) compound interest is paid for part of a period?

Solution a   First we find an equivalent payment R per quarter by accumulating monthly payments to the end of a 3-month period at a simple interest rate of 8%.

$$R = 200[1 + (.08)(\tfrac{2}{12})] + 200[1 + (.08)(\tfrac{1}{12})] + 200$$
$$= 600 + 200(.08)(\tfrac{3}{12}) = \$604.00$$

The accumulated value S equals

$$S = 604\, s_{\overline{20}|.02} = \$14\ 675.61$$

Solution b   First, find the rate i per month equivalent to 2% per quarter-year, such that

$$(1 + i)^{12} = (1.02)^4$$
$$1 + i = (1.02)^{1/3}$$
$$i = (1.02)^{1/3} - 1$$
$$i = .00662271$$

Second, find the accumulated value S of an ordinary simple annuity with $R = 200, n = 60, i = .00662271$

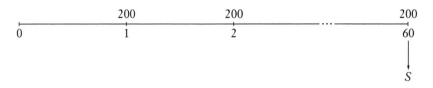

$$S = 200s_{\overline{60}|i} = \$14\ 675.18$$

∎

OBSERVATION: Not many people are aware of the fact that simple interest brings more money for a part of a conversion period than compound interest. The difference is not large; however, it is important to know the rules that apply to the particular transaction.

Exercise 4.1

Part A

1. George deposits $200 at the beginning of each year in a bank account that earns interest at $j_4 = 6\%$. How much money will be in his bank account at the end of 5 years?

2. A car is purchased by paying $2000 down and then $300 each quarter-year for 3 years. If the interest on the loan was $j_2 = 14\%$, what did the car sell for?

3. An insurance policy requires premium payments of $15 at the beginning of each month for 20 years. Find the discounted value of these payments at $j_4 = 5\%$.

4. Payments of $1000 are to be made at the end of each half-year for the next ten years. Find their discounted value if the interest rate is 12% per annum compounded a) half-yearly; b) quarterly; c) annually; d) continuously.

5. Deposits of $100 are made at the end of each quarter to a bank account for 5 years. Find the accumulated value of these payments if a) $j_{12} = 6\%$; b) $j_4 = 6\%$; c) $j_1 = 6\%$; d) $j_\infty = 6\%$.

6. Find the monthly payment on a $3000 loan for 4 years at a) $j_4 = 9\%$; b) $j_\infty = 9\%$.

7. Find the discounted value at $j_2 = 10\%$ of 20 annual payments of $200 each, the first one due five years hence.

8. Julia borrows $10 000. The loan is to be repaid with equal payments at the end of each month for the next 5 years. Find the size of these payments if a) $j_4 = 12\%$; b) $j_1 = 12\%$.

9. Upon graduation, Scott determines that he has borrowed $8000 from the provincial loan plan over his three years of university. This loan must be repaid with monthly payments (first payment at the end of the first month) over the next 5 years. If $j_2 = 11\%$, find the monthly payment.

10. How much must be deposited in a bank account at the end of each quarter for 4 years to accumulate $4000 if a) $j_{12} = 6\%$; b) $j_2 = 6\%$; c) $j_{365} = 6\%$?

11. Upon her husband's death, a widow finds that her husband had a $30 000 life insurance policy. One option available to her is a monthly annuity over a 10-year period. If the insurance company pays 6% per annum, what monthly income (at the end of each month) would this provide?

12. A city wants to accumulate $500 000 over the next 20 years to redeem an issue of bonds. What payment will be required at the end of each 6 months to accumulate this amount if interest is earned at 7% per annum, compounded monthly?

13. Deposits of $200 are made at the end of each month for 5 years into an account where interest is paid at $j_4 = 9\%$. Find the accumulated value of the account at the end of 5 years if
    a) simple interest is paid for part of a period;
    b) compound interest is paid for part of a period.

14. Find the accumulated value of $100 deposits made at the end of each month for 5 years at $j_1 = 5\%$ if a) compound interest; b) simple interest is paid for part of an interest period.

15. Which is cheaper?
    a) Buy a car for $14 000 and after three years trade it in for $4000.
    b) Rent a car for $350 a month payable at the end of each month for three years. Assume maintenance and license costs are identical and $j_1 = 7.9\%$.

16. An insurance company pays 5% per annum on money left with them. What would be the cost of an annuity certain paying $250 at the end of each month for 10 years if compound interest is paid for each part of a period?

17. What rate of interest j_4 must be earned for deposits of $100 at the end of each month to accumulate to $2000 in 18 months?

18. A furniture suite listing for $2100 may be purchased for $300 down and 18 monthly payments of $100. If cash is paid, a 10% discount is given. Find the highest interest rate j_2 at which the buyer can borrow money in order to take advantage of the cash discount.

19. Amanda's New Year's resolution is to make regular weekly deposits of $25 into her savings account earning $j_{12} = 6\%$. Find the number of full deposits and the size of the smaller concluding deposit necessary to accumulate $3000.

20. Jack has made deposits of $250 at the end of each month for 2 years into an investment fund paying interest at $j_4 = 8\%$. What monthly deposits for the next 12 months will bring the fund up to $10 000?

21. An annuity consists of 40 payments of $300 each made at intervals of three months. Interest is at $j_1 = 12\%$. Find the value of this annuity at each of the following times:
    a) three months before the time of the first payment;
    b) at the time of the last payment;
    c) at the time of the first payment;
    d) three months after the last payment;
    e) four years and three months before the first payment.

22. A couple made monthly deposits of $400 into an account paying interest at 8.25% compounded quarterly. The first deposit was made on August 1, 2000; the last deposit was made on December 1, 2003.
    a) How much money will be in the account on January 1, 2004?
    b) How much money will they be able to withdraw monthly, starting on February 1, 2004, and ending February 1, 2005, to pay for their expenses during their planned trip around the world?

23. Melvin borrows $1000 at 16% compounded semi-annually. The loan is to be repaid with 10 equal monthly payments, the first due one year from today. Determine the size of the payments.

24. A university graduate must repay the government $4200 for outstanding student loans. The graduate can afford to pay $150 monthly. If the first payment is made at the end of the first year and money is worth 11% compounded quarterly, find the number of full payments required and the size of the smaller concluding payment.

25. At age 65 a man takes his life savings of $96 000 and buys a 20-year annuity with monthly payments. Find the size of these payments if interest is at 4% per annum compounded a) daily; b) continuously.

Part B

1. Using the "replacement of rate" method, fill in the table below to show the accumulated value of an ordinary annuity of $1, p times per year, for 10 years at $j_m = 12\%$.

	$m = 2$	$m = 4$	$m = 12$
$p = 2$			
$p = 4$			
$p = 12$	$i = (1.06)^{1/6} - 1$ $s_{\overline{120}|i}$		

2. Develop the replacement formula $R = \dfrac{W}{s_{\overline{m/p}|i}}$ to replace an ordinary general annuity with payments of W made p times a year by an equivalent ordinary simple annuity with payments of R made m times a year.

3. Using the "replacement of payment" method, fill in the table below to show the discounted value of an ordinary annuity of $1, p times per year, for 10 years at $j_m = 12\%$.

	$m = 2$	$m = 4$	$m = 12$
$p = 2$			$\dfrac{a_{\overline{120}|.01}}{s_{\overline{6}|.01}}$
$p = 4$			
$p = 12$			

4. Convert an annuity with semi-annual payments of $500 into an equivalent annuity with
   a) annual payments if money is worth $j_1 = 8\%$;
   b) quarterly payments if money is worth 10% per annum compounded semi-annually.

5. Convert an annuity with quarterly payments of $1000 into an equivalent annuity with
   a) monthly payments if money is worth $j_{12} = 16\%$;
   b) annual payments if money is worth 13% per annum compounded quarterly.

6. Develop the following formulas for the accumulated value S and the discounted value A of an ordinary general annuity with payments W, p times per year, for k years at rate j_m.

$$S = W\frac{(1 + i)^{km} - 1}{(1 + i)^{m/p} - 1}$$

$$A = W\frac{1 - (1 + i)^{-km}}{(1 + i)^{m/p} - 1} \text{ where } i = \frac{j_m}{m}$$

7. Develop the following formulas for the accumulated value S and the discounted value A of a general annuity due with payments W, p times per year, for k years at rate j_m.

$$S = W\frac{s_{\overline{km}|i}}{s_{\overline{m/p}|i}}(1 + i)^{m/p}$$

$$A = W\frac{a_{\overline{km}|i}}{s_{\overline{m/p}|i}}(1 + i)^{m/p} \text{ where } i = \frac{j_m}{m}$$

8. A father has saved money in a fund that was set up to help his son to pay for his 4-year university program. The fund will pay $300 at the beginning of each month for eight months (September through April) plus an extra $2000 each September 1st for four years. If $j_4 = 8\%$, what is the value of the fund on the first day of university (before any withdrawals)?

9. A used car may be purchased for $7600 cash or $600 down and 20 monthly payments of $400 each, the first payment to be made in 6 months. What annual effective rate of interest does the installment plan use?

10. A $5000 loan is repaid by 24 monthly payments of $175 each, followed by 24 monthly payments of $160 each. What interest rate j_1 is being charged?

11. Three years from now, a condominium owners' association will need $100 000 to be used for major renovations. For the last 2 years they made deposits of $4000 at the end of each quarter into a fund that earns interest at $j_{365} = 12\%$. What quarterly deposits to the fund will be needed to reach the goal of $100 000, 3 years from now?

12. In a recent court case, the ABC Development Company sued the XYZ Trust Company. The ABC Development Company had borrowed $8.2 million from the XYZ Trust Company at 13%. The loan was paid off over 3 years with monthly payments. The ABC Development Company thought the rate of interest was $j_1 = 13\%$. The XYZ Trust Company was using $j_{365} = 13\%$. The court sided with the Development Company and awarded them the total dollar difference (without interest). Find the size of the award.

13. To prepare for early retirement, a self-employed businesswoman makes RRSP deposits of $11 500 each year for 20 years, starting on her 32nd birthday. When she is 55, she plans to make the first of 30 equal annual withdrawals. What will be the size of each withdrawal if interest is at 10% compounded a) semi-annually; b) continuously?

14. A deposit of $2000 is made to open an account on April 1, 2002. Quarterly deposits of $300 are then made for 5 years, starting July 1, 2002. Starting October 1, 2008, the first of a sequence of $1000 quarterly withdrawals is made. Assuming an interest rate of 10% compounded continuously, find the balance in the account a) on October 1, 2005; b) on October 1, 2011.

15. Show that the accumulated value S of an ordinary annuity of n payments of $R each, p payments per year over t years, at rate j_∞ is

$$S = R\,s_{\overline{n}|j_\infty} = R\frac{e^{j_\infty t} - 1}{e^{j_\infty/p} - 1}$$

16. Show that the discounted value A of an ordinary annuity of n payments of $\$R$ each, p payments per year over t years, at rate j_∞ is

$$A = R \, a_{\overline{n}|j_\infty} = R\frac{1 - e^{-j_\infty t}}{e^{\,j_\infty/p} - 1}$$

Section 4.2   Mortgages in Canada

During a lifetime, individuals or households may borrow money for a range of purposes, including cars, education, holidays, and personal equipment. However, the biggest loan that most individuals take out is to enable them to buy their home. These housing loans are called mortgages, as the institution lending money (say, the bank) normally requires the borrower to deliver the title of the property as security for the loan. That is, should the individual be unable to repay the loan, the lender has the right to sell the property and use the proceeds to repay the loan. It is also worth noting that many housing loans are initially set for a term of 20 to 25 years, as individuals expect to repay them throughout their working lifetime.

Canadian mortgage regulations require that the interest can be compounded, at most, semi-annually, whereas mortgage payments are usually made monthly. Thus mortgage amortizations in Canada are, in effect, general annuities. There are no legal ceilings on mortgage interest rates in Canada; the rates fluctuate with the price of money on the open market.*

While the monthly payment on a Canadian mortgage is usually determined using a repayment period of 20, 25 or 30 years, the interest rate stated in the mortgage is not guaranteed for that length of time. Rather, the interest rate will change after (usually) one, three or five years depending on the mortgage chosen. At the time the interest rate is renegotiated, the mortgage is open and can be repaid in full. Refinancing of a mortgage will be discussed in Chapter 5.

Note: In the United States, the interest on mortgages is compounded monthly and payments are usually made monthly.

EXAMPLE 1   In July 1979, mortgage rates in Canada averaged around $j_2 = 11\%$. Two years later, mortgage rates reached the unprecedented level of $j_2 = 22\%$. Given a $\$100\ 000$ mortgage to be repaid over 25 years, find the required monthly payment at the two different rates of interest.

Solution a   At $j_2 = 11\%$, first find the rate i per month equivalent to $5\frac{1}{2}\%$ per half-year, such that

$$(1 + i)^{12} = (1.055)^2$$
$$1 + i = (1.055)^{1/6}$$
$$i = (1.055)^{1/6} - 1$$
$$i = .008963394$$

* For the current mortgage rates offered by major Canadian banks, see **www.royalbank.com**, **www.bmc.ca**, **www.tdbank.com**, **www.scotiabank.ca**, **www.cibc.com**.

Second, find the monthly payment R of an ordinary simple annuity with $A = 100\ 000$, $n = 300$, $i = .008963394$.

$$R = \frac{100\ 000}{a_{\overline{300}|i}} = \$962.53$$

At $j_2 = 11\%$ the monthly mortgage payment required is $962.53.

Solution b   At $j_2 = 22\%$, first find the rate i per month equivalent to 11% per half-year such that

$$(1 + i)^{12} = (1.11)^2$$
$$1 + i = (1.11)^{1/6}$$
$$i = (1.11)^{1/6} - 1$$
$$i = .017545481$$

Second, find the monthly payment R of an ordinary simple annuity with $A = 100\ 000$, $n = 300$, and $i = .017545481$.

$$R = \frac{100\ 000}{a_{\overline{300}|i}} = \$1764.11$$

At $j_2 = 22\%$ the monthly mortgage payment required is $1764.11. Notice the significant effect the rate of interest has on a mortgage payment.

■

EXAMPLE 2   A couple is considering a mortgage loan for $130 000 at 6.5% compounded semi-annually.
  a) Calculate the couple's monthly mortgage payment based on the following repayment periods: 25 years, 20 years, 15 years and 10 years.
  b) Suppose that the couple could arrange a mortgage loan where only interest due is paid in monthly payments and the principal as a lump sum is returned at the time of the resale of the home or at the end of 10 years, whichever comes first. Find the monthly interest payment.
  c) If the couple can afford to pay $1300 monthly on their mortgage loan, how many full payments will be required and what will be the smaller concluding payment 1 month after the last full payment?

Solution a   First, find the rate i per month equivalent to 3.25% per half-year, such that

$$(1 + i)^{12} = (1.0325)^2$$
$$1 + i = (1.0325)^{1/6}$$
$$i = (1.0325)^{1/6} - 1$$
$$i = .00534474$$

Second, find the monthly payment R of an ordinary simple annuity with $A = 130\ 000$, $i = .00534474$.

For $n = 300$ (25-year period) : $R = \dfrac{130\ 000}{a_{\overline{300}|i}} = \$870.77$

For $n = 240$ (20-year period) : $R = \dfrac{130\ 000}{a_{\overline{240}|i}} = \$962.65$

For $n = 180$ (15-year period) : $R = \dfrac{130\ 000}{a_{\overline{180}|i}} = \$1126.28$

For $n = 120$ (10-year period) : $R = \dfrac{130\ 000}{a_{\overline{120}|i}} = \$1470.42$

OBSERVATION: Borrowers are advised to choose the term of the loan best suited to their particular situation.

Solution b Interest rate i per month calculated in **Solution a** is $i = .00534474$ and the monthly interest payment is then $130\ 000 \times .00534474 = \$694.82$.

OBSERVATION: Compare the above interest payment with the monthly mortgage payment using a 25-year term to see that most of the mortgage payment pays interest on the outstanding balance and only a small portion of the mortgage payment is used to reduce the principal. (This is true in the early part of the term of a mortgage.) Detailed discussion of the amortization of a mortgage loan is presented in Chapter 5.

Solution c We have an ordinary simple annuity with $A = 130\ 000$, $R = 1300$, $i = .00534474$ and we want to calculate n from equation **(11)**.

$$1300\, a_{\overline{n}|i} = 130\ 000$$

$$a_{\overline{n}|i} = 100$$

$$\frac{1 - (1.00534474)^{-n}}{.00534474} = 100$$

$$1 - (1.00534474)^{-n} = .534474008$$

$$(1.00534474)^{-n} = .465525992$$

$$-n \log 1.00534474 = \log .465525992$$

$$n = 143.4361224$$

Thus there will be 143 full monthly payments and a smaller concluding payment $\$X$ 1 month after the last full payment.

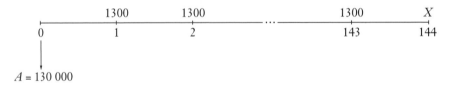

$$A = 130\ 000$$

Using 144 (12 years) as a focal date, we obtain the equation of value for unknown X at rate $i = .00534474$.

$$1300\, s_{\overline{143}|i}\, (1 + i) + X = 130\ 000(1 + i)^{144}$$

$$279\ 526.83 + X = 280\ 094.64$$

$$X = \$567.81$$

■

Exercise 4.2

Part A

In each question listed below, find the monthly mortgage payment.

No.	Mortgage Loan	Interest Rate	Repayment Period
1.	$105 000	$j_2 = 6.25\%$	25 years
2.	$ 93 000	$j_2 = 14.75\%$	20 years
3.	$200 000	$j_2 = 7.50\%$	22 years
4.	$135 000	$j_2 = 9.75\%$	12 years

5. A $100 000 mortgage is to be paid off by monthly payments over 25 years. Find the monthly payment if a) $j_2 = 5\%$; b) $j_2 = 7\%$; c) $j_2 = 9\%$.

6. A $150 000 mortgage is obtained at a rate $j_2 = 10\frac{1}{4}\%$. The mortgage can be paid over over 20, 25 or 30 years. Find the monthly payments necessary under each of these three options.

7. Politicians have sometimes suggested putting a ceiling on mortgage interest rates. Mr. and Mrs. Gordon want to buy a house that would require a $160 000 mortgage to be paid off in 25 years. Interest is at rate $j_2 = 9\%$. If the government guaranteed lower mortgage interest rates, then we would expect the demand for housing to increase, thus forcing house prices upward (given a constant supply). Would Mr. and Mrs. Gordon be better off if they could get a government mortgage at $j_2 = 6\%$ but the price of the house rose such that it required a $180 000 mortgage?

8. A couple is looking at buying one of two houses. The smaller house would require a $130 000 mortgage, the larger house a $160 000 mortgage. If $j_2 = 7.5\%$, what difference would there be in their monthly payments between these two houses (assume a 25-year repayment period is used in both cases)?

9. Mr. and Mrs. Battiston are considering the purchase of a house. They would require a mortgage loan of $120 000 at rate $j_2 = 6.75\%$. If they can afford to pay $900 monthly, how many full payments will be required and what will be the smaller concluding payment?

10. The Belangers can buy a certain house listed at $200 000 and pay $35 000 down. They can get a mortgage for $165 000 at $j_2 = 7\%$ payable monthly over 25 years. The seller of the house offers them the house for $210 000 and he will give them a $175 000 mortgage at $j_2 = 6\%$ with a 25-year repayment period. If these interest rates are guaranteed for 25 years, what should they do?

11. A couple requires a $120 000 mortgage loan at $j_2 = 7.6\%$ to buy a new house.
    a) Find the couple's monthly mortgage payment based on the following repayment periods: 25 years, 20 years and 15 years.
    b) If the couple can afford to pay $1000 monthly on their mortgage loan, what repayment period (in years) should they request?

Part B

In each question listed below, find the unknown part.

No.	Mortgage Loan	Interest Rate	Repayment Period	Monthly Payment
1.	?	$j_2 = 6.75\%$	25 years	$920
2.	$100 000	$j_2 = ?$	20 years	$828
3.	$ 89 000	$j_2 = 6.45\%$	?	$750
4.	$ 80 000	$j_2 = 5.62\%$	17 years	?

5. The Friedmans have a $135 000 mortgage with monthly payments over 25 years at $j_2 = 7\%$. Because they each get paid weekly, they decide to switch to weekly payments (still at $j_2 = 7\%$ and paid over 25 years). Compare their weekly payments to their monthly payments.

6. A couple must decide between a red house and a blue house. To purchase the red house they must take a $155 000 mortgage to be repaid over 20 years at 7.8% compounded semi-annually. To purchase the blue house, they must take a mortgage to be repaid over 25 years at 7.1% compounded semi-annually. If the monthly mortgage payments are equal, determine the value of the mortgage on the blue house.

7. You want to take a $105 000 mortgage at $j_2 = 9\%$ and can afford to pay up to $300 per week. What repayment period in years should you request and what will be your weekly payment?

Section 4.3   Perpetuities

Illustration: Let us consider a person who invests $10 000 at rate $j_2 = 10\%$, keeps the original investment intact and collects $500 interest at the end of each half-year. As long as the interest rate does not change and the original principal of $10 000 is kept intact, interest payments of $500 can be collected forever. We say the interest payments of $500 form a perpetuity.

The present value or discounted value of this infinite series of payments is $10 000, as shown on the diagram below.

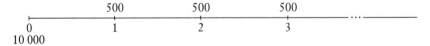

A **perpetuity** is an annuity whose payments begin on a fixed date and continue forever. Examples of perpetuities are the series of interest payments from a sum of money invested permanently at a certain interest rate, a scholarship paid from an endowment on a perpetual basis, and the dividends on a share of preferred stock.

It is meaningless to speak about the accumulated value of a perpetuity, since there is no end to the term of a perpetuity. The discounted value, however, is well defined as the equivalent dated value of the set of payments at the beginning of the term of the perpetuity.

Discounted values of perpetuities are very useful in capitalization problems and will be discussed in Chapter 7, Section 7.3.

The terminology defined for annuities applies to a perpetuity as well. First we shall discuss an **ordinary simple perpetuity**, that is, a series of level periodic payments, made at the ends of interest periods, which continues forever. The other simple perpetuities, i.e., perpetuities due and deferred, may be handled using the concept of an equation of value.

Let A be the discounted value of an ordinary simple perpetuity; let $i > 0$ be the interest rate per period; and let R be the periodic payment of the perpetuity.

Then A must be equivalent to the set of payments R as shown on the diagram below.

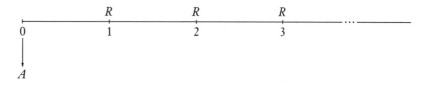

From first principles,

$$A = R(1 + i)^{-1} + R(1 + i)^{-2} + R(1 + i)^{-3} + \ldots \to \infty$$

We know that the sum of an infinite geometric progression can be expressed as

$$\frac{t_1}{1 - r} \quad \text{if} \quad -1 < r < 1$$

In our infinite geometric progression $t_1 = R(1 + i)^{-1}$ and $r = (1 + i)^{-1}$. Obviously, $0 < (1 + i)^{-1} < 1$ for $i > 0$. Therefore,

$$A = \frac{R(1 + i)^{-1}}{1 - (1 + i)^{-1}} = \frac{R}{(1 + i) - 1} = \frac{R}{i}$$

Alternatively, it is evident that A will perpetually provide $R = Ai$ as interest payments on the invested capital A at the end of each interest period as long as it remains invested at rate i.

From $R = Ai$ we obtain the discounted value A of an ordinary simple perpetuity

$$\boxed{A = \frac{R}{i}} \tag{12}$$

OBSERVATION: Formula **(12)** for the discounted value A of an ordinary simple perpetuity can also be obtained as the limit of the discounted value of an ordinary simple annuity

$$A = \lim_{n \to \infty} Ra_{\overline{n}|i} = \lim_{n \to \infty} R\frac{1 - (1 + i)^{-n}}{i} = \frac{R}{i}$$

Actuaries use the notation $a_{\overline{\infty}|i} = \lim_{n \to \infty} a_{\overline{n}|i} = \frac{1}{i}$ when referring to the discounted value of a \$1 ordinary simple perpetuity.

EXAMPLE 1   How much money is needed to establish a scholarship fund paying scholarships of $1000 each half-year if the endowment can be invested at $j_2 = 10\%$ and if the first scholarships will be provided
   a) a half-year from now;
   b) immediately;
   c) 4 years from now?

Solution a   The payments form an ordinary simple perpetuity and we have $R = 1000$ and $i = .05$; and using **(12)** we calculate

$$A = \frac{1000}{.05} = \$20\ 000$$

Solution b   An extra $1000 is needed immediately. Thus the endowment is the sum of the above perpetuity and $1000, i.e., $21 000.

Solution c   If the first scholarships are awarded 4 years from now, the fund will have to contain $21 000 at that time. We have to find the discounted value of $21 000 for 4 years at $j_2 = 10\%$. We obtain

$$21\ 000(1.05)^{-8} = \$14\ 213.63$$

∎

OBSERVATION: Since a perpetuity is an annuity with $n \to \infty$, we may develop formulas for the discounted value A of other simple perpetuities as in Section 3.4:

$$A = \frac{R}{i}(1 + i) = \frac{R}{i} + R \qquad \text{for a simple perpetuity due}$$

$$A = \frac{R}{i}(1 + i)^{-k} \qquad \text{for a simple perpetuity deferred } k \text{ periods.}$$

EXAMPLE 2   A company is expected to pay $0.90 every 3 months on a share of its preferred stock. What should a share of the stock be selling for, if money is worth a) $j_4 = 6\%$; b) $j_4 = 8\%$?

Solution a   We have $R = 0.90$ and $i = .015$; and using **(12)** we calculate

$$A = \frac{0.90}{.015} = \$60$$

Solution b   Using $i = .02$, we calculate $A = \frac{.90}{.02} = \$45$

Note the effect of interest rates on preferred stock values.

∎

When the payment interval and the interest period do not coincide, we have a **general perpetuity.** The simplest way to solve a general perpetuity problem is to find the equivalent rate of interest per payment interval and then use Formula **(12)**.

EXAMPLE 3   What should a share of the stock in Example 2 be selling for if the money is worth a) $j_{12} = 6\%$; b) $j_1 = 6\%$?

Solution a   First we find the rate i per quarter-year equivalent to $\frac{.06}{12} = .005$ per month, such that

$$(1 + i)^4 = (1 + .005)^{12}$$
$$1 + i = (1.005)^3$$
$$i = (1.005)^3 - 1$$
$$i = .015075125$$

We now have an ordinary simple perpetuity with $R = .90$ and $i = .015075125$; and using **(12)** we calculate

$$A = \frac{.90}{i} = \$59.70$$

Solution b   First we find the rate i per quarter-year equivalent to .06 per year, such that

$$(1 + i)^4 = 1.06$$
$$1 + i = (1.06)^{1/4}$$
$$i = (1.06)^{1/4} - 1$$
$$i = .014673846$$

We now have an ordinary simple perpetuity with $R = .90$ and $i = .014673846$; and using **(12)** we calculate

$$A = \frac{.90}{i} = \$61.33$$

■

Exercise 4.3

Part A

1. Find the discounted value of an ordinary simple perpetuity paying $50 a month if a) $j_{12} = 9\%$; b) $j_{12} = 12\%$; c) $j_{12} = 15\%$.
2. Find the discounted value of an ordinary simple perpetuity paying $400 a year if interest is a) $j_1 = 8\%$; b) $j_1 = 12.48\%$.
3. How much money is needed to establish a scholarship fund paying $1500 annually if the fund will earn interest at $j_1 = 14\%$ and the first payment will be made a) at the end of the first year; b) immediately; c) 5 years from now?
4. On September 1, 2003, a philanthropist gives a university a fund of $50 000, which is invested at $j_2 = 10\%$. If semi-annual scholarships are awarded for 20 years from this grant, what is the size of each scholarship if the first one is awarded on a) September 1, 2003; b) September 1, 2005?

5. If the semi-annual scholarships in question 4 were to be awarded indefinitely, what would be the size of the payments for the two starting dates listed above?

6. It costs the C.N.R. $100 at the end of each month to maintain a level-crossing gate system. How much can the company contribute toward the cost of an underpass that will eliminate the level-crossing system if money is worth 15% per annum payable monthly?

7. How much money is needed to establish a research fund paying $2000 semi-annually forever (first payment at the end of six months) if money is worth
a) $j_1 = 12\frac{1}{2}\%$; b) $j_2 = 12\frac{1}{2}\%$; c) $j_{12} = 9\%$?

8. A family is considering putting aluminum siding on their house as it needs painting immediately. Painting the house costs $4200 and must be done every four years. What price can the family afford for the aluminum siding if they earn interest at $j_1 = 8.75\%$ on their savings?

9. The XYZ company has a stock that pays a semi-annual dividend of $4. If the stock sells for $64, what yield j_2 did the investor desire? What is the equivalent rate j_1?

10. The discounted value of a perpetuity is $20 000 and the yield on the fund is $j_2 = 8\%$. What payment will this fund provide a) at the end of each month; b) at the beginning of each year?

11. Deposits of $1000 are placed into a fund at the beginning of each year for the next 20 years. At the end of the 20th year, annual payments from the fund commence and continue forever. If $j_1 = 12\%$, find the value of these payments.

12. On the basis of an unspecified interest rate, i per annum, a perpetuity paying $330 at the end of each year forever may be purchased for $3000. Find i.

13. If money is worth $j_1 = 4\%$, find the present value of a deferred perpetuity of $1000 per year if the first payment is due at the end of 6 years.

14. What annual deposits are needed for 15 years to provide for a perpetuity of $2000 per year, with the first payment in 10 years? Assume interest is at $j_1 = 8\%$.

Part B

1. Use an infinite geometric progression to derive the formula for the discounted value of a simple perpetuity due.

2. Derive $a_{\overline{n}|i} = \frac{1-(1+i)^{-n}}{i}$ as the difference between the discounted value of an ordinary simple perpetuity of $1 per period and the discounted value of an ordinary simple perpetuity of $1 per period deferred for n periods.

3. In 1997, a research foundation was established by a fund of $2 000 000 invested at a rate that would provide $240 000 payments at the end of each year, forever.
a) What interest rate was being earned in the fund?
b) After the payment in 2002, the foundation learned that the rate of interest earned on the fund was being changed to $j_1 = 10\%$. If the foundation wants to continue annual payments forever, what size will the new payments be?
c) If the foundation continues with the $240 000 payments annually, how many full payments can be made at the new rate?

4. A university estimates that its new campus centre will require $3000 for upkeep at the end of each year for the next 5 years and $5000 at the end of each year thereafter indefinitely. If money is worth $j_1 = 12\%$, how large an endowment is necessary for the future upkeep of the campus centre?

5. You take out a loan for L at $j_{12} = 18\%$ and repay $300 at the end of each month for as long as necessary. This loan is invested at $j_1 = 10\%$ and provides for a *perpetuity due* that pays the prize in the annual "Liar's Contest."

The prize is $200 for the 1st year and increases by $150 each year until it reaches $500. From then on the prize remains $500. Find the time and amount of the final repayment on the loan.

6. A university receives a certain sum as a bequest and invests it to earn $j_1 = 8\%$. The fund can be used to pay for a lecturer at $60 000 payable at the end of each year forever, or the money can be used to pay for a new building that the university is planning to erect. The building will be paid for with 25 equal annual payments, the first of which is due 4 years from today when the building will be occupied. Find the amount of each building payment.

7. How much must you deposit at the end of each year for 10 years to fund a perpetuity of $2500 per year with the 1st payment in 15 years? Interest rate is $j_1 = 6\%$ for 15 years, then $j_{12} = 9\%$.

8. A perpetuity paying $1000 at the end of each year is replaced with an annuity paying $X at the end of each month for 10 years. Find X if $j_4 = 15\%$ in both cases.

9. If, in question eight, the annuity paid $250 at the end of each month, how long would the annuity last and what would be the size of the final smaller payment?

10. On September 1, 2000, a wealthy industrialist gives a university a fund of $100 000 which is invested at 8% compounded daily. If semi-annual scholarships are awarded for 20 years from this grant, what is the size of each scholarship if the first one is awarded on a) September 1, 2000; b) September 1, 2002?

11. If the semi-annual scholarships in question 10 were awarded indefinitely, what would be the size of the payments for the two starting dates listed above?

12. Find the present value of a perpetuity of $362.99 payable every 4 years, the first payment 3 years hence, at the annual effective rate of 4%.

13. Find the monthly deposit needed for 5 years to provide for a perpetuity of $400 per month. The 1st perpetuity payment is made 2 years after the last deposit, and interest changes from $j_{12} = 8\%$ to $j_{12} = 9\%$ on that date.

14. You deposit $1000 per year for 10 years at $j_1 = 8\%$. This fund then provides for a perpetuity of $3000 per year, with the first payment made n years after the final deposit. At the time of the first perpetuity payment, interest rates fall to $j_4 = 7\%$. Find n.

15. A perpetuity pays $4000 per year, as follows:
    a) in odd-numbered years, a payment of $4000 is made at the end of the year;
    b) in even-numbered years, a payment of $1000 is made at the end of each quarter.

    Interest is at an annual effective rate of 8%. At the beginning of an odd-numbered year, this perpetuity is exchanged for another of equal value which provides semi-annual payments, the first payment due six months hence. What is the semi-annual payment of the new perpetuity?

<hr>

Section 4.4   ## Annuities Where Payments Vary

Thus far, all the annuities considered have had level payments. Unfortunately, this is not always the case in real life. Thus, it is necessary to be able to handle situations where the size of the payments may vary.

First we consider situations where **payments vary in geometric progression**.

EXAMPLE 1 Mr. Fung wants to buy an annuity of $2000 a year for 10 years that is protected against inflation. The XYZ Trust Company offers to sell him an annuity where payments increase each year by exactly 5%. Find the cost of this annuity if $j_1 = 6\%$. (Assume the payments are at the end of each year and the first payment is $2000.)

Solution We arrange the data on a time diagram below.

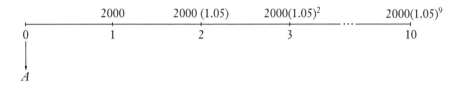

Using 0 as a focal date, we write the equation of value for the discounted value A of these payments at $j_1 = 6\%$

$$A = 2000(1.06)^{-1} + 2000(1.05)(1.06)^{-2} + ... + 2000(1.05)^9(1.06)^{-10}$$

The expression on the right-hand side of the equal sign of the above equation is the sum of 10 terms of a geometric progression (see Appendix 2) with first term $t_1 = 2000(1.06)^{-1}$ and common ratio $r = (1.05)(1.06)^{-1}$. Thus, applying the formula for the sum of n terms of a geometric progression $S_n = t_1\dfrac{1-r^n}{1-r}$ we obtain

$$A = 2000(1.06)^{-1}\left[\frac{1 - (1.05)^{10}(1.06)^{-10}}{1 - (1.05)(1.06)^{-1}}\right] = \$18\ 086.75$$

The cost of the annuity is $18 086.75.

It is worth noting that a 10-year $2000 annuity purchased at $j_1 = 6\%$ with no inflation factor would only cost $2000a_{\overline{10}|6\%} = \$14\ 720.17$. ∎

Second, we consider situations where **payments vary in arithmetic progression.** We illustrate a popular method of solution in the following two examples.

EXAMPLE 2 Mrs. Soros has a job that pays $25 000 a year. Each year she gets a $1000 raise. What is the discounted value of her income for the next 15 years at $j_1 = 7\%$? (Assume payments are at the end of each year with a first payment of $25 000.)

Solution: We arrange the data on a time diagram below.

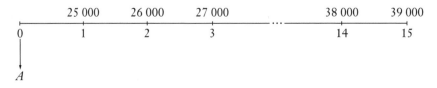

Using 0 as the focal date we write the equation of value for the discounted value A of her income

$$A = 25\ 000(1.07)^{-1} + 26\ 000(1.07)^{-2} + 27\ 000(1.07)^{-3} + \dots$$
$$+38\ 000(1.07)^{-14} + 39\ 000(1.07)^{-15}$$

If we multiply A by (1.07) we obtain

$$(1.07)A = 25\ 000 + 26\ 000(1.07)^{-1} + 27\ 000(1.07)^{-2} + \dots$$
$$+38\ 000(1.07)^{-13} + 39\ 000(1.07)^{-14}$$

Subtracting the first equation from the second gives

$$.07A = 25\ 000 + 1000[(1.07)^{-1} + (1.07)^{-2} + \dots + (1.07)^{-14}] - 39\ 000(1.07)^{-15}$$
$$.07A = 25\ 000 + 1000a_{\overline{14}|.07} - 39\ 000(1.07)^{-15}$$
$$A = \frac{1}{.07}[25\ 000 + 1000a_{\overline{14}|.07} - 39\ 000(1.07)^{-15}]$$
$$A = \$280\ 143.90$$

∎

EXAMPLE 3   Consider an annuity with a term of n periods in which payments begin at P at the end of the first period and change by Q each period thereafter. Assuming an interest rate i per period, and $P > 0, P + (n-1)Q > 0$ (to avoid negative payments if $Q < 0$), show that

a) the discounted value of the annuity is given by

$$A = Pa_{\overline{n}|i} + Q\frac{a_{\overline{n}|i} - n(1+i)^{-n}}{i}$$

b) the accumulated value of the annuity is given by

$$S = Ps_{\overline{n}|i} + Q\frac{s_{\overline{n}|i} - n}{i}$$

Solution a   We arrange the data on a time diagram below.

Using 0 as the focal date, we write the equation of value for the discounted value A of the annuity

$$A = P(1+i)^{-1} + (P+Q)(1+i)^{-2} + (P+2Q)(1+i)^{-3} + \dots$$
$$+ [P + (n-2)Q](1+i)^{-(n-1)} + [P + (n-1)Q](1+i)^{-n}$$

If we multiply A by $(1+i)$ we obtain

$$(1+i)A = P + (P+Q)(1+i)^{-1} + (P+2Q)(1+i)^{-2} + (P+3Q)(1+i)^{-3} + \dots$$
$$+ [P + (n-2)Q](1+i)^{-(n-2)} + [P + (n-1)Q](1+i)^{-(n-1)}$$

Subtracting the first equation from the second we obtain

$$iA = P + Q[(1+i)^{-1} + (1+i)^{-2} + ... + (1+i)^{-(n-1)}] - P(1+i)^{-n} - (n-1)Q(1+i)^{-n}$$
$$= P[1 - (1+i)^{-n}] + Q[(1+i)^{-1} + (1+i)^{-2} + ... + (1+i)^{-n}] - nQ(1+i)^{-n}$$
$$= P[1 - (1+i)^{-n}] + Qa_{\overline{n}|i} - nQ(1+i)^{-n}$$

Thus

$$A = P\frac{1-(1+i)^{-n}}{i} + Q\frac{a_{\overline{n}|i} - n(1+i)^{-n}}{i} = Pa_{\overline{n}|i} + Q\frac{a_{\overline{n}|i} - n(1+i)^{-n}}{i}$$

Solution b   The accumulated value

$$S = A(1+i)^n = [Pa_{\overline{n}|i} + Q\frac{a_{\overline{n}|i} - n(1+i)^{-n}}{i}](1+i)^n = Ps_{\overline{n}|i} + Q\frac{s_{\overline{n}|i} - n}{i}$$

∎

OBSERVATION: Using the formula of Example 3 (part a) to solve Example 2, we substitute $P = 25\ 000, Q = 1000, n = 15, i = .07$ to obtain
$$A = 25\ 000a_{\overline{15}|.07} + 1000\frac{a_{\overline{15}|.07} - 15(1.07)^{-15}}{.07} = \$280\ 143.90$$

Exercise 4.4

Part A

1. Find the discounted value of a series of 20 annual payments of $500 if $j_1 = 5\%$ and we want to allow for an inflation factor of $j_1 = 2\%$. (Assume the payments are at the end of each year and the first payment is $500.)

2. A court is trying to determine the discounted value of the future income of a man paralyzed in a car accident. At the time of the accident the man was earning $45 000 a year and anticipated getting a 4% raise each year. He is 30 years away from retirement. If money is worth $j_1 = 5\%$, what is the discounted value of his future income? (Assume the payments are at the end of each year and the first payment is 45 000(1.04) = $46 800.)

3. Find the discounted value of a series of 15 payments made at the end of each year at $j_1 = 6\%$ if the first payment is $300, the second payment is $600, the third $900 and so on.

4. Mrs. Tong invests $10 000 in a preferred stock that pays an annual dividend at 12%. (That is $1200 at the end of each year.) Mrs. Tong invests the dividend payment in her investment account, which pays annual interest at $j_1 = 10\%$. What is the total accumulated value of her assets at the end of 5 years? Show that if the two rates of interest (12% and 10%) had been equal (at rate i), the answer would have been $10\ 000(1+i)^5$.

5. An investor deposits $1000 at the beginning of each year in a special fund paying interest at $j_1 = 10\%$. These interest payments are then deposited in a bank account paying interest at $j_1 = 6\%$. How much money has been accumulated at the end of 6 years?

6. A certain site is returning an annual rent of $5000 per year payable at the beginning of the year. It is expected that the rent will increase, on the average, 6% per year. Calculate the present value of the site at $j_1 = 10\%$.

7. Mr. Martin needs to have his house painted immediately. Painting the house will cost $1200, so he is trying to decide if he should put on aluminum siding. If we assume the house must be painted every 5th year (forever) and that the cost of painting will rise by 3% per year (forever), how much should Mr. Martin be willing to pay for aluminum siding if he can earn $j_1 = 6\%$ on his money?

8. Find the accumulated value of quarterly payments of $100, $110, $120,... for 20 years if interest is at $j_{12} = 12\%$ and payments are at the end of each quarter.

9. What is the present value of quarterly payments of $100, $110, $120, $130,..., for 20 years, if $j_2 = 8\%$ and payments are at the end of each quarter?

10. An annuity provides for 30 annual payments. The first payment of $100 is made immediately and the remaining payments increase by 8% per annum. Interest is calculated at $j_1 = 13.4\%$. Calculate the present value of this annuity.

11. Find the discounted value one year before the first payment of a series of 20 annual payments the first of which is $500. The payments inflate at $j_1 = 6\%$ and interest is at $j_1 = 8\%$.

12. What is the accumulated value of quarterly payments of $100, $110, $120,... for 15 years if interest is $j_2 = 10\%$ and payments are at the end of each quarter?

13. A decreasing annuity will pay $800 at the end of 6 months, $750 at the end of 1 year etc., until a final payment of $350 is made at the end of 5 years. Find the present value of the payments if money is worth 8% compounded monthly.

14. Find the present value of monthly payments of $20, $25, $30, $35,... for 100 months if interest is $j_\infty = \delta = 10\%$ and payments are at the end of each month.

Part B

1. Consider an annuity where payments vary as represented in the following diagram $(0 < i_1 < 1)$:

Further, assume the interest rate is i_2 per interest period. Let i be the rate such that
$$(1 + i) = \frac{1 + i_2}{1 + i_1}$$
Show that the discounted value A of the above annuity at rate i_2 is
$$A = R a_{\overline{n}|i}$$

2. Derive an algebraic proof of the following identities and interpret by means of time diagrams.

a) $a_{\overline{1}|i} + a_{\overline{2}|i} + a_{\overline{3}|i} + ... + a_{\overline{n}|i} = \dfrac{n - a_{\overline{n}|i}}{i}$

b) $s_{\overline{1}|i} + s_{\overline{2}|i} + s_{\overline{3}|i} + ... + s_{\overline{n-1}|i} = \dfrac{s_{\overline{n}|i} - n}{i}$

3. Show that the discounted value A of a decreasing annuity whose n payments at the end of each year are $nR, (n-1)R, (n-2)R, ..., 2R, R$ is $A = \dfrac{R}{i}(n - a_{\overline{n}|i})$ at rate i per year.

4. The principal on a loan of $5000 is to be repaid in equal annual instalments of $1000 each at the end of each of the next 5 years. Interest will be paid at the end of each year at $j_1 = 4\frac{1}{2}\%$ on the entire outstanding balance, including the $1000 payment then due. Find the purchase price of the loan to yield $j_1 = 5\%$.

5. Find the present value of an annuity where payments are $200 per month at the end of each month during the first year; $195 per month during the second year; $190 per month in the third year, etc., with monthly payments decreasing by $5 after the end of each year. Five dollars will be paid at the end of each month during the 40th year and nothing thereafter. Interest is at $j_{12} = 12\%$.

6. Mrs. Rider has just retired and is trying to decide between two retirement income options as to where she should place her life savings. Fund A will pay her quarterly payments for 25 years starting at $3000 at the end of the first quarter. Fund A will increase her payments each quarter thereafter using an inflation factor equivalent to 4% per annum compounded quarterly. Fund B pays $4500 at the end of each quarter for 25 years with no inflation factor. Which fund should Mrs. Rider choose if she compares the funds using 6% per annum compounded quarterly?

7. Find the present value of a perpetuity under which an amount p is paid at the end of the second year, $p + q$ at the end of the fourth year, $p + 2q$ at the end of the sixth year, $p + 3q$ at the end of the eighth year, etc., if interest is at rate i per year.

8. Consider the perpetuity whose payments at the end of each year are $R, R + p, R + 2p, ..., R + (n - 1)p, R + np, R + np, ...$. The payments increase by a constant amount p until they reach $R + np$, after which they continue without change. Show that the discounted value A of such a perpetuity at rate i per annum is given by

$$A = \frac{R + p\, a_{\overline{m}|i}}{i}$$

9. A man deposits $100 at the beginning of each year into a fund paying interest to him in cash at $j_1 = 6\%$ each year on the principal in the fund. At the end of each year he deposits the interest payments into a second fund earning interest at $j_1 = 4\%$. At the end of which year will the interest fund first exceed the principal fund?

10. For the same lump-sum payment, an insurance company will provide a choice of
    a) an ordinary annuity of $1000 per year for 20 years; or
    b) an ordinary increasing annuity of X per year initially, inflating at 10% per year, for 20 years.
    Find X if interest is at $j_1 = 8\%$. What is the final payment of the increasing annuity?

11. Jeanne has won a lottery that pays $1000 per month in the first year, $1100 per month in the second year, $1200 per month in the third year, etc. Payments are made at the end of each month for 10 years. Using an annual effective interest rate of 3%, calculate the present value of this prize.

12. Consider the increasing $1 annuity in the diagram below

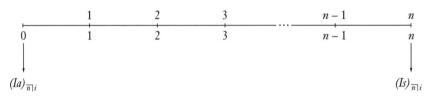

where $(Ia)_{\overline{n}|i}$ denotes the discounted value, $(Is)_{\overline{n}|i}$ denotes the accumulated value of this annuity and i is the interest rate per period. Show that

$$(Ia)_{\overline{n}|i} = a_{\overline{n}|i} + \frac{a_{\overline{n}|i} - n(1 + i)^{-n}}{i} = \frac{\ddot{a}_{\overline{n}|i} - nv^n}{i}$$

$$(Is)_{\overline{n}|i} = s_{\overline{n}|i} + \frac{s_{\overline{n}|i} - n}{i} = \frac{\ddot{s}_{\overline{n}|i} - n}{i}$$

13. Consider the decreasing $1 annuity in the diagram below

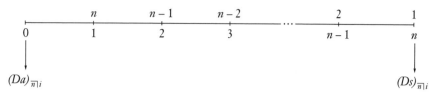

where $(Da)_{\overline{n}|i}$ denotes the discounted value, $(Ds)_{\overline{n}|i}$ denotes the accumulated value of this annuity and i is the interest rate per period. Show that

$$(Da)_{\overline{n}|i} = na_{\overline{n}|i} - \frac{a_{\overline{n}|i} - n(1+i)^{-n}}{i} = \frac{n - a_{\overline{n}|i}}{i}$$

$$(Ds)_{\overline{n}|i} = ns_{\overline{n}|i} - \frac{s_{\overline{n}|i} - n}{i} = \frac{n(1+i)^n - s_{\overline{n}|i}}{i}$$

14. Consider a perpetuity in which payments begin at P at the end of the first period and increase by Q per period thereafter. Assuming the interest rate i per period and $P > 0$, $Q > 0$ (to avoid negative payments), show that the discounted value of this perpetuity is given by

$$\frac{P}{i} + \frac{Q}{i^2}$$

(Hint: take the limit of the formulas in Example 3a as n approaches infinity.)

15. Using the result of Problem 14,

a) show that the discounted value of a $1 increasing perpetuity, denoted by $(Ia)_{\overline{\infty}|i}$, is

$$(Ia)_{\overline{\infty}|i} = \frac{1}{i} + \frac{1}{i^2}$$

b) calculate the present value of a perpetuity whose payments start at $100 at the end of the first month and increase by $2 per month thereafter, assuming $j_{12} = 6\%$.

16. Find the accumulated value of deposits of $1, $2, $3, ..., $98, $99, $100, $99, $98, ..., $3, $2, $1 if interest is 2% per period and deposits are at the end of periods.

17. It is desired to accumulate a fund of $18 000 at the end of 3 years by equal deposits at the beginning of each month. If the deposits earn interest at $j_{12} = 9\%$ but the interest can be reinvested only at $j_{12} = 6\%$, find the size of the necessary deposit.

18. Find the discounted value of payments made at the beginning of each year indefinitely at $j_2 = 12\%$ if the first payment is $100; the second is $200; the third is $300 and so on.

Section 4.5   Summary and Review Exercises

- Change of rate approach to general annuity problems: Find the equivalent interest rate per payment period and then solve a simple annuity problem using the methods of Chapter 3.

- Discounted value of a simple perpetuity is given by

$$A = \frac{R}{i} \qquad \text{for an ordinary perpetuity}$$

$$A = \frac{R}{i}(1+i) \qquad \text{for a perpetuity due}$$

$$A = \frac{R}{i}(1+i)^{-k} \qquad \text{for a deferred perpetuity}$$

- To calculate the discounted value A (or the accumulated value S) of an annuity with payments varying in a constant ratio, we write A as a sum of discounted values of individual payments (or S as a sum of accumulated values of individual payments) and then find the sum of n terms of the resulting geometric progression.

- To calculate the discounted value A of an annuity with payments varying by a constant difference, we first write A as a sum of discounted values of individual payments. Then we multiply the equation for A by $(1 + i)$. Subtracting the equation for A from the equation for $A(1 + i)$ and simplifying the resulting equation for $A(1 + i) - A = Ai$, we obtain the value of A. Similarly we can calculate the accumulated value S.

- For an annuity with a term of n periods in which payments begin at P at the end of the first period and change by Q each period thereafter (assuming $i > 0, P > 0, P + (n - 1)Q > 0$), the discounted value A is given by

$$A = Pa_{\overline{n}|i} + Q\frac{a_{\overline{n}|i} - n(1 + i)^{-n}}{i}$$

and the accumulated value S is given by

$$S = Ps_{\overline{n}|i} + Q\frac{s_{\overline{n}|i} - n}{i}$$

Review Exercises 4.5

1. Replace an annuity with quarterly payments of $300 by an equivalent annuity with
   a) semi-annual payments if money is worth 5% compounded semi-annually;
   b) monthly payments if money is worth 14% compounded quarterly.

2. Find the accumulated and the discounted value of payments of $100 at the end of each quarter-year for 10 years at $j_{12} = 4\%$.

3. Deposits of $100 are made at the end of each month to an account paying interest at 8% compounded semi-annually. How much money will be accumulated in the account at the end of 3 years if
   a) simple interest is paid for part of a conversion period;
   b) compounded interest is paid for part of a conversion period?

4. How many monthly deposits of $100 each and what final deposit one month later will be necessary to accumulate $3000 if interest is
   a) at $6\frac{1}{2}\%$ compounded semi-annually;
   b) at 15% compounded quarterly?

5. A company wishes to have $150 000 in a fund at the end of 8 years. What deposit at the end of each month must they make if the fund pays interest at 5% compounded daily?

6. Steve buys a car worth $15 000. He pays $3000 down and agrees to pay $500 at the end of each month as long as necessary. Find the number of full payments and the final payment one month later if interest is at $j_1 = 14.2\%$.

7. If it takes $50 per month for 18 months to repay a loan of $800, what nominal rate compounded semi-annually is being charged? What annual effective rate is being charged?

8. A lot is sold for $8000 down and 6 semi-annual payments of $3000, the first due at the end of 2 years. Find the cash value of the lot if money is worth 6% compounded daily.

9. Find the accumulated and discounted value of payments of $100 made at the beginning of each month for 5 years at $j_4 = 14\%$.

10. An annuity consists of 60 payments of $200 at the end of each month. Interest is at $j_1 = 11.05\%$. Find the value of this annuity at each of the following times:
    a) at the time of the first payment;
    b) two years before the first payment;
    c) at the time of the last payment.

11. Which is cheaper?
    a) Buy a car for $13 600 and after three years trade it in for $3600.
    b) Lease a car for $318 a month payable at the beginning of each month for three years.
    Assume maintenance costs are identical and that money is worth $j_1 = 8\%$.

12. On November 1, 2000, a research fund of $300 000 was established to provide for equal annual grants for 10 years. What will be the size of each grant if
    a) the fund earns interest at 8% compounded daily and the first grant is awarded on November 1, 2000;
    b) the fund earns interest at 10% compounded monthly and the first grant is awarded on November 1, 2003?

13. A couple requires a $95 000 mortgage loan at $j_2 = 6\frac{3}{4}\%$ to buy a new house.
    a) Find the couple's monthly mortgage payment based on the following repayment periods: 30 years, 20 years and 10 years.
    b) If the couple can afford to pay up to $850 monthly on their mortgage loan, what repayment period in years should they request? What will be their monthly payment?

14. An annuity paying $200 at the end of each year for 20 years is replaced by another annuity paying $X at the end of every 6 months for 12 years. Find X if $j_{12} = 12\%$ in both cases.

15. Starting on her 36th birthday, Ms. Gagnon deposits $2000 a year into an investment account. Her last deposit is at age 65. Starting one month later, she makes monthly withdrawals for 15 years. If $j_4 = 8\%$ throughout, find the size of the monthly withdrawals.

16. Danielle and Sue would like to open their own business in 3 years. They estimate they will need $40 000 at that time. How much should they deposit at the end of each month into their account if they have $5000 in the account now and interest is at $j_4 = 7\%$?

17. What sum of money should be set aside to provide an income of $500 a month for a period of 3 years if money earns interest at $j_{12} = 6\%$ and the first payment is to be received a) one month from now; b) immediately; c) two years from now?

18. The proceeds of a $100 000 death benefit are left on deposit with an insurance company for seven years at an annual effective interest rate of 5%. The balance at the end of seven years is paid to the beneficiary in 120 equal monthly payments of X, with the first payment made immediately. During the payout period, interest is credited at an annual effective interest rate of 3%. Calculate X.

19. A certain stock is expected to pay a dividend of $4 at the end of each quarter for an indefinite period in the future. If an investor wishes to realize an annual effective yield of 12%, how much should he pay for the stock?

20. How much money is needed to establish a scholarship fund paying $1000 a year indefinitely if
    a) the fund earns interest at $j_1 = 12\%$ and the first scholarship is provided at the end of 3 years?

b) the fund earns interest at $j_{12} = 12\%$ and the first scholarship is provided immediately?

21. On the assumption that a farm will net $15 000 annually indefinitely, what is a fair price for it if money is worth a) $j_1 = 8\%$, b) $j_{12} = 15\%$?

22. Mr. Jenkins invests $1000 at the end of each year for 10 years in an investment fund which pays $j_1 = 8\%$. The fund pays the interest out in cheque form at the end of each year and does not allow deposits of less than $1000. Mr. Jenkins deposits his annual interest payment from the fund into his bank account, which pays interest at $j_1 = 4\%$. How much money does he have at the end of 10 years?

23. Show that in Question 22, if both rates of interest had been equal to $i\%$, the answer would have been $1000\, s_{\overline{10}|i}$.

24. Find the discounted and the accumulated value of a decreasing annuity of 20 payments at the end of each year at $j_1 = 12\%$, if the first payment is $2000, the second $1900 and so on, the last payment being $100.

25. Find the discounted value of a series of payments that start at $18 000 at the end of year one and then increase by $2000 each year forever (i.e., $20 000, $22 000, etc.) if interest is at $j_1 = 10\%$.

26. How much a month for 4 years at 12% compounded continuously would you have to save in order to receive $500 a month for 3 years afterward?

27. Under the terms of a contract, the following payments are guaranteed:
a) $100 000 payable immediately.
b) $75 000 per year for 5 years, payable in equal monthly instalments of $6250 at the end of each month.
c) $50 000 per year, payable annually at the end of the 6th through the 10th years.
If interest is at 5% compounded annually, find the discounted value at the beginning of the 4th year of the remaining payments due under the contract.

28. Every two years Mrs. Furtado deposits $1000 into a fund that pays interest at $j_2 = 10\%$. The first deposit is on her 53rd birthday and the last deposit is on her 65th birthday. Beginning three months after her 65th birthday and continuing every three months thereafter, she withdraws equal amounts X, which will exactly exhaust the fund on her 79th birthday. Find X.

29. Find the present value of an ordinary annuity paid annually for 25 years if payments are $1000 per year for the first 7 years, $5000 for the following 8 years, and $2000 in the final 10 years. Interest is $j_{12} = 6\%$ throughout this period.

30. A $5000 loan at $j_1 = 14\%$ is repaid by 10 quarterly payments, as shown. Find X.

	X	X	X	$2X$	$2X$	$2X$		$2X$
0	1	2	3	4	5	6	...	10

31. $100 is deposited in an account paying $j_{12} = 6\%$ every month for 20 years. Exactly 6 months after the last deposit the rate changes to $j_4 = 5\%$. On this date the first withdrawal is made in a series of quarterly withdrawals for 15 years. Find the amount of each withdrawal.

32. Find the present value of future contributions to a pension plan, for a person aged 35 earning $30 000 per year and expecting to retire at age 65. The pension plan requires contributions of 5% of salary and the employee expects to receive average annual salary increases of 3%. Use an annual effective rate of interest of 7% and assume contributions are at the end of each year.

CASE STUDY I

Canadian mortgages

Consider a purchase of a $120 000 house with a down payment of $20 000 and a $100 000 mortgage at $j_2 = 7\%$.

a) Find the monthly payment based on the following repayment periods: 25 years, 20 years and 15 years.

b) Using a 20-year repayment period, find the concluding payment, the total cost of financing, and the interest and principal parts of the first payment.

c) If you can afford to pay up to $1000 a month on the mortgage, what repayment period (in full years) should you request?

d) Compare the total payouts using monthly payments of $1000 with weekly payments of $250.

Chapter Five

Amortization Method and Sinking Funds

Section 5.1   ## Amortization of a Debt

In this chapter we shall discuss different methods of repaying interest-bearing loans, which is one of the most important applications of annuities in business transactions.

The first and most common method is the **amortization method**. When this method is used to liquidate an interest-bearing debt, a series of periodic payments, usually equal, is made. Each payment pays the interest on the unpaid balance and also repays a part of the outstanding principal. As time goes on, the outstanding principal is gradually reduced and interest on the unpaid balance decreases.

When a debt is amortized by equal payments at equal payment intervals, the debt becomes the discounted value of an annuity. The size of the payment is determined by the methods used in the annuity problems of the preceding chapters. The common commercial practice is to round the payment up to the next cent. This practice will be used in this textbook unless specified otherwise. Instead of rounding up to the next cent, the lender may round up to the next dime or dollar. In any case the rounding of the payment up to the cent, to the dime or to the dollar will result in a smaller concluding payment. An equation of value at the time of the last payment will give the size of the smaller concluding payment.

EXAMPLE 1   A loan of $20 000 is to be amortized with equal monthly payments over a period of 10 years at $j_{12} = 12\%$. Find the concluding payment if the monthly payment is rounded up to a) the cent; b) the dime.

Solution   First we find the monthly payment R, given $A = 20\ 000$, $n = 120$ and $i = .01$. Using equation **(11)** of Section 3.3 we calculate

$$R = \frac{20\ 000}{a_{\overline{120}|.01}} = 286.9418968$$

and set up an equation of value for X at 120,

$$X + Rs_{\overline{119}|.01}(1.01) = 20\ 000(1.01)^{120}$$

Solution a   If the monthly payment is rounded up to the next cent, we have $R = \$286.95$ and calculate

$$X = 20\ 000(1.01)^{120} - 286.95s_{\overline{119}|.01}(1.01)$$
$$= 66\ 007.74 - 65\ 722.65$$
$$= \$285.09$$

Solution b   If the monthly payment is rounded up to the next dime, we have $R = \$287$ and calculate

$$X = 20\ 000(1.01)^{120} - 287s_{\overline{119}|.01}(1.01)$$
$$= 66\ 007.74 - 65\ 734.10$$
$$= \$273.64$$

■

When interest-bearing debts are amortized by means of a series of equal payments at equal intervals, it is important to know how much of each payment goes for interest and how much goes for the reduction of principal. For example, this may be a necessary part of determining one's taxable income or tax deductions. We construct an **amortization schedule**, which shows the progress of the amortization of the debt.

> **CALCULATION TIP:** The common commercial practice for loans that are amortized by equal payments is to round the payment up to the next cent (sometimes up to the next nickel, dime or dollar if so specified). For each loan you should calculate the reduced final payment. For all other calculations (other than the regular amortization payment), use the normal round-off procedure. Do not round off the interest rate per conversion period. Use all the digits provided by your calculator (store the rate in a memory of the calculator) to avoid significant round-off errors. When calculating the total payout of the loan (also called the total debt in the sum of digits method of section 5.4), find the total sum of all regular (rounded-up) payments and the final reduced (rounded-off) payment.

EXAMPLE 2   A debt of $22 000 with interest at $j_4 = 10\%$ is to be amortized by payments of $5000 at the end of each quarter for as long as necessary. Make out an amortization schedule showing the distribution of the payments as to interest and the repayment of principal.

Solution   The interest due at the end of the first quarter is $2\frac{1}{2}\%$ of $22 000 or $550.00. The first payment of $5000 at this time will pay the interest and will also reduce the **outstanding principal balance*** by $4450. Thus the outstanding principal after the first payment is reduced to $17 550. The interest due at the end of the second quarter is $2\frac{1}{2}\%$ of $17 550 or $438.75. The second payment of $5000 pays the interest and reduces the indebtedness by $4561.25. The outstanding principal now becomes $12 988.75. This procedure is repeated and the results are tabulated below in the amortization schedule.

Payment Number	Monthly Payment	Interest Payment	Principal Payment	Outstanding Balance
				$22 000.00
1	5 000.00	550.00	4 450.00	17 550.00
2	5 000.00	438.75	4 561.25	12 988.75
3	5 000.00	324.72	4 675.28	8 313.47
4	5 000.00	207.84	4 792.16	3 521.31
5	3 609.34	88.03	3 521.31	0
Totals	23 609.34	1609.34	22 000.00	

It should be noted that the 5th payment is only $3609.34, which is the sum of the outstanding principal at the end of the 4th quarter plus the interest due at $2\frac{1}{2}\%$. The totals at the bottom of the schedule are for checking purposes. The total amount of principal repaid must equal the original debt. Also, the total of the periodic payments must equal the total interest plus the total principal returned. Note that the entries in the principal repaid column (except the final payment) are in the ratio $(1 + i)$. That is

$$\frac{4561.25}{4450.00} \doteq \frac{4675.28}{4561.25} \doteq \frac{4792.16}{4675.28} \doteq 1.025.$$

∎

EXAMPLE 3   Consider a $6000 loan that is to be repaid over 5 years with monthly payments at $j_{12} = 6\%$. Show the first three lines of the amortization schedule.

Solution   First, we find the monthly payment R given $A = 6000$, $n = 60$ and $i = .005$.

$$R = \frac{6000}{a_{\overline{60}|.005}} = \$116.00$$

To make out the amortization schedule for the first three months, we follow the method outlined in Example 2 with $i = .005$.

---

*__Outstanding principal balance__ is also referred to as outstanding principal or outstanding balance.

Payment Number	Monthly Payment	Interest Payment	Principal Payment	Outstanding Balance
				$6000.00
1	116.00	30.00	86.00	5914.00
2	116.00	29.57	86.43	5827.57
3	116.00	29.14	86.86	5740.71

As before, the entries in the principal repaid column are in the ratio $1 + i$. That is,

$$\frac{86.43}{86.00} \doteq \frac{86.86}{86.43} \doteq 1.005$$

∎

EXAMPLE 4 Prepare a spreadsheet and show the first 3 months and the last 3 months of a complete amortization schedule for the loan of Example 3 above. Also show total payments, total interest and total principal paid.

Note: There are several spreadsheets available (such as EXCEL, LOTUS, QUATTRO PRO) and all can generate required schedules very well. In this text we will present Excel spreadsheets.

Solution We summarize the entries in an Excel spreadsheet and their interpretation in the table below. In any amortization schedule we reserve cells A1 through E1 for headings. The same headings will be used in all future Excel spreadsheets showing amortization schedules.

	CELL	ENTER	INTERPRETATION
Headings	A1	Pmt #	'Payment number'
	B1	Payment	'Monthly payment'
	C1	Interest	'Interest payment'
	D1	Principal	'Principal payment'
	E1	Balance	'Outstanding principal balance'
Line 0	A2	0	Time starts
	E2	6,000	Mortgage amount
Line 1	A3	=A2+1	End of month 1
	B3	116.00	Monthly payment (up to the next cent)
	C3	=E2*0.005	Interest 1 = (Balance 0) (0.005)
	D3	=B3–C3	Principal 1 = Payment – Interest 1
	E3	=E2–D3	Balance 1 = Balance 0 – Principal 1

To generate the complete schedule copy A3.E3 to A4.E62

To get the last payment adjust B62 = E61+C62

To get totals apply Σ to B3.D62

Below are the first 3 months and the last 3 months of a complete amortization schedule together with the required totals.

	A	B	C	D	E
1	Pmt#	Payment	Interest	Principal	Balance
2	0				6,000.00
3	1	116.00	30.00	86.00	5,914.00
4	2	116.00	29.57	86.43	5,827.57
5	3	116.00	29.14	86.86	5,740.71
⋮	⋮	⋮	⋮	⋮	⋮
60	58	116.00	1.72	114.28	230.05
61	59	116.00	1.15	114.85	115.20
62	60	115.78	0.58	115.20	0.00
63		6,959.78	959.78	6,000.00	

■

EXAMPLE 5 A couple purchases a home and signs a mortgage contract for $80 000 to be paid in equal monthly payments over 25 years with interest at $j_2 = 10\frac{1}{2}\%$. Find the monthly payment and make out a partial amortization schedule showing the distribution of the first payments as to interest and repayment of principal.

Solution Since the interest is compounded semi-annually and the payments are paid monthly we have a general annuity problem. First we calculate rate i per month equivalent to $5\frac{1}{4}\%$ per half-year and store it in a memory of the calculator.

$$(1 + i)^{12} = (1.0525)^2$$
$$(1 + i) = (1.0525)^{1/6}$$
$$i = (1.0525)^{1/6} - 1$$
$$i = .008564515$$

The monthly payment $R = \dfrac{80\ 000}{a_{\overline{300}|i}} = \$742.67$.

To make out the amortization schedule for the first 6 months, we follow the same method as in the previous two examples with $i = .008564515$.

Payment Number	Monthly Payment	Interest Payment	Principal Payment	Outstanding Balance
				$80 000.00
1	742.67	685.16	57.51	79 942.49
2	742.67	684.67	58.00	79 884.49
3	742.67	684.17	58.50	79 825.99
4	742.67	683.67	59.00	79 766.99
5	742.67	683.17	59.50	79 707.49
6	742.67	682.66	60.01	79 647.48
Totals	4456.02	4103.50	352.52	

During the first 6 months only $352.52 of the original $80 000 debt is repaid. It should be noted that over 92% of the first six payments goes for interest and less than 8% for the reduction of the outstanding balance.

Again, the entries in the principal repaid column are in the ratio of $(1 + i)$. That is:

$$\frac{58.00}{57.51} \doteq \frac{58.50}{58.00} \doteq \cdots \doteq \frac{60.01}{59.50} \doteq 1 + i$$

∎

EXAMPLE 6 Prepare an Excel spreadsheet and show the first 6 months and the last 6 months of a complete amortization schedule for the mortgage loan of Example 5 above. Also show total payments, total interest and total principal paid.

Solution: We summarize the entries for line 0 and line 1 of an Excel spreadsheet.

CELL	ENTER
A2	0
E2	80,000.00
A3	=A2+1
B3	742.67
C3	= E2*((1 + 0.105/2)^(1/6)-1)
D3	=B3–C3
E3	=E2–D3

To generate the complete schedule copy A3.E3 to A4.E302

To get the last payment adjust B302 =E301+C302

To get the totals apply Σ to B3.D302

Below are the first 6 months and the last 6 months of a complete amortization schedule together with the required totals.

	A	B	C	D	E
1	Pmt#	Payment	Interest	Principal	Balance
2	0				80,000.00
3	1	742.67	685.16	57.51	79,942.49
4	2	742.67	684.67	58.00	79,884.49
5	3	742.67	684.17	58.50	79,825.99
6	4	742.67	683.67	59.00	79,766.99
7	5	742.67	683.17	59.50	79,707.49
8	6	742.67	682.66	60.01	79,647.47

	A	B	C	D	E
⋮	⋮	⋮	⋮	⋮	⋮
297	295	742.67	36.97	705.70	3,611.48
298	296	742.67	30.93	711.74	2,899.74
299	297	742.67	24.83	717.84	2,181.91
300	298	742.67	18.69	723.98	1,457.92
301	299	742.67	12.49	730.18	727.74
302	300	733.97	6.23	727.74	–
303		222,792.30	142,792.30	80,000.00	

∎

EXAMPLE 7   General amortization schedule

Consider a loan of $\$A$ to be repaid with level payments of $\$R$ at the end of each period for n periods, at rate i per period. Show that in the kth line of the amortization schedule ($1 \leq k \leq n$)

a) Interest payment $I_k = iRa_{\overline{n-k+1}|i} = R[1 - (1 + i)^{-(n-k+1)}]$

b) Principal payment $P_k = R(1 + i)^{-(n-k+1)}$

c) Outstanding balance $B_k = Ra_{\overline{n-k}|i}$

Verify that

d) The sum of the principal payments equals the original value of the loan.

e) The sum of the interest payments equals the total payments less the original value of the loan.

f) The principal payments are in the ratio $(1 + i)$.

Solution a   The outstanding balance after the $(k - 1)$st payment is the discounted value of the remaining $n - (k - 1) = n - k + 1$ payments, that is

$$B_{k-1} = Ra_{\overline{n-k+1}|i}$$

Interest paid in the k-th payment is

$$I_k = iB_{k-1} = i\,Ra_{\overline{n-k+1}|i} = i\,R\frac{1 - (1 + i)^{-(n-k+1)}}{i} = R[1 - (1 + i)^{-(n-k+1)}]$$

Solution b   Principal repaid in the k-th payment is

$$P_k = R - I_k = R - R[1 - (1 + i)^{-(n-k+1)}] = R(1 + i)^{-(n-k+1)}$$

Solution c   The outstanding balance after the k-th payment is the discounted value of the remaining $n - k$ payment, that is

$$B_k = Ra_{\overline{n-k}|i}$$

Solution d   The sum of the principal payments is

$$\sum_{k=1}^{n} P_k = \sum_{k=1}^{n} R(1 + i)^{-(n-k+1)} = R[(1 + i)^{-n} + \ldots + (1 + i)^{-1}] = Ra_{\overline{n}|i} = A$$

Solution e   The sum of the interest payment is

$$\sum_{k=1}^{n} I_k = \sum_{k=1}^{n} R[1 - (1 + i)^{-(n-k+1)}] = n\,R - Ra_{\overline{m}|i} = n\,R - A$$

Solution f $$\dfrac{P_{k+1}}{P_k} = \dfrac{R(1+i)^{-[n-(k+1)+1]}}{R(1+i)^{-(n-k+1)}} = (1+i)^{(-n+k)+(n-k+1)} = (1+i)$$

$\blacksquare$

Exercise 5.1

Part A

1. A loan of \$5000 is to be amortized with equal quarterly payments over a period of 5 years at $j_4 = 12\%$. Find the concluding payment if the quarterly payment is rounded up to a) the cent; b) the dime.

2. A loan of \$20 000 is to be amortized with equal monthly payments over a 3-year period at $j_{12} = 8\%$. Find the concluding payment if the monthly payment is rounded up to a) the cent; b) the dime.

3. A \$5000 loan is to be amortized with 8 equal semi-annual payments. If interest is at $j_2 = 14\%$, find the semi-annual payment and construct an amortization schedule.

4. A loan of \$900 is to be amortized with 6 equal monthly payments at $j_{12} = 12\%$. Find the monthly payment and construct an amortization schedule.

5. A loan of \$1000 is to be repaid over 2 years with equal quarterly payments. Interest is at $j_{12} = 6\%$. Find the quarterly payment required and construct an amortization schedule to show the interest and principal portion of each payment.

6. A debt of \$50 000 with interest at $j_4 = 8\%$ is to be amortized by payments of \$10 000 at the end of each quarter for as long as necessary. Create an amortization schedule.

7. A \$10 000 loan is to be repaid with semi-annual payments of \$2500 for as long as necessary. If interest is at $j_{12} = 12\%$, do a complete amortization schedule showing the distribution of each payment into principal and interest.

8. A debt of \$2000 will be repaid by monthly payments of \$500 for as long as necessary, the first payment to be made at the end of 6 months. If interest is at $j_{12} = 9\%$, find the size of the debt at the end of 5 months and make out the complete schedule starting at that time.

9. A couple buys some furniture for \$1500. They pay off the debt at $j_{12} = 18\%$ by paying \$200 a month for as long as necessary. The first payment is at the end of 3 months. Do a complete amortization schedule for this loan showing the distribution of each payment into principal and interest.

10. A \$16 000 car is purchased by paying \$1000 down and then equal monthly payments for 3 years at $j_{12} = 15\%$. Find the size of the monthly payments and complete the first three lines of the amortization schedule.

11. A mobile home worth \$46 000 is purchased with a down payment of \$6000 and monthly payments for 15 years. If interest is $j_2 = 10\%$, find the monthly payment required and complete the first 6 lines of the amortization schedule.

12. A couple purchases a home worth \$256 000 by paying \$86 000 down and then taking out a mortgage at $j_2 = 7\%$. The mortgage will be amortized over 25 years with equal monthly payments. Find the monthly payment and do a partial amortization schedule showing the distribution of the first 6 payments as to interest and principal. How much of the principal is repaid during the first 6 months?

13. In September 1981, mortgage interest rates in Canada peaked at $j_2 = 21\frac{1}{2}\%$. Redo Question 12 using $j_2 = 21\frac{1}{2}\%$.

Part B

1. On a loan with $j_{12} = 12\%$ and monthly payments, the amount of principal in the 6th payment is $40.
   a) Find the amount of principal in the 15th payment.
   b) If there are 36 equal payments in all, find the amount of the loan.

2. A loan is being repaid over 10 years with equal annual payments. Interest is at $j_1 = 10\%$. If the amount of principal repaid in the third payment is $100, find the amount of principal repaid in the 7th payment.

3. The ABC Bank develops a special scheme to help their customers pay their loans off quickly. Instead of making payments of $$X$ once a month, mortgage borrowers are asked to pay $$\frac{X}{4}$ once a week (52 times a year).

   The Gibsons are buying a house and need a $180 000 mortgage. If $j_2 = 8\%$, determine
   a) the monthly payment required to amortize the debt over 25 years;
   b) the weekly payment $\frac{X}{4}$ suggested in the scheme;
   c) the number of weeks it will take to pay off the debt using the suggested scheme.

   Compare these results and comment.

4. A loan is being repaid by monthly instalments of $100 at $j_{12} = 18\%$. If the loan balance after the fourth month is $1200, find the original loan value.

5. A loan is being repaid with 20 annual instalments at $j_1 = 15\%$. In which instalment are the principal and interest portions most nearly equal to each other?

6. A loan is being repaid with 10 annual instalments. The principal portion of the seventh payment is $110.25 and the interest portion is $39.75. What annual effective rate of interest is being charged?

7. A loan at $j_1 = 9\%$ is being repaid by monthly payments of $750 each. The total principal repaid in the twelve monthly instalments of the 8th year is $400. What is the total interest paid in the 12 instalments of the 10th year?

8. Below is part of a mortgage amortization schedule.

Payment	Interest	Principal	Balance
	$243.07	$31.68	
	242.81	31.94	

Determine
   a) the monthly payment;
   b) the effective rate of interest per month;
   c) the nominal rate of interest j_2, rounded to nearest $\frac{1}{8}\%$ (Note: Use this rounded nominal rate from here on);
   d) the outstanding balance just after the first payment shown above;
   e) the remaining period of the mortgage if interest rates don't change.

9. On mortgages repaid by equal annual payments covering both principal and interest, a mortgage company pays as a commission to its agents 10% of the portion of each scheduled instalment, which represents interest. What is the total commission paid per $1000 of original mortgage loan if it is repaid with n annual payments at rate i per year?

10. Fred buys a personal computer from a company whose head office is in the United States. The computer includes software designed to calculate mortgage amortization schedules. Fred has a $75 000 mortgage with 20-year amortization at $j_2 = 6\frac{1}{2}\%$. According to his bank statement, his monthly payment is $555.38. When Fred uses his software, he enters his original principal balance of $75 000, the amortization period of 20 years and the nominal annual rate of interest of $6\frac{1}{2}\%$. The computer produces output that says the monthly payment should be $559.18. Which answer is correct? Can you explain the error?

11. As part of the purchase of a home on January 1, 2000, you have just negotiated a mortgage in the amount of $150 000. The amortization period for calculation of the level monthly payments (principal and interest) has been set at 25 years and the interest rate is 9% per annum compounded semi-annually.
   a) What level monthly payment is required, assuming the first one to be made at February 1, 2000?
   b) It has been suggested that if you multiply the monthly payment [calculated in a)] by 12, divide by 52 then pay the resulting amount each week, with the first payment at January 8, 2000, then the amortization period will be shortened. If the lender agrees to this, at what time in the future will the mortgage be fully paid?
   c) What will be the size of the smaller, final weekly payment, made 1 week after the last regular payment [as calculated for part b)]?

12. Part of an amortization schedule shows

Payment	Interest	Principal	Balance
	440.31	160.07	
	438.71	161.67	

Complete the next 2 lines of this schedule.

13. You lend a friend $15 000 to be amortized by semi-annual payments for 8 years, with interest at $j_2 = 9\%$. You deposit each payment in an account paying $j_{12} = 7\%$. What annual effective rate of interest have you earned over the entire 8-year period?

14. A loan of $10 000 is being repaid by semi-annual payments of $1000 on account of principal. Interest on the outstanding balance at j_2 is paid in addition to the principal repayments. The total of all payments is $12 200. Find j_2.

15. A loan of A is to be repaid by 16 equal semi-annual instalments, including principal and interest, at rate i per half year. The principal in the first instalment (six months hence) is $30.83. The principal in the last is $100. Find the annual effective rate of interest.

16. A loan is to be repaid by 16 quarterly payments of $50, $100, $150,...,$800, the first payment due three months after the loan is made. Interest is at a nominal annual rate of 8% compounded quarterly. Find the total amount of interest contained in the payments.

17. A loan of $20 000 with interest at $j_{12} = 6\%$ is amortized by equal monthly payments over 15 years. In which payment will the interest portion be less than the principal portion, for the first time?

18. You are choosing between two mortgages for $160 000 with a 20-year amortization period. Both charge $j_2 = 7\%$ and permit the mortgage to be paid off in less than 20 years. Mortgage A allows you to make weekly payments, with each payment being $\frac{1}{4}$ of the normal monthly payment. Mortgage B allows you to make double the usual monthly payment every 6 months. Assuming that you will take advantage of the mortgage provisions, calculate the total interest charges over the life of each mortgage to determine which mortgage costs less.

19. In the United States, in an effort to advertise low rates of interest, but still achieve high rates of return, lenders sometimes charge *points*. Each point is a 1% discount from the face value of the loan. Suppose a home is being sold for $220 000 and that the buyer pays $60 000 down and gets a $160 000 15-year mortgage at $j_{12} = 8\%$. The lender charges 5 points, that is 5% of 160 000 = $8000, so the loan is $152 000, but $160 000 is repaid. What is the true interest rate, j_{12} on the loan?

Section 5.2   **Outstanding Balance**

It is quite important to know the amount of principal remaining to be paid at a certain time. The borrower may want to pay off the outstanding balance of the debt in a lump sum, or the lender may wish to sell the contract.

One could find the outstanding balance by making out an amortization schedule. This becomes rather tedious without a spreadsheet when a large number of payments are involved. In this section we shall calculate the outstanding balance directly from an appropriate equation of value.

Let B_k denote the outstanding balance immediately after the kth payment has been made.

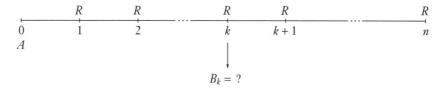

Two methods for finding B_k are available.

The Retrospective Method: This method uses the past history of the debt—the payments that have been made already. The outstanding balance B_k is calculated as the difference between the accumulated value of the debt and the accumulated value of the payments already made.

Thus,

$$\boxed{B_k = A(1 + i)^k - Rs_{\overline{k}|i}} \tag{13}$$

Equation **(13)** always gives the correct value of the outstanding balance B_k and can be used in all cases, even if the number of payments is not known or the last payment is an irregular one.

The Prospective Method: This method uses the future prospects of the debt—the payments yet to be made. The outstanding balance B_k is calculated as the discounted value of the $(n - k)$ payments yet to be made.

If all the payments, including the last one, are equal, we obtain

$$\boxed{B_k = Ra_{\overline{n-k}|i}} \tag{14}$$

While equations **(13)** and **(14)** are algebraically equivalent (see Problem B9), equation **(14)** cannot be used when the concluding payment is an irregular one. However, it is useful if you don't know the original value of the loan. When the concluding payment is an irregular one, equation **(14)** is still applicable if suitably modified, but it is usually simpler to use the retrospective method than to discount the $(n - k)$ payments yet to be made.

EXAMPLE 1 A loan of $2000 with interest at $j_{12} = 12\%$ is to be amortized by equal payments at the end of each month over a period of 18 months. Find the outstanding balance at the end of 8 months.

Solution First we calculate the monthly payment R, given $A = 2000$, $n = 18$, $i = .01$,

$$R = \frac{2000}{a_{\overline{18}|.01}} = \$121.97$$

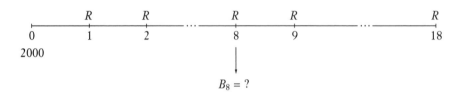

The retrospective method We have $A = 2000$, $R = 121.97$, $k = 8$, $i = .01$ and calculate B_8 using equation **(13)**

$$B_8 = 2000(1.01)^8 - 121.97s_{\overline{8}|.01}$$
$$= 2165.71 - 1010.60 = \$1155.11$$

The prospective method We have $R = 121.97$, $n - k = 10$, $i = .01$ and calculate B_8 using equation **(14)**

$$B_8 = 121.97a_{\overline{10}|.01} = \$1155.21$$

The difference of 10 cents is due to rounding the monthly payment R up to the next cent. The concluding payment is, in fact, slightly smaller than the regular payment $R = 121.97$, so the value of B_8 under the prospective method is not accurate.

◼

EXAMPLE 2 On July 15, 2000, a couple borrowed $10 000 at $j_{12} = 15\%$ to start a business. They plan to repay the debt in equal monthly payments over 8 years with the first payment on August 15, 2000. a) How much principal did they repay during 2000? b) How much interest can they claim as a tax deduction during 2000?

Solution We arrange our data on a time diagram below.

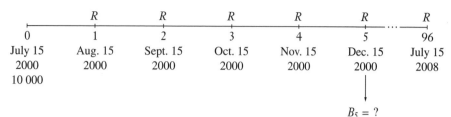

First we calculate the monthly payment R, given $A = 10\,000$, $n = 96$, $i = .0125$

$$R = \frac{10\,000}{a_{\overline{96}|.0125}} = \$179.46$$

Then we calculate the outstanding balance B_5 on December 15, 2000, after the 5th payment has been made. We have $A = 10\,000$, $R = 179.46$, $k = 5$, $i = .0125$, and using equation **(13)** we calculate

$$B_5 = 10\,000(1.0125)^5 - 179.46 s_{\overline{5}|.0125}$$
$$= 10\,640.82 - 920.01$$
$$= \$9720.81$$

The total reduction in principal in 2000 is the difference between the outstanding balance on December 15, 2000, after the 5th payment has been made, and the original debt of $10\,000. Thus, they repaid $10\,000 - 9720.81 = \$279.19$ of principal during 2000.

To get the total interest paid in 2000, we subtract the amount they repaid on principal from the total of the 5 payments, i.e., total interest $= 5 \times 179.46 - 279.19 = \$618.11$. They can deduct $618.11 as an expense on their 2000 income tax return.

∎

This method of amortization is quite often used to pay off loans incurred in purchasing a property. In such cases, the outstanding balance is called the **seller's equity**. The amount of principal that has been paid already plus the down payment is called the **buyer's equity**, or **owner's equity**. At any point in time we have the following relation:

Buyer's equity + Seller's equity = Original selling price

The buyer's equity starts with the down payment and is gradually increased with each periodic payment by the part of the payment which is applied to reduce the outstanding balance. It should be noted that the buyer's equity, as defined above, does not make any allowance for increases or decreases in the value of the property.

EXAMPLE 3 The Mancinis buy a cottage worth $178\,000 by paying $38\,000 down and the balance, with interest at $j_2 = 11\%$, in monthly instalments of $2000 for as long as necessary. Find the Mancinis' equity at the end of 5 years.

Solution The monthly payments form a general annuity. First, we calculate a monthly rate of interest i equivalent to $5\frac{1}{2}\%$ each half-year.

$$(1 + i)^{12} = (1.055)^2$$
$$(1 + i) = (1.055)^{\frac{1}{6}}$$
$$i = (1.055)^{\frac{1}{6}} - 1$$
$$i = .008963394$$

Using the retrospective method, we calculate the seller's equity as the outstanding balance at the end of 60 months.

$$B_{60} = 140\ 000(1 + i)^{60} - 2000s_{\overline{60}|i}$$
$$= 239\ 140.20 - 158\ 008.10 = \$81\ 132.10$$

The Mancinis' equity is then

$$\text{Buyer's equity} = 178\ 000 - 81\ 132.10 = \$96\ 867.90.$$

■

Exercise 5.2

Part A

1. To pay off the purchase of a car, Chantal got a $15 000, 3-year bank loan at $j_{12} = 12\%$. She makes monthly payments. How much does she still owe on the loan at the end of two years (24 payments)? Use both the retrospective and prospective methods.

2. A debt of $10 000 will be amortized by payments at the end of each quarter of a year for 10 years. Interest is at $j_4 = 10\%$. Find the outstanding balance at the end of 6 years.

3. On July 1, 2000, Brian borrowed $30 000 to be repaid with monthly payments (first payment August 1, 2000) over 3 years at $j_{12} = 8\%$. How much principal did he repay in 2000? How much interest?

4. A couple buys a house worth $156 000 by paying $46 000 down and taking out a mortgage for $110 000. The mortgage is at $j_2 = 11\%$ and will be repaid over 25 years with monthly payments. How much of the principal does the couple pay off in the first year?

5. On May 1, 2000, the Morins borrow $4000 to be repaid with monthly payments over 3 years at $j_{12} = 9\%$. The 12 payments made during 2001 will reduce the principal by how much? What was the total interest paid in 2001?

6. To pay off the purchase of home furnishings, a couple takes out a bank loan of $2000 to be repaid with monthly payments over 2 years at $j_{12} = 15\%$. What is the outstanding debt just after the 10th payment? What is the principal portion of the 11th payment?

7. A couple buys a piece of land worth $200 000 by paying $50 000 down and then taking a loan out for $150 000. The loan will be repaid with quarterly payments over 15 years and is at $j_4 = 12\%$. Find the couple's equity at the end of 8 years.

8. A family buys a house worth $326 000. They pay $110 000 down and then take out a 5-year mortgage for the balance at $j_2 = 6\frac{1}{2}\%$ to be amortized over 20 years. Payments will be made monthly. Find the outstanding balance at the end of 5 years and the owners' equity at that time.

9. Land worth $80 000 is purchased by a down payment of $12 000 and the balance in equal monthly instalments for 15 years. If interest is at $j_{12} = 9\%$, find the buyer's and seller's equity in the land at the end of 9 years.

10. A loan of $10 000 is being repaid by instalments of $200 at the end of each month for as long as necessary. If interest is at $j_4 = 8\%$, find the outstanding balance at the end of 1 year.

11. A debt is being amortized at an annual effective rate of interest of 15% by payments of $500 made at the end of each year for 11 years. Find the outstanding balance just after the 7th payment.

12. A loan is to be amortized by semi-annual payments of $802, which includes principal and interest at 8% compounded semi-annually. What is the original amount of the loan if the outstanding balance is reduced to $17 630 at the end of 3 years?

13. A loan is being repaid with semi-annual instalments of $1000 for 10 years at 10% per annum convertible semi-annually. Find the amount of principal in the 6th instalment.

14. Jones purchased a cottage, paying $10 000 down and agreeing to pay $500 at the end of every month for the next ten years. The rate of interest is $j_2 = 12\%$. Jones discharges the remaining indebtedness without penalty by making a single payment at the end of 5 years. Find the extra amount that Jones pays in addition to the regular payment then due.

Part B

1. With mortgage rates at $j_2 = 11\%$, the XYZ Trust Company makes a special offer to its customers. It will lend mortgage money and determine the monthly payment as if $j_2 = 9\%$. The mortgage will be carried at $j_2 = 11\%$ and any deficiency that results will be added to the outstanding balance. If the Moras are taking out a $100 000 mortgage to be repaid over 25 years under this scheme, what will their outstanding balance be at the end of 5 years?

2. The Hwangs can buy a home for $190 000. To do so would require taking out a $125 000 mortgage from a bank at $j_2 = 11\%$. The loan will be repaid over 25 years with the rate of interest fixed for 5 years.
    The seller of the home is willing to give the Hwangs a mortgage at $j_2 = 10\%$. The monthly payment will be determined using a 25-year repayment schedule. The seller will guarantee the rate of interest for 5 years at which time the Hwangs will have to pay off the seller and get a mortgage from a bank. If the Hwangs accept this offer, the seller wants $195 000 for the house, forcing the Hwangs to borrow $130 000. If the Hwangs can earn $j_{12} = 8\%$ on their money, what should they do?

3. The Smiths buy a home and take out a $160 000 mortgage on which the interest rate is allowed to float freely. At the time the mortgage is issued, interest rates are $j_2 = 10\%$ and the Smiths choose a 25-year amortization schedule. Six months into the mortgage, interest rates rise to $j_2 = 12\%$. Three years into the mortgage (after 36 payments) interest rates drop to $j_2 = 11\%$ and four years into the mortgage, interest rates drop to $j_2 = 9\frac{1}{2}\%$. Find the outstanding balance of the mortgage after 5 years. (The monthly payment is set at issue and does not change.)

4. A young couple buys a house and assumes a $90 000 mortgage to be amortized over 25 years. The interest rate is guaranteed at $j_2 = 8\%$. The mortgage allows the couple to make extra payments against the outstanding principal each month. By saving carefully the couple manages to pay off an extra $100 each month. Because of these extra payments, how long will it take to pay off the mortgage and what will be the size of the final smaller monthly payment?

5. A loan is made on January 1, 1985, and is to be repaid by 25 level annual instalments. These instalments are in the amount of $3000 each and are payable on December 31 of the years 1985 through 2009. However, just after the December 31, 1989, instalment has been paid, it is agreed that, instead of continuing the annual instalments on the basis just described, henceforth instalments will be payable quarterly with the first such quarterly instalment being payable on March 31, 1990, and the last one on December 31, 2009. Interest is at an annual effective rate of 10%. By changing from the old repayment schedule to the new one, the borrower will reduce the total amount of payments made over the 25-year period. Find the amount of this reduction.

6. The ABC Trust Company issues loans where the monthly payments are determined by the rate of interest that prevails on the day the loan is made. After that, the rate of interest varies according to market forces but the monthly payments do not change in dollar size. Instead, the length of time to full repayment is either lengthened (if interest rates rise) or shortened (if interest rates fall).

    Medhaf takes out a 10-year, $20 000 loan at $j_2 = 8\%$. After exactly 2 years (24 payments) interest rates change. Find the duration of the remaining loan and the final smaller payment if the new interest rate is a) $j_2 = 9\%$; b) $j_2 = 7\%$.

7. Big Corporation built a new plant in 1997 at a cost of $1 700 000. It paid $200 000 cash and assumed a mortgage for $1 500 000 to be repaid over 10 years by equal semi-annual payments due each June 30 and December 31, the first payment being due on December 31, 1997. The mortgage interest rate is 11% per annum compounded semi-annually and the original date of the loan was July 1, 1997.
    a) What will be the total of the payments made in 1999 on this mortgage?
    b) Mortgage interest paid in any year (for this mortgage) is an income tax deduction for that year. What will be the interest deduction on Big Corporation's 1999 tax form?
    c) Suppose the plant is sold on January 1, 2001. The buyer pays $650 000 cash and assumes the outstanding mortgage. What is Big Corporation's capital gain (amount realized less original price) on the investment in the building?

8. Given a loan L to be repaid at rate i per period with equal payments R at the end of each period for n periods, give the Retrospective and Prospective expressions for the outstanding balance of the loan at time k (after the kth payment is made) and prove that the two expressions are equal.

9. A 5-year loan is being repaid with level monthly instalments at the end of each month, beginning with January 2000 and continuing through December 2004. A 12% nominal annual interest rate compounded monthly was used to determine the amount of each monthly instalment. On which date will the outstanding balance of this loan first fall below one-half of the original amount of the loan?

10. A debt is amortized at $j_4 = 10\%$ by payments of $300 per quarter. If the outstanding principal is $2853.17 just after the kth payment, what was it just after the $(k-1)$st payment?

11. A loan of $3000 is to be repaid by annual payments of $400 per annum for the first 5 years and payments of $450 per year thereafter for as long as necessary. Find the total number of payments and the amount of the smaller final payment made one year after the last regular payment. Assume an annual effective rate of 7%.

12. Five years ago, Justin deposited $1000 into a fund out of which he draws $100 at the end of each year. The fund guarantees interest at 5% on the principal on deposit during the year. If the fund actually earns interest at a rate in excess of 5%, the excess interest earned during the year is paid to Justin at the end of the year in addition to the regular $100 payment. The fund has been earning 8% each year for the past 5 years. What is the total payment Justin now receives?

13. An advertisement by the Royal Trust said,

> Our new Double-Up mortgage can be paid off faster and that dramatically reduces the interest you pay. You can double your payment any or every month, with no penalty. Then your payment reverts automatically to its normal amount the next month. What this means to you is simple. You will pay your mortgage off sooner. And that's good news because you can save thousands of dollars in interest as a result.

Consider a $120 000 mortgage at 8% per annum compounded semi-annually, amortized over 25 years with no anniversary pre payments.
a) Find the required monthly payment for the above mortgage.
b) What is the total amount of interest paid over the full amortization period (assuming $j_2 = 8\%$)?
c) Suppose that payments were "Doubled-Up," according to the advertisement, every 6th and 12th month
  i) How many years and months would be required to pay off the mortgage?
  ii) What would be the total amount of interest paid over the full amortization period?
  iii) How much of the loan would still be outstanding at the end of 3 years, just after the Doubled-Up payment then due?
  iv) How much principal is repaid in the payment due in 37 months?

Section 5.3   Refinancing a Loan—The Amortization Method

It is common to want to renegotiate a long-term loan after it has been partially paid off.

If the loan is refinanced at a lower rate, the discounted value of the savings due to lower interest charges must be compared with the cost of refinancing to decide whether the refinancing would be a profitable one.

EXAMPLE 1 Mr. Bouchard buys $5000 worth of home furnishings from the ABC Furniture Mart. He pays $500 down and agrees to pay the balance with monthly payments over 5 years at $j_{12} = 18\%$. The contract he signs stipulates that if he pays off the contract early there is a penalty equal to three months' payments. After two years (24 payments) Mr. Bouchard realizes that he could borrow the money from the bank at $j_{12} = 12\%$. He realizes that to do so means he will have to pay the three-month penalty on the ABC Furniture Mart contract. Should he refinance?

Solution   First, find the monthly payments required under the original contract.

$$R_1 = \frac{4500}{a_{\overline{60}|.015}} = \$114.28$$

Now, find the outstanding balance B_{24} on the original contract at the end of two years.

$$B_{24} = 4500(1.015)^{24} - 114.28s_{\overline{24}|0.015} = \$3160.52$$

If Mr. Bouchard repays the loan he will also pay a penalty equal to $3 \times R = \$342.84$. Therefore, the amount he must borrow from the bank is

$$3160.52 + 342.84 = \$3503.36$$

Thus, the new monthly payments on the bank loan are

$$R_2 = \frac{3503.36}{a_{\overline{36}|.01}} = \$116.37$$

Therefore, he should not refinance since \$116.37 is larger than \$114.28.   ∎

The largest single loan the average Canadian is likely to make will be the mortgage on one's home. At one time, interest rates on mortgages were guaranteed for the life of the repayment schedule, which could be as long as 25 or 30 years. Now, the longest period of guaranteed interest rates available is usually 5 years, although homeowners may choose mortgages where the interest rate is adjusted every three years, every year or even daily to current interest rates.

A homeowner who wishes to repay the mortgage in full before the defined renegotiation date will have to pay a penalty. In some cases, this penalty is defined as three months of interest on the amount prepaid. In other cases the penalty varies according to market conditions and can be determined by the lender at the time of prepayment (the formula that the lender must use may be defined in the mortgage contract, however). This situation is illustrated in Example 3.

With the high interest rates of the late 1980s and the drop in rates in the early 1990s, many new and different mortgages are being offered to the prospective borrower. It is important that students are capable of analyzing these contracts fully. Several examples are illustrated in the exercises contained in this chapter.

EXAMPLE 2 A couple purchased a home and signed a mortgage contract for \$180 000 to be paid with monthly payments calculated over a 25-year period at $j_2 = 10\%$. The interest rate is guaranteed for 5 years. After 5 years, they renegotiate the interest rate and refinance the loan at $j_2 = 6\frac{1}{2}\%$. There is no penalty if a mortgage is refinanced at the end of an interest rate guarantee period. Find
  a) the monthly payment for the initial 5-year period;
  b) the new monthly payments after 5 years;
  c) the accumulated value of the savings for the second 5-year period (at the end of that period) at $j_{12} = 4\frac{1}{2}\%$; and
  d) the outstanding balance at the end of 10 years.

Solution a   First we calculate a monthly rate of interest i equivalent to $j_2 = 10\%$

$$(1 + i)^{12} = (1.05)^2$$
$$(1 + i) = (1.05)^{\frac{1}{6}}$$
$$i = (1.05)^{\frac{1}{6}} - 1$$
$$i = .008164846$$

Now $A = 180\ 000$, $n = 300$, $i = .008164846$ and we calculate the monthly payment R_1 for the initial 5-year period.

$$R_1 = \frac{180\ 000}{a_{\overline{300}|i}} = \$1610.08$$

Solution b   First we calculate the outstanding balance of the loan after 5 years.

$$180\ 000(1 + i)^{60} - 1610.08s_{\overline{60}|i}$$
$$= 293\ 201.03 - 124\ 015.89 = \$169\ 185.12$$

This outstanding balance is refinanced at $j_2 = 6\frac{1}{2}\%$ over a 20-year period. First, find the rate i per month equivalent to $j_2 = 6\frac{1}{2}\%$.

$$(1 + i)^{12} = (1.0325)^2$$
$$(1 + i) = (1.0325)^{\frac{1}{6}}$$
$$i = (1.0325)^{\frac{1}{6}} - 1$$
$$i = .00534474$$

Now $A = 169\ 185.12$, $n = 240$, $i = .00534474$ and we calculate the new monthly payment R_2.

$$R_2 = \frac{169\ 185.12}{a_{\overline{240}|i}} = \$1252.82$$

Solution c   The monthly savings after the loan is renegotiated after 5 years at $j_2 = 6\frac{1}{2}\%$ is

$$1610.08 - 1252.82 = \$357.26$$

If these savings are deposited in an account paying $j_{12} = 4\frac{1}{2}\%$, the accumulated value of the savings at the end of the 5-year period is:

$$357.26s_{\overline{60}|.00375} = \$23\ 988.42$$

Solution d   The outstanding balance of the loan at the end of 10 years at $i = .00534474$ is

$$169\ 185.12(1 + i)^{60} - 1252.82s_{\overline{60}|i}$$
$$= 232\ 950.03 - 88\ 344.94 = \$144\ 605.09$$

The outstanding balance at the end of 10 years is \$144 605.09. That means that only $180\ 000 - 144\ 605.09 = \$35\ 394.91$ of the loan was repaid during the first 10 years (they still owe 80.3% of the original \$180 000 loan). This is despite the fact that payments in the first 10 years totalled $(60 \times 1610.08 + 60 \times 1252.82)$ or \$171 774.

■

EXAMPLE 3 The Knapps buy a house and borrow $145\ 000$ from the ABC Insurance Company. The loan is to be repaid with monthly payments over 30 years at $j_2 = 9\%$. The interest rate is guaranteed for 5 years. After exactly two years of making payments, the Knapps see that interest rates have dropped to $j_2 = 7\%$ in the market place. They ask to be allowed to repay the loan in full so they can refinance. The Insurance Company agrees to renegotiate but sets a penalty exactly equal to the money the company will lose over the next 3 years. Find the value of the penalty.

Solution First, find the monthly payments R required on the original loan at $j_2 = 9\%$. Find i such that

$$(1 + i)^{12} = (1.045)^2$$
$$(1 + i) = (1.045)^{\frac{1}{6}}$$
$$i = (1.045)^{\frac{1}{6}} - 1$$
$$i = .007363123$$

Now $A = 145\ 000$, $n = 360$, $i = .007363123$, and

$$R = \frac{145\ 000}{a_{\overline{360}|i}} = \$1149.61$$

Next we find the outstanding balance after 2 years

$$= 145\ 000(1 + i)^{24} - 1149.61s_{\overline{24}|i}$$
$$= 172\ 915.20 - 30\ 058.08 = \$142\ 857.12$$

Also, we find the outstanding balance after 5 years if the loan is not renegotiated.

$$145\ 000(1 + i)^{60} - 1149.61s_{\overline{60}|i}$$
$$= 225\ 180.57 - 86\ 335.54 = \$138\ 845.03$$

If the insurance company renegotiates, it will receive $\$142\ 857.12$ plus the penalty $\$X$ now. If they do not renegotiate, they will receive $\$1149.61$ a month for 3 years plus $\$138\ 845.03$ at the end of 3 years. These two options should be equivalent at the current interest rate $j_2 = 7\%$.

```
              +X
Option 1:  142 857.12
           ├────────┼────────┼─── ··· ───┼────────┤
           0        1        2           35       36
Option 2:           1149.61  1149.61      1149.61  1149.61
                                                   +138 845.03
```

Find the monthly rate i equivalent to $j_2 = 7\%$.

$$(1 + i)^{12} = (1.035)^2$$
$$(1 + i) = (1.035)^{\frac{1}{6}}$$
$$i = (1.035)^{\frac{1}{6}} - 1$$
$$i = .00575004$$

Using 0 as the focal date we solve an equation of value for X.

$$X + 142\ 857.12 = 1149.61a_{\overline{36}|i} + 138\ 845.03(1 + i)^{-36}$$
$$X + 142\ 857.12 = 37\ 286.97 + 112\ 950.52$$
$$X = \$7380.37$$

Thus the penalty at the end of 2 years is \$7380.37. If the penalty had been three times the monthly interest on the outstanding balance at the end of 2 years, it would have been

$$3 \times 142\ 857.12 \times .007363123 = \$3155.62$$

■

EXAMPLE 4   The Wongs borrow \$10 000 from a bank to buy some home furnishings. Interest is $j_{12} = 9\%$ and the term of the loan is 3 years. After making 15 monthly payments, the Wongs miss the next two monthly payments. The bank forces them to renegotiate the loan at a new, higher interest rate $j_{12} = 10\frac{1}{2}\%$. Determine

a) the original monthly payments R_1;

b) the outstanding balance of the loan at the time the 17th monthly payment would normally have been made.

c) the new monthly payments R_2 if the original length of the loan term is not extended.

Solution a   Using $A = 10\ 000$, $n = 36$, $i = \frac{.09}{12} = .0075$ we calculate the original monthly payment

$$R_1 = \frac{10\ 000}{a_{\overline{36}|i}} = \$318.00$$

Solution b   The first 15 monthly payments of \$318 were made. Therefore at the end of 15 months the outstanding balance was

$$10\ 000(1 + i)^{15} - 318s_{\overline{15}|i}$$
$$= 11\ 186.03 - 5028.75 = \$6157.28$$

Then no payments were made for 2 months. Therefore the outstanding balance at the end of 17 months is

$$\$6157.28(1 + i)^2 = \$6249.99$$

Solution c   We have $A = 6249.99$, $n = 19$, $i = \frac{.105}{12} = .00875$ and we calculate the new monthly payment

$$R_2 = \frac{6249.99}{a_{\overline{19}|i}} = \$358.49$$

■

Exercise 5.3

Part A

1. A borrower is repaying a $5000 loan at $j_{12} = 15\%$ with monthly payments over 3 years. Just after the 12th payment (at the end of 1 year) he has the balance refinanced at $j_{12} = 12\%$. If the number of payments remains unchanged, what will be the new monthly payment and what will be the monthly savings in interest?

2. A 5-year, $6000 loan is being amortized with monthly payments at $j_{12} = 18\%$. Just after making the 30th payment, the borrower has the balance refinanced at $j_{12} = 12\%$ with the term of the loan to remain unchanged. What will be the monthly savings in interest?

3. A borrower has a $5000 loan with the "Easy-Credit" Finance Company. The loan is to be repaid over 4 years at $j_{12} = 24\%$. The contract stipulates an early repayment penalty equal to 3 months' payments. Just after the 20th payment, the borrower determines that his local bank would lend him money at $j_{12} = 16\%$. Should he refinance?

4. The Jones family buys a fridge and stove totalling $1400 from their local appliance store. They agree to pay off the total amount with monthly payments over 3 years at $j_{12} = 15\%$. If they wish to pay off the contract early they will experience a penalty equal to 3 months' interest on the effective outstanding balance. After 12 payments they see that interest rates at their local bank are $j_{12} = 11\%$. Should they refinance?

5. Consider a couple who bought a house in Canada in 1976. Assume they needed a $60 000 mortgage, which was to be repaid with monthly payments over 25 years. In 1976, interest rates were $j_2 = 10\frac{1}{2}\%$. What was their monthly payment? In 1981 (on the 5th anniversary of their mortgage) their mortgage was renegotiated to reflect current market rates. The repayment schedule was to cover the remaining 20 years and interest rates were now $j_2 = 22\%$. What was the new monthly payment? What effect might this have on homeowners?

6. The Steins buy a house and take out an $85 000 mortgage. The mortgage is amortized over 25 years at $j_2 = 9\%$. After $3\frac{1}{2}$ years, the Steins sell their house and the buyer wants to set up a new mortgage better tailored to his needs. The Steins find out that in addition to repaying the principal balance on their mortgage, they must pay a penalty equal to three months' interest on the outstanding balance. What total amount must they repay?

7. A couple borrows $15 000 to be repaid with monthly payments over 48 months at $j_4 = 10\%$. They make the first 14 payments and miss the next three. What new monthly payments over 31 months would repay the loan on schedule?

8. A loan of $20 000 is to be repaid in 13 equal annual payments, each payable at year-end. Interest is at $j_1 = 8\%$. Because the borrower is having financial difficulties, the lender agrees that the borrower may skip the 5th and 6th payments. Immediately after the 6th payment would have been paid, the loan is renegotiated to yield $j_1 = 10\%$ for the remaining 7 years. Find the new level annual payment for each of the remaining 7 years.

9. A loan of $50 000 was being repaid by monthly level instalments over 20 years at $j_{12} = 9\%$ interest. Now, when 10 years of the repayment period are still to run, it is proposed to increase the interest rate to $j_{12} = 10\frac{1}{2}\%$. What should the new level payment be so as to liquidate the loan on its original due date?

10. The Mosers buy a camper trailer and take out a $15 000 loan. The loan is amortized over 10 years with monthly payments at $j_2 = 18\%$.
    a) Find the monthly payment needed to amortize this loan.
    b) Find the amount of interest paid by the first 36 payments.
    c) After 3 years (36 payments) they could refinance their loan at $j_2 = 16\%$ provided they pay a penalty equal to three months' interest on the outstanding balance. Should they refinance? (Show the difference in their monthly payments.)

Part B

1. A couple buys a home and signs a mortgage contract for $120 000 to be paid with monthly payments over a 25-year period at $j_2 = 10\frac{1}{2}\%$. After 5 years, they renegotiate the interest rate and refinance the loan at $j_2 = 7\%$; find
    a) the monthly payment for the initial 5-year period;
    b) the new monthly payment after 5 years;
    c) the accumulated value of the savings for the second 5-year period at $j_{12} = 3\%$ valued at the end of the second 5-year period;
    d) the outstanding balance at the end of 10 years.

2. Mrs. McDonald is repaying a debt with monthly payments of $100 over 5 years. Interest is at $j_{12} = 12\%$. At the end of the second year she makes an extra payment of $350. She then shortens her payment period by 1 year and renegotiates the loan without penalty and without an interest change. What are her new monthly payments over the remaining two years?

3. Mr. Fisher is repaying a loan at $j_{12} = 15\%$ with monthly payments of $1500 over 3 years. Due to temporary unemployment, Mr. Fisher missed making the 13th through the 18th payments inclusive. Find the value of the revised monthly payments needed starting in the 19th month if the loan is still to be repaid at $j_{12} = 15\%$ by the end of the original 3 years.

4. Mrs. Metcalf purchased property several years ago, (but she cannot remember exactly when). She has a statement in her possession, from the mortgage company, which shows that the outstanding loan amount on January 1, 2000, is $28 416.60 and that the current interest rate is 11% per annum compounded semi-annually. It further shows that monthly payments are $442.65 (Principal and Interest only).

    Assuming continuation of the interest rate until maturity and that monthly payments are due on the 1st of every month
    a) What is the final maturity date of the mortgage?
    b) What will be the amount of the final payment?
    c) Show the amortization schedule entries for January 1 and February 1, 2000.
    d) Calculate the new monthly payment required (effective after December 31, 1999) if the interest rate is changed to 8% per annum compounded semi-annually and the other terms of the mortgage remain the same.
    e) Calculate the new monthly payment if Mrs. Metcalf decides that she would like to have the mortgage completely repaid by December 31, 2009 (i.e., last payment at December 1, 2009). Assume the 11% per annum compounded semi-annually interest rate, and that new payments will start on February 1, 2000.

5. A loan effective January 1, 1995, is being amortized by equal monthly instalments over 5 years using interest at a nominal annual rate of 12% compounded monthly. The first such instalment was due February 1, 1995, and the last such instalment was to be due January 1, 2000. Immediately after the 24th instalment was made on January 1, 1997, a new level monthly

instalment is determined (using the same rate of interest) in order to shorten the total amortization period to $3\frac{1}{2}$ years, so the final instalment will fall due on July 1, 1998. Find the ratio of the new monthly instalment to the original monthly instalment.

6. A \$150 000 mortgage is to be amortized by monthly payments for 25 years. Interest is $j_2 = 9\frac{1}{2}\%$ for 5 years and could change at that time. No penalty is charged for full or partial payment of the mortgage after 5 years.
   a) Calculate the regular monthly payment and the reduced final mortgage payment assuming the $9\frac{1}{2}\%$ rate continues for the entire 25 years.
   b) What is the outstanding balance after
       i) 2 years; ii) 5 years?
   c) During the 5-year period, there is a penalty of 6 months' interest on any principal repaid early. After 2 years, interest rates fall to $j_2 = 8\%$ for 3-year mortgages. Calculate the new monthly payment if the loan is refinanced and set up to have the same outstanding balance at the end of the initial 5-year period. Would it pay to refinance? (Exclude consideration of any possible refinancing costs other than the penalty provided in the question.)

7. A \$200 000 mortgage is taken out at $j_2 = 6\frac{1}{2}\%$, to be amortized over 25 years by monthly payments. Assume that the $6\frac{1}{2}\%$ rate continues for the entire life of the mortgage.
   a) Find the regular monthly payment and the reduced final payment.
   b) Find the accumulated value of all the interest payments on the mortgage, at the time of the final payment.
   c) Find the outstanding principal after 5 years.
   d) If an extra \$5000 is paid off in 5 years (no penalty) and the monthly payments continue as before, how much less will have to be paid over the life of the mortgage?

8. Two years ago the Tongs took out a \$175 000 mortgage that was to be amortized over a 25-year period with monthly payments. The initial interest rate was set at $j_2 = 11\frac{1}{4}\%$ and guaranteed for 5 years. After exactly two years of payments, the Tongs see that mortgage interest rates for a 3-year term are $j_2 = 7\frac{1}{2}\%$. They ask the bank to let them pay off the old mortgage in full and take out a new mortgage with a 23-year amortization schedule with $j_2 = 7\frac{1}{2}\%$ guaranteed for 3 years. The bank replies that the Tongs must pay a penalty equal to the total dollar difference in the interest payments over the next three years. The mortgage allows the Tongs to make a 10% lump-sum principal repayment at any time (i.e., 10% of the remaining outstanding balance). The Tongs argue that they should be allowed to make this 10% lump-sum payment first and then determine the interest penalty. How many dollars will they save with respect to the interest penalty if they are allowed to make the 10% lump-sum repayment?

9. Mr. Adams has just moved to Waterloo. He has been told by his employer that he will be transferred out of Waterloo again in exactly three years. Mr. Adams is going to buy a house and requires a \$150 000 mortgage. He has his choice of two mortgages with monthly payments and 25-year amortization period.

   Mortgage A is at $j_2 = 10\%$. This mortgage stipulates, however, that if you pay off the mortgage any time before the fifth anniversary, you will have to pay a penalty equal to three months' interest on the outstanding balance at the time of repayment.

   Mortgage B is at $j_2 = 10\frac{1}{2}\%$ but can be paid off at any time without penalty. Given that Mr. Adams will have to repay the mortgage in three years and that he can save money at $j_1 = 6\%$, which mortgage should he choose?

Section 5.4   Refinancing a Loan – The Sum-of-Digits Method

A few Canadian lending institutions, when determining the outstanding principal on a consumer loan, do not use the amortization method outlined in the previous two sections, but rather they use an approximation to the amortization method called the **Sum-of-Digits Method** or the **Rule of 78**.

We know that under the amortization method, each payment made on a loan is used partly to pay off interest owing and partly to pay off principal owing. We also know that the interest portion of each payment is largest at the early instalments and gets progressively smaller over time. Thus, if a consumer borrowed \$1000 at $j_{12} = 12\%$ and repaid the balance over 12 months, we would get the following amortization schedule:

Payment Number	Monthly Payment	Interest Payment	Principal Payment	Outstanding Balance
				\$1000.00
1	88.85	10.00	78.85	921.15
2	88.85	9.21	79.64	841.51
3	88.85	8.42	80.43	761.08
4	88.85	7.61	81.24	679.84
5	88.85	6.80	82.05	597.79
6	88.85	5.98	82.87	514.92
7	88.85	5.15	83.70	431.22
8	88.85	4.31	84.54	346.68
9	88.85	3.47	85.38	261.30
10	88.85	2.61	86.24	175.06
11	88.85	1.75	87.10	87.96
12	88.84	0.88	87.96	0
Totals	1066.20	66.20	1000.00	

Under the sum-of-digits approximation, we find the interest portion of each payment by taking a defined proportion of the total interest paid—that is, \$66.20. The proportion used is found as follows.

The loan is paid off over 12 months. If we sum the digits from 1 to 12 we get 78. We now say that the amount of interest assigned to the first payment is $\frac{12}{78} \times 66.20 = \$10.18$. The amount of interest assigned to the second payment is $\frac{11}{78} \times 66.20 = \$9.34$. This process continues until the amount of interest assigned to the last payment is $\frac{1}{78} \times \$66.20 = \$0.85$. In general, the interest portion of the kth payment, I_k, is given by:

$$\boxed{I_k = \frac{n - k + 1}{s_n}\, I}$$

where n = number of payments
$\quad s_n$ = sum of digits from 1 to $n = \dfrac{n(n + 1)}{2}$
$\quad I$ = total interest

OBSERVATION: Notice that the total of the interest column is still $66.20 since $(\frac{12}{78} + \frac{11}{78} + ... + \frac{1}{78}) = \frac{78}{78}$. Since many consumer loans are paid off over 12 months, the alternative name, the Rule of 78, is sometimes used.

If we apply the sum-of-digits method to the loan described above, we will get the repayment schedule shown below.

Payment Number	Monthly Payment	Interest Payment	Principal Payment	Outstanding Balance
				$1000.00
1	88.85	10.18	78.67	921.33
2	88.85	9.34	79.51	841.82
3	88.85	8.49	80.36	761.46
4	88.85	7.64	81.21	680.25
5	88.85	6.79	82.06	598.19
6	88.85	5.94	82.91	515.28
7	88.85	5.09	83.76	431.52
8	88.85	4.24	84.61	346.91
9	88.85	3.39	85.46	261.45
10	88.85	2.55	86.30	175.15
11	88.85	1.70	87.15	88.00
12	88.85	0.85	88.00	0
Totals	1066.20	66.20	1000.00	

OBSERVATION: One important point to notice here is that while the error is small, the use of the sum-of-digits approximation leads to an outstanding balance that is always larger than that obtained by using the amortization method. Thus, if you want to refinance your loan to repay the balance early, there is a penalty attached because of the use of the sum-of-digits approximation.

In the above illustration we can see that if the consumer paid his loan off in full after one month the lending institution would have realized an interest return of $10.18 on $1000 over one month, or $i = 1.018\%$. This corresponds to $j_{12} = 12.216\%$ as opposed to $j_{12} = 12\%$.

EXAMPLE 1 Consider a $6000 loan that is to be repaid over 5 years at $j_{12} = 6\%$. Show the first three lines of the repayment schedule using: a) the amortization method; b) the sum-of-digits method.

Solution First we calculate the monthly payment R required

$$R = \frac{6000}{a_{\overline{60}|.005}} = \$116.00$$

Solution a   From Example 3 in Section 5.1 we have the repayment schedule using the amortization method as shown below.

Payment Number	Monthly Payment	Interest Payment	Principal Payment	Outstanding Balance
				$6000.00
1	116.00	30.00	86.00	5914.00
2	116.00	29.57	86.43	5827.57
3	116.00	29.14	86.86	5740.71

Solution b   Using the sum-of-digits approach we first calculate the total dollar value of all interest payments. We can find the smaller concluding payment X at the end of 5 years using the method of Section 5.1,

$$X = 6000(1.005)^{60} - 116.00 s_{\overline{59}|.005}(1.005)$$
$$= 8093.10 - 7977.32 = \$115.78$$

and then calculate the total interest in the loan as the difference between the total dollar value of all payments and the value of the loan; i.e.,

$$(59 \times 116.00 + 115.78) - 6000 = \$959.78$$

The sum of the digits from 1 to 60 is 1830. Thus,

$$\text{Interest in first payment} = \frac{60}{1830} \times 959.78 = \$31.47$$

$$\text{Interest in second payment} = \frac{59}{1830} \times 959.78 = \$30.94$$

$$\text{Interest in third payment} = \frac{58}{1830} \times 959.78 = \$30.42$$

This leads to the following entries in our repayment schedule:

Payment Number	Monthly Payment	Interest Payment	Principal Payment	Outstanding Balance
				$6000.00
1	116.00	31.47	84.53	5915.47
2	116.00	30.94	85.06	5830.41
3	116.00	30.42	85.58	5744.83

This illustrates the penalty involved in refinancing a loan, where the lending institution uses the sum-of-digits approximation. The penalty can be quite significant on long-term loans at high interest rates. In fact, it is possible under the sum-of-digits approximation for the interest portion of early payments to exceed the dollar size of these payments, which leads to an outstanding balance that is actually larger than the original loan (see Problem 7 in Exercise 5.4, Part A).

In the above illustration, if the consumer repaid the loan after one month the lending institution would earn $31.47 of interest on their $6000 over one month, or $i = 0.5245\%$. This corresponds to $j_{12} = 6.294\%$ as opposed to $j_{12} = 6\%$.

We pointed out in Section 3.6 that most provinces in Canada have "Truth-in-Lending" Acts to protect the consumer. Unfortunately, these acts apply only if the loan is not renegotiated before the normal maturity date. Paying off a loan early constitutes "breaking the contract" and the Truth-in-Lending laws do not apply once the contract is broken.

If you have a consumer loan from a Canadian lending institution using the Rule of 78, to find the outstanding balance, the loans officer will not go through an entire repayment schedule but rather will use the method outlined in Example 2.

EXAMPLE 2 To pay off the purchase of a mobile home, a couple takes out a loan of $45 000 to be repaid with monthly payments over 10 years at $j_{12} = 9\%$. Using the sum-of-digits method find the outstanding debt at the end of 2 years.

Solution The regular monthly payment R and the final payment X are

$$R = \frac{45\ 000}{a_{\overline{120}|.0075}} = \$570.05$$

$$X = 45\ 000(1.0075)^{120} - 570.05 s_{\overline{119}|.0075}(1.0075)$$
$$= 110\ 311.07 - 109\ 742.76 = \$568.31$$

Total interest $= (119 \times 570.05 + 568.31) - 45\ 000 = \$23\ 404.26$

The total interest to be repaid over the remaining 96 months is called the **interest rebate** and is calculated as

$$\frac{\text{sum of digits 1 to 96}}{\text{sum of digits 1 to 120}} \times 23\ 404.26 = \frac{4656}{7260} \times 23\ 404.26 = \$15\ 009.67$$

The couple would be given the following figures:

Original total debt $= 119R + X$	$= 68\ 404.26$
Less interest rebate	$= 15\ 009.67$
Less payments to date $= 24 \times 570.05$	$= 13\ 681.20$
Outstanding balance at the end of 2 years	$= \$39\ 713.39$

It is interesting to note that the outstanding balance at the end of 2 years under the amortization method is

$$45\ 000(1.0075)^{24} - 570.05 s_{\overline{24}|.0075} = 53\ 838.61 - 14\ 928.74$$
$$= \$38\ 909.87$$

----

* As of January 1, 1983, the Rule of 78 Method became illegal for consumer loans issued by federally chartered Canadian banks only.

EXAMPLE 3 Prepare an Excel spreadsheet and show the first three months and last three months of a complete repayment schedule for the $6000 loan of Example 1 using the sum-of-digits method. Also show total payments, total interest, and total principal paid.

Solution: We use the same headings in cells A1 through E1 as in the amortization schedule of Example 1 and we summarize the entries for line 0 and line 1 of an Excel spreadsheet.

CELL	ENTER
A2	0
E2	6,000
A3	=A2+1
B3	=116.00
C3	=((60−A3+1)/1830)*959.78
D3	=B3−C3
E3	=E2−D3

To generate the complete schedule copy A3.E3 to A4.E62

To get the last payment adjust B62 = E61+C62

To get the totals apply Σ to B3.D62

Below are the first 3 months and the last 3 months of a complete repayment schedule with required totals, using the sum-of-digits method.

	A	B	C	D	E
1	Pmt#	Payment	Interest	Principal	Balance
2	0				6,000.00
3	1	116.00	31.47	84.53	5,915.47
4	2	116.00	30.94	85.06	5,830.41
5	3	116.00	30.42	85.58	5,744.83
⋮	⋮	⋮	⋮	⋮	⋮
60	58	116.00	1.57	114.43	230.31
61	59	116.00	1.05	114.95	115.26
62	60	115.78	0.52	115.26	–
63		6,959.78	959.78	6,000.00	

Exercise 5.4

Part A

1. A loan of $900 is to be repaid with 6 equal monthly payments at $j_{12} = 12\%$. Find the monthly payment and construct the repayment schedule using the sum-of-digits method. (Compare this answer to Exercise 5.1, Part A, Question 4.)

2. A loan of $1000 is to be repaid over one year with equal monthly payments at $j_{12} = 9\%$. Find the monthly payment and construct the repayment schedule using the sum-of-digits method.

3. To pay off the purchase of a car, Derek got a $15 000 3-year loan at $j_{12} = 9\%$. He makes equal monthly payments. Find the outstanding balance on the loan just after the 24th payment using the sum-of-digits method.

4. Christine borrows $20 000 to be repaid by monthly payments over 3 years at $j_{12} = 10\frac{1}{2}\%$. Find the outstanding balance at the end of 16 months using the sum-of-digits method.

5. Raymond is repaying a $5000 loan at $j_{12} = 15\%$ with monthly payments over 3 years. Just after the 12th payment he has the balance refinanced at $j_{12} = 12\%$. The balance is determined by the sum-of-digits method. If the number of payments remains unchanged, what will be the new monthly payments and what will be the monthly savings in interest? (Compare this to Exercise 5.3, Part A, Question 1.)

6. A 5-year $6000 loan is to be repaid with monthly payments at $j_{12} = 18\%$. Just after making the 30th payment, the borrower has the balance refinanced at $j_{12} = 12\%$ with the term of the loan to remain unchanged. If the balance is determined by the sum-of-digits method, what will be the monthly savings in interest? (Compare this to Exercise 5.3, Part A, Question 2.)

7. Consider a $10 000 loan being repaid with monthly payments over 15 years at $j_{12} = 15\%$. Find the outstanding balance at the end of 2 years and at the end of 5 years using both the sum-of-digits method and the amortization method.

8. Michelle has a $5000 loan that is being repaid by monthly payments over 4 years at $j_{12} = 9\%$. The lender uses the sum-of-digits method to determine outstanding balances. After 1 year of payments the lender's interest rate on new loans has dropped to $j_{12} = 6\%$. Will Michelle save money by refinancing the loan? (The term of the loan will not be changed.)

Part B

1. Matthew can borrow $15 000 at $j_4 = 15\%$ and repay the loan with monthly payments over 10 years. If he wants to pay the loan off early, the outstanding balance will be determined using the sum-of-digits method.

    He can also borrow $15 000 with monthly payments over 10 years at $j_4 = 16\%$ and pay the loan off at any time without penalty. The outstanding balance will be determined using the amortization method.

    Matthew has an endowment insurance policy coming due in 4 years that could be used to pay off the outstanding balance at that time in full. Which loan should he take if he earns $j_{12} = 6\%$ on his savings?

2. A loan of $18 000 is to be repaid with monthly payments over 10 years at $j_2 = 8\frac{1}{2}\%$. Using the sum-of-digits method,

    a) construct the first two and the last two lines of the repayment schedule;
    b) find the interest and the principal portion of the 10th payment;

    c) find the outstanding balance at the end of 2 years and compare it with the outstanding balance at the same time calculated by the amortization method;

    d) advise whether the loan should be refinanced at the end of 2 years at current rate $j_2 = 8\%$ with the term of the loan unchanged.

3. A \$20 000 home renovation loan is to be amortized over 10 years by monthly payments, with each regular payment rounded up to the next dollar, and the last payment reduced accordingly. Interest on the loan is at $j_4 = 10\%$. After 4 years the loan is fully paid off with an extra payment. Find the amount of this final payment if the sum-of-digits method is used to calculate the outstanding principal.

Sinking Funds

When a specified amount of money is needed at a specified future date, it is a good practice to accumulate systematically a fund by means of equal periodic deposits. Such a fund is called a **sinking fund**. Sinking funds are used to pay off debts (see Section 5.6), to redeem bond issues, to replace worn-out equipment, to buy new equipment, or in one of the depreciation methods (see Section 7.4).

Since the amount needed in the sinking fund, the time the amount is needed and the interest rate that the fund earns are known, we have an annuity problem in which the size of the payment, the sinking-fund deposit, is to be determined. A schedule showing how a sinking fund accumulates to the desired amount is called a **sinking-fund schedule**.

EXAMPLE 1 An eight-storey condominium apartment building consists of 146 two-bedroom apartment units of equal size. The Board of Directors of the Homeowners' Association estimated that the building will need new carpeting in the halls at a cost of \$25 800 in 5 years.

Assuming that the association can invest their money at $j_{12} = 8\%$, what should be the monthly sinking-fund assessment per unit?

Solution The sinking-fund deposits form an ordinary simple annuity with $S = 25\ 800$, $i = \frac{2}{3}\%$, $n = 60$. We calculate the total monthly sinking-fund deposit

$$R = \frac{25\ 800}{s_{\overline{60}|2/3\%}} = \$351.13 \text{ (rounded off)}$$

Per-unit assessment should be

$$\frac{351.13}{146} = \$2.41$$

■

EXAMPLE 2 Show the first three lines and the last two lines of the sinking-fund schedule, explaining the growth of the fund in Example 1.

Solution At the end of the first month, a deposit of \$351.13 is made and the fund contains \$351.13. This amount earns interest at $\frac{2}{3}\%$ for 1 month, i.e., $351.13 \times \frac{2}{300} = \$2.34$. Thus the total increase at the end of the second month is the second payment plus interest on the amount in the fund, i.e., $351.13 + 2.34 = \$353.47$, and the fund will contain $\$704.60$. This procedure may be repeated to complete the entire schedule.

In order to complete the last two lines of the sinking-fund schedule without running the complete schedule, we may calculate the amount in the fund at the end of the 58th month as the accumulated value of 58 payments, i.e.,

$$351.13 s_{\overline{58}|2/3\%} = \$24\,764.04$$

and complete the schedule from that point. The calculations are tabulated below.

End of the Month	Interest on Fund at $\frac{2}{3}$ %	Deposit	Increase in Fund	Amount in Fund
1	-	351.13	351.13	351.13
2	2.34	351.13	353.47	704.60
3	4.70	351.13	355.83	1060.43
⋮	⋮	⋮		
58				24 764.04
59	165.09	351.13	516.22	25 280.26
60	168.54	351.20*	519.67	25 800.00

* The last deposit is adjusted to have the final amount in the fund equal $\$25\,800$.

■

EXAMPLE 3 Prepare an Excel spreadsheet and show the first 3 months and the last 3 months of the sinking-fund schedule of Example 2 above.

Solution: We reserve cells A1 through E1 for headings and summarize the entries in the table below.

	CELL	ENTER	INTERPRETATION
Headings	A1	Deposit #	'Deposit number' or 'End of the month #'
	B1	Interest	'Interest on fund'
	C1	Deposit	'Monthly deposit'
	D1	Increase	'Increase in fund'
	E1	Amount	'Amount in fund'
Line 0	A2	0	Time starts
	E2	0	Amount in fund at the beginning of month 1
Line 1	A3	=A2+1	End of month 1
	B3	=E2*(0.08/12)	Interest on fund at the end of month 1
	C3	351.13	Monthly deposit
	D3	=B3+C3	Increase in fund at the end of month 1
	E3	=E2+D3	Amount in fund at the end of month 1

To generate the complete schedule copy A3.E3 to A4.E62

To get the last deposit adjust C62 = E62–E61–B62

Below are the first 3 months and the last 3 months of a sinking-fund schedule

	A	B	C	D	E
1	Deposit#	Interest	Deposit	Increase	Amount
2	0				0
3	1	0	351.13	351.13	351.13
4	2	2.34	351.13	353.47	704.60
5	3	4.70	351.13	355.83	1,060.43
⋮	⋮	⋮	⋮	⋮	⋮
60	58	161.67	351.13	512.80	24,764.04
61	59	165.09	351.13	516.22	25,280.26
62	60	168.54	351.20	519.74	25,800.00

■

Exercise 5.5

Part A

1. A couple is saving a down payment for a home. They want to have $15 000 at the end of 4 years in an account paying interest at $j_1 = 6\%$. How much must be deposited in the fund at the end of each year? Make out a schedule showing the growth of the fund.

2. A company wants to save $100 000 over the next 5 years so they can expand their plant facility. How much must be deposited at the end of each year if their money earns interest at $j_1 = 8\%$? Make out a schedule for this problem.

3. What quarterly deposit is required in a bank account to accumulate $2000 at the end of 2 years if interest is at $j_4 = 4\%$? Prepare a schedule for this problem.

4. A sinking fund earning interest at $j_4 = 6\%$ now contains $1000. What quarterly deposits for the next 5 years will cause the fund to grow to $10 000? How much is in the fund at the end of 3 years?

5. A cottagers' association decides to set up a sinking fund to save money to have their cottage road widened and paved. They want to have $250 000 at the end of 5 years and they can earn interest at $j_1 = 9\%$. What annual deposit is required per cottager if there are 30 cottages on the road? Show the complete schedule.

6. Find the quarterly deposits necessary to accumulate $10 000 over 10 years in a sinking fund earning interest at $j_4 = 6\%$. Find the amount in the fund at the end of 9 years and complete the rest of the schedule.

7. A city needs to have $200 000 at the end of 15 years to retire a bond issue. What annual deposits will be necessary if their money earns interest at $j_1 = 7\%$? Make out the first three and last three lines of the schedule.

8. What monthly deposit is required to accumulate $3000 at the end of 2 years in a bank account paying interest at $j_4 = 10\%$?

9. A couple wants to save $200 000 to buy some land. They can save $3500 each quarter-year in a bank account paying $j_4 = 9\%$. How many years (to the nearest quarter) will it take them and what is the size of the final deposit?

10. In its manufacturing process, a company uses a machine that costs $75 000 and is scrapped at the end of 15 years with a value of $5000. The company sets up a sinking fund to finance the replacement of the machine, assuming no change in price, with level payments at the end of each year. Money can be invested at an annual effective interest rate of 4%. Find the value of the sinking fund at the end of the 10th year.

Part B

1. A homeowners' association decided to set up a sinking fund to accumulate $50 000 by the end of 3 years to improve recreational facilities. What monthly deposits are required if the fund earns 5% compounded daily? Show the first three and the last two lines of the sinking-fund schedule.

2. Consider an amount that is to be accumulated with equal deposits R at the end of each interest period for 5 periods at rate i per period. Hence, the amount to be accumulated is $Rs_{\overline{5}|i}$. Do a complete schedule for this sinking fund. Verify that the sum-of-the-interest column plus the sum-of-the-deposit column equals the sum of the increase-in-the-fund column, and both sums equal the final amount in the fund.

3. A sinking fund is being accumulated at $j_{12} = 6\%$ by deposits of $200 per month. If the fund contains $5394.69 just after the kth deposit, what did it contain just after the $(k - 1)$st deposit?

Section 5.6 The Sinking-Fund Method of Retiring a Debt

A common method of paying off long-term loans is to pay the interest on the loan at the end of each interest period and create a sinking fund to accumulate the principal at the end of the term of the loan. Usually, the deposits into the sinking fund are made at the same times as the interest payments on the debt are made to the lender. The sum of the interest payment and the sinking-fund payment is called the **periodic expense** or **cost of the debt**. It should be noted that the sinking fund remains under the control of the borrower. At the end of the term of the loan, the borrower returns the whole principal as a lump-sum payment by transferring the accumulated value of the sinking fund to the lender.

When the sinking-fund method is used, we define the **book value** of the borrower's debt at any time as the original principal minus the amount in the sinking fund. The book value of the debt may be considered as the outstanding balance of the loan.

EXAMPLE A city issues $1 000 000 of bonds paying interest at $j_2 = 9\frac{1}{8}\%$, and by law it is required to create a sinking fund to redeem the bonds at the end of 8 years. If the fund is invested at $j_2 = 8\%$, find a) the semi-annual expense of the debt; b) the book value of the city's indebtedness at the beginning of the 7th year.

Solution a

Semi-annual interest payment on the debt: $1\ 000\ 000 \times 0.045625 = \$45\ 625$

Semi-annual deposit into the sinking fund: $\qquad R = \dfrac{1\ 000\ 000}{s_{\overline{16}|.04}} = \$45\ 820$

Semi-annual expense of the debt $\qquad\qquad\qquad\qquad = \$91\ 445$

Solution b The amount in the sinking fund at the end of the 6th year is the accumulated value of the deposits; i.e.,

$$45\ 820 s_{\overline{12}|.04} = \$688\ 482.41$$

The book value of the city's indebtedness at the beginning of the 7th year is then

$$1\ 000\ 000 - 688\ 482.41 = \$311\ 517.59$$

■

Exercise 5.6

Part A

1. A borrower of $5000 agrees to pay interest semi-annually at $j_2 = 10\%$ on the loan and to build up a sinking fund, which will repay the loan at the end of 5 years. If the sinking fund accumulates at $j_2 = 4\%$, find his total semi-annual expense. How much is in the sinking fund at the end of 4 years?

2. A city borrows $250\ 000$, paying interest annually on this sum at $j_1 = 9\frac{1}{2}\%$. What annual deposits must be made into a sinking fund earning interest at $j_1 = 3\frac{1}{2}\%$ in order to pay off the entire principal at the end of 15 years? What is the total annual expense of the debt?

3. A company issues $500\ 000$ worth of bonds, paying interest at $j_2 = 8\%$. A sinking fund with semi-annual deposits accumulating at $j_2 = 4\%$ is established to redeem the bonds at the end of 20 years. Find
   a) the semi-annual expense of the debt;
   b) the book value of the company's indebtedness at the end of the 15th year.

4. A city borrows $2\ 000\ 000$ to build a sewage treatment plant. The debt requires interest at $j_2 = 10\%$. At the same time, a sinking fund is established, which earns interest at $j_2 = 4\frac{1}{2}\%$ to repay the debt in 25 years. Find
   a) the semi-annual expense of the debt;
   b) the book value of the city's indebtedness at the beginning of the 16th year.

5. On a debt of 4000, interest is paid monthly at $j_{12} = 12\%$ and monthly deposits are made into a sinking fund to retire the debt at the end of 5 years. If the sinking fund earns interest at $j_4 = 3.6\%$, what is the monthly expense of the debt?

6. On a debt of $10\ 000$, interest is paid semi-annually at $j_2 = 10\%$ and semi-annual deposits are made into a sinking fund to retire the debt at the end of 5 years. If the sinking fund earns interest at $j_{12} = 6\%$, what is the semi-annual expense of the debt?

7. Interest at $j_2 = 12\%$ on a loan of 3000 must be paid semi-annually as it falls due. A sinking fund accumulating at $j_4 = 8\%$ is established to enable the debtor to repay the loan at the end of 4 years.
   a) Find the semi-annual sinking fund deposit and construct the last two lines of the sinking fund schedule, based on semi-annual deposits.
   b) Find the semi-annual expense of the loan.
   c) What is the outstanding principal (book value of the loan) at the end of 2 years?

8. A 10-year loan of \$10 000 at $j_1 = 11\%$ is to be repaid by the sinking-fund method, with interest and sinking fund payments made at the end of each year. The rate of interest earned in the sinking fund is $j_1 = 5\%$. Immediately after the 5th year's payment, the lender requests that the outstanding principal be repaid in one lump sum. Calculate the amount of extra cash the borrower has to raise in order to extinguish the debt.

Part B

1. A company issues \$2 000 000 worth of bonds paying interest at $j_{12} = 10\frac{1}{2}\%$. A sinking fund accumulating at $j_4 = 6\%$ is established to redeem the bonds at the end of 15 years. Find
a) the monthly expense of the debt;
b) the book value of the company's indebtedness at the beginning of the 6th year.

2. A man is repaying a \$10 000 loan by the sinking-fund method. His total monthly expense is \$300. Out of this \$300, interest is paid to the lender at $j_{12} = 12\%$ and a deposit is made to a sinking fund earning $j_{12} = 9\%$. Find the duration of the loan and the final smaller payment.

3. A \$100 000 loan is to be repaid in 15 years, with a sinking fund accumulated to repay principal *plus interest*. The loan charges $j_2 = 12\%$, while the sinking fund earns $j_2 = 5\%$. What semi-annual sinking fund deposit is required?

4. A loan of \$20 000 bears interest on the amount outstanding at $j_1 = 10\%$. A deposit is to be made in a sinking fund earning interest at $j_1 = 4\%$, which will accumulate enough to pay one-half of the principal at the end of 10 years. In addition, the debtor will make level payments to the creditor, which will pay interest at $j_1 = 10\%$ on the outstanding balance first and the remainder will repay the principal. What is the total annual payment, including that made to the creditor and that deposited in the sinking fund, if the loan is to be completely retired at the end of 10 years?

5. John borrows \$10 000 for 10 years and uses a sinking fund to repay the principal. The sinking-fund deposits earn an annual effective interest rate of 5%. The total required payment for both the interest and the sinking-fund deposit made at the end of each year is \$1445.05. Calculate the annual effective interest rate charged on the loan.

6. A company borrows \$10 000 for five years. Interest of \$600 is paid semi-annually. To repay the principal of the loan at the end of 5 years, equal semi-annual deposits are made into a sinking fund that credits interest at a nominal rate of 8% compounded quarterly. The first payment is due in 6 months. Calculate the annual effective rate of interest that the company is paying to service and retire the debt.

7. On August 1, 1993, Mrs. Chan borrows \$20 000 for 10 years. Interest at 11% per annum convertible semi-annually must be paid as it falls due. The principal is replaced by means of level deposits on February 1 and August 1 in years 1994 to 2003 (inclusive) into a sinking fund earning $j_1 = 7\%$ in 1994 through December 31, 1998, and $j_1 = 6\%$ January 1, 1999, through 2003.
a) Find the semi-annual expense of the loan.
b) How much is in the sinking fund just after the August 1, 2002, deposit?
c) Show the sinking-fund schedule entries at February 1, 2003, and August 1, 2003.

8. Mr. White borrows \$15 000 for 10 years. He makes total payments, annually, of \$2000. The lender receives $j_1 = 10\%$ on his investment each year for the first 5 years and $j_1 = 8\%$ for the second 5 years. The balance of each payment is invested in a sinking fund earning $j_1 = 7\%$.

a) Find the amount by which the sinking fund is short of repaying the loan at the end of 10 years.
b) By how much would the sinking-fund deposit (in each of the first 5 years only) need to be increased so that the sinking fund at the end of 10 years will be just sufficient to repay the loan?

Section 5.7 Comparison of Amortization and Sinking-Fund Methods

We have discussed the two most common methods of paying off long-term loans: the amortization method and the sinking-fund method. When there are several sources available from which to borrow money, it is important to know how to compare the available loans and choose the cheapest one. The borrower should choose that source for which the periodic expense of the debt is the lowest. When the amortization method is used, the periodic expense of the debt is equal to the periodic amortization payment. When the sinking-fund method is used, the periodic expense of the debt is the sum of the interest payment and the sinking-fund deposit.

To study the relationship between the amortization and sinking-fund methods, we define the following:

L = principal of the loan

n = number of interest periods during the term of the loan

i_1 = loan rate per interest period using amortization

i_2 = loan rate per interest period using the sinking fund

i_3 = sinking-fund rate per interest period

E_A = periodic expense using amortization = $\dfrac{L}{a_{\overline{n}|i_1}} = L\left(\dfrac{1}{s_{\overline{n}|i_1}} + i_1\right)$
     (See Exercise 3.3B 1 b.)

E_S = periodic expense using sinking fund = $\dfrac{L}{s_{\overline{n}|i_3}} + L i_2 = L\left(\dfrac{1}{s_{\overline{n}|i_3}} + i_2\right)$

We shall examine the relationship between E_A and E_S for different levels of rates i_1, i_2, i_3.

a) Let $i_1 = i_2 = i_3 \equiv i$, then $E_A = E_S$

b) Let $i_3 < i_1 = i_2$, then $i_3 < i_1$ implies
$$s_{\overline{n}|i_3} < s_{\overline{n}|i_1}$$
$$\frac{1}{s_{\overline{n}|i_1}} < \frac{1}{s_{\overline{n}|i_3}}$$
$$\frac{1}{s_{\overline{n}|i_1}} + i_1 < \frac{1}{s_{\overline{n}|i_3}} + i_2$$
$$E_A < E_S$$

c) Let $i_3 > i_1 = i_2$, then $i_3 > i_1$ implies
$$s_{\overline{n}|i_3} > s_{\overline{n}|i_1}$$
$$\frac{1}{s_{\overline{n}|i_1}} > \frac{1}{s_{\overline{n}|i_3}}$$
$$\frac{1}{s_{\overline{n}|i_1}} + i_1 > \frac{1}{s_{\overline{n}|i_3}} + i_2$$
$$E_A > E_S$$

d) Let $i_3 < i_1 < i_2$, then $i_3 < i_1$ implies

$$s_{\overline{m}|i_3} < s_{\overline{m}|i_1}$$
$$\frac{1}{s_{\overline{m}|i_1}} < \frac{1}{s_{\overline{m}|i_3}}$$
$$\frac{1}{s_{\overline{m}|i_1}} + i_1 < \frac{1}{s_{\overline{m}|i_3}} + i_2$$
$$E_A < E_S$$

e) Let $i_3 < i_2 < i_1$. In this case, which is the most common, we can't tell which method is cheaper. We must calculate the actual periodic costs to determine the cheaper source of money, i.e., the source with the least periodic expense.

EXAMPLE 1 A company wishes to borrow $500 000 for 5 years. One source will lend the money at $j_2 = 12\%$ if it is amortized by semi-annual payments. A second source will lend the money at $j_2 = 11\%$ if only the interest is paid semi-annually and the principal is returned in a lump sum at the end of 5 years. If the second source is used, a sinking fund will be established by semi-annual deposits that accumulate at $j_{12} = 6\%$. How much can the company save semi-annually by using the better plan?

Solution   When the first source is used, the semi-annual expense of the debt is

$$E_A = \frac{500\ 000}{a_{\overline{10}|0.06}} = \$67\ 933.98$$

When the second source is used, the interest on the debt paid semi-annually is $5\frac{1}{2}\%$ of $500\ 000 = \$27\ 500.00$.

To calculate the semi-annual deposit into the sinking fund, we must first calculate the semi-annual rate i equivalent to $j_{12} = 6\%$

$$(1+i)^2 = (1.005)^{12}$$
$$i = (1.005)^6 - 1$$
$$i = .030377509$$

Now we calculate the semi-annual deposit R into the sinking fund

$$R = \frac{500\ 000}{s_{\overline{10}|i}} = \$43\ 539.48$$

The semi-annual expense of the debt using the second source is

$$E_S = 27\ 500 + 43\ 539.48 = \$71\ 039.48$$

Thus, the first source, using amortization, is cheaper and the company can save
$$71\ 039.48 - 67\ 933.98 = \$3105.50 \text{ semi-annually.}$$

■

EXAMPLE 2 A firm wants to borrow $500 000. One source will lend the money at $j_4 = 10\%$ if interest is paid quarterly and the principal is returned in a lump sum at the end of 10 years. The firm can set up a sinking fund at $j_4 = 7\%$. At what rate j_4 would it be less expensive to amortize the debt over 10 years?

Solution We calculate the quarterly expense of the debt.

Interest payment: $\qquad$ $500\ 000 \times .025 = \$12\ 500.00$

Sinking-fund deposit: $\qquad \dfrac{500\ 000}{s_{\overline{40}|.0175}} = \$\ 8\ 736.05$

Quarterly expense: $\qquad\qquad\qquad = \$21\ 236.05$

The amortization method will be as expensive if the quarterly amortization payment is equal to \$21 236.05. Thus, we want to find the interest rate i per quarter (and then j_4) given $A = 500\ 000$, $R = 21\ 236.05$, $n = 40$.
We have

$$500\ 000 = 21\ 236.05 a_{\overline{40}|i}$$
$$a_{\overline{40}|i} = 23.544868$$

We want to find the rate $j_4 = 4i$ such that $a_{\overline{40}|i} = 23.5449$. A starting value to solve $a_{\overline{40}|i} = 23.5449$ is

$$i = \frac{1 - (\frac{23.5449}{40})^2}{23.5449}$$

or $j_4 = 4i = 11.10\%$. Using linear interpolation, we calculate

	$a_{\overline{40}|i}$	j_4	
$.9633\begin{cases} .5332\begin{cases} 24.0781 \\ 23.5449 \end{cases} \\ 23.1148 \end{cases}$		$\begin{array}{l} 11\% \\ j_4 \\ 12\% \end{array}\left.\begin{array}{l} \\ \end{array}\right\}d\ \Big\}1\%$	

$$\frac{d}{1\%} = \frac{.5332}{.9663}$$
$$d \doteq .55\%$$
$$\text{and } j_4 = 11.55\%$$

If the firm can borrow the money and amortize the debt at less than $j_4 = 11.55\%$, then it will be less expensive than a straight loan at $j_4 = 10\%$ with a sinking fund at $j_4 = 7\%$.

■

Exercise 5.7

Part A

1. A company borrows \$50 000 to be repaid in equal annual instalments at the end of each year for 10 years. Find the total annual cost under the following conditions:
   a) the debt is amortized at $j_1 = 9\%$;
   b) interest at 9% is paid on the debt and a sinking fund is set up at $j_1 = 9\%$;
   c) interest at 9% is paid on the debt and a sinking fund is set up at $j_1 = 6\%$.

2. A company can borrow \$180 000 for 15 years. They can amortize the debt at $j_1 = 10\%$, or they can pay interest on the loan at $j_1 = 9\%$ and set up a sinking fund at $j_1 = 7\%$ to repay the loan. Which plan is cheaper and by how much per annum?

3. A firm wants to borrow \$60 000 to be repaid over 5 years. One source will lend them the money at $j_2 = 10\%$ if it is amortized by semi-annual payments. A second source will lend them money at $j_2 = 9\frac{1}{2}\%$ if only the interest is paid semi-annually and the principal is returned in a lump sum at the end of 5 years. The firm can earn $j_2 = 4\%$ on their savings. Which source should be used for the loan and how much will be saved each half-year?

4. A company can borrow $100 000 for 10 years by paying the interest as it falls due at $j_2 = 9\%$ and setting up a sinking fund at $j_2 = 7\%$ to repay the debt. At what rate j_2 would an amortization plan have the same semi-annual cost?

5. A city can borrow $500 000 for 20 years by issuing bonds on which interest will be paid semi-annually at $j_2 = 9\frac{1}{8}\%$. The principal will be paid off by a sinking fund consisting of semi-annual deposits invested at $j_2 = 8\%$. Find the nominal rate j_2 at which the loan could be amortized at the same semi-annual cost.

6. A firm can borrow $200 000 at $j_1 = 9\%$ and amortize the debt for 10 years. From a second source, the money can be borrowed at $j_1 = 8\frac{1}{2}\%$ if the interest is paid annually and the principal is repaid in a lump sum at the end of 10 years. What yearly rate j_1 must the sinking fund earn for the annual expense to be the same under the two options?

7. A company wants to borrow $500 000. One source of funds will agree to lend the money at $j_4 = 8\%$ if interest is paid quarterly and the principal is paid in a lump sum at the end of 15 years. The firm can set up a sinking fund at $j_4 = 6\%$ and will make quarterly deposits.
   a) What is the total quarterly cost of the loan?
   b) At what rate j_4 would it be less expensive to amortize the debt over 15 years?

8. You are able to repay an $80 000 loan by either a) amortization at $j_{12} = 7\%$ with 12 months payments; or b) at $j_{12} = 6\frac{1}{2}\%$ using a sinking fund earning $j_{12} = 4\%$, and paid off in 1 year. Which method is cheaper?

Part B

1. A company needs to borrow $200 000 for 6 years. One source will lend them the money at $j_2 = 10\%$ if it is amortized by monthly payments. A second source will lend the money at $j_4 = 9\%$ if only the interest is paid monthly and the principal is returned in a lump sum at the end of 6 years. The company can earn interest at $j_{365} = 6\%$ on the sinking fund. Which source should be used for the loan and how much will be saved monthly?

2. Tanya can borrow $10 000 by paying the interest on the loan as it falls due at $j_2 = 12\%$ and by setting up a sinking fund with semi-annual deposits that accumulate at $j_{12} = 9\%$ over 10 years to repay the debt. At what rate j_4 would an amortization scheme have the same semi-annual cost?

3. A loan of $100 000 at 8% per annum is to be repaid over 10 years; $20 000 by the amortization method and $80 000 by the sinking-fund method, where the sinking fund can be accumulated with annual deposits at $j_4 = 5\%$. What extra annual payment does the above arrangement require as compared to repayment of the whole loan by the amortization method?

4. A company wants to borrow a large amount of money for 15 years. One source would lend the money at $j_2 = 9\%$, provided it is amortized over 15 years by monthly payments. The company could also raise the money by issuing bonds paying interest semi-annually at $j_2 = 8\frac{1}{2}\%$ and redeemable at par in 15 years. In this case, the company would set up a sinking fund to accumulate the money needed for the redemption of the bonds at the end of 15 years. What rate j_{12} on the sinking fund would make the monthly expense the same under the two options?

5. A $10 000 loan is being repaid by the sinking-fund method. Total annual outlay (each year) is $1400 for as long as necessary, plus a smaller final payment made 1 year after the last regular payment. If the lender receives $j_1 = 8\%$ and the sinking fund accumulates at $j_1 = 6\%$, find the time and amount of the last irregular final payment.

6. A \$5000 loan can be repaid quarterly for 5 years using amortization and an interest rate of $j_{12} = 10\%$ or by a sinking fund to repay both principal and accumulated interest. If paid by a sinking fund, the interest on the loan will be $j_{12} = 9\%$. What annual effective rate must the sinking fund earn to make the quarterly cost the same for both methods?

Section 5.8

Summary and Review Exercises

- Outstanding balance B_k (immediately after the kth payment has been made) by the retrospective method (looking back)
$$B_k = A(1 + i)^k - Rs_{\overline{k}|i}$$
by the prospective method (looking ahead) assuming all payments equal
$$B_k = Ra_{\overline{n-k}|i}$$

- Total interest = Total payments – Amount of Loan

- For a loan of \$$A$ to be amortized with level payments of \$$R$ at the end of each period for n periods, at rate i per period, in the kth line of the amortization schedule $(1 \le k \le n)$:

  Interest payment $I_k = R[1 - (1 + i)^{-(n-k+1)}]$

  Principal payment $P_k = R(1 + i)^{-(n-k+1)}$

  Outstanding balance $B_k = Ra_{\overline{n-k}|i}$

  Successive principal payments are in the ratio $1 + i$, that is $\dfrac{P_{k+1}}{P_k} = (1 + i)$.

- In the sum-of-digits approximation to the amortization method, the interest portion of the kth payment is given by
$$I_k = \frac{n - k + 1}{s_n} I$$
where n = number of payments

  s_n = sum of digits from 1 to $n = \dfrac{n(n + 1)}{2}$

  I = total interest

- For a sinking fund designed to accumulate a specified amount of \$$S$ by equal deposits of \$$R$ at the end of each period for n periods, at rate i per period, in the k-th line of the sinking-fund schedule $(1 \le k \le n)$:

  Interest on fund $= iRs_{\overline{k-1}|i} = R[(1 + i)^{k-1} - 1]$

  Increase in fund $= R + R[(1 + i)^{k-1} - 1] = R(1 + i)^{k-1}$

  Amount in fund $= Rs_{\overline{k}|i}$

- When a loan is paid off by the sinking-fund method, the borrower pays interest on the loan at the end of each period and accumulates the principal of the loan in a sinking fund. The principal of the loan is repaid at the end of the term of the loan as a lump sum from the sinking fund.

  Periodic expense of the loan = Interest payment + Sinking-fund deposit.

  Book value of the loan after k periods = Principal of the loan – Amount in the sinking fund.

Review Exercises 5.8

1. The Roberts borrow $15 000 to be repaid with monthly payments over 10 years at $j_{12} = 15\%$.
   a) Find the monthly payment required.
   b) Find the outstanding balance of the loan after three years (36 payments) and split the 37th payment into principal and interest under
      i) the amortization method;
      ii) the sum-of-digits method.

2. A loan of $20 000 is to be amortized by 20 quarterly payments over 5 years at $j_{12} = 7\frac{1}{2}\%$. Split the 9th payment into principal and interest.

3. A loan of $10 000 is repaid by 5 equal annual payments at $j_2 = 14\%$. What is the total amount of interest paid?

4. A company wants to borrow a large sum of money to be repaid over 10 years. The company can issue bonds paying interest at $j_2 = 8\frac{1}{2}\%$ redeemable at par in 10 years. A sinking fund earning $j_{12} = 7\frac{1}{2}\%$ can be used to accumulate the amount needed in 10 years to redeem the bonds. At what rate j_2 would the semi-annual cost be the same if the debt were amortized over 10 years?

5. Interest at $j_2 = 12\%$ on a debt of $3000 must be paid as it falls due. A sinking fund accumulating at $j_4 = 8\%$ is established to enable the debtor to repay the loan at the end of four years. Find the semi-annual sinking-fund deposit and construct the last two lines of the sinking-fund schedule.

6. For a $60 000 mortgage at $j_2 = 10\%$ amortized over 25 years, find
   a) the level monthly payment required;
   b) the outstanding balance just after the 48th payment;
   c) the principal portion of the 49th payment; and
   d) the total interest paid in the first 48 payments.

7. A $100 000 loan is being amortized over 12 years with monthly payments at $j_2 = 8\%$. After 5 years you want to renegotiate the loan. You discover that the outstanding balance will be calculated using the sum-of-digits method. You can refinance by borrowing the money needed to pay off the outstanding balance at $j_2 = 7\%$ and repay this new loan by accumulating the total principal and interest that will be due in 7 years in a sinking fund earning $j_{12} = 6\%$. Should you refinance?

8. As part of the purchase of a home on January 1, 1996, you negotiated a mortgage in the amount of $110 000. The amortization period for calculation of the level payments (principal and interest) was 25 years and the initial interest rate was 6% per annum compounded semi-annually.
   a) What was the initial monthly payment?
   b) During 1996-2000 inclusive (and January 1, 2001) all monthly mortgage payments were made as they became due. What was the balance of the loan owing just after the payment made January 1, 2001?
   c) At January 1, 2001 (just after the payment then due) the loan was renegotiated at 8% per annum compounded semi-annually (with the end date of the amortization period unchanged). What was the new monthly payment?
   d) All payments, as above, have been faithfully made. How much of the September 1, 2001, payment will be principal and how much represents interest?

9. A couple has a $150 000, 5-year mortgage at $j_2 = 9\%$ with a 20-year amortization period. After exactly 3 years (36 payments) they could renegotiate a new mortgage at $j_2 = 7\%$. If the bank charges an interest penalty of 3 times the monthly interest due on the outstanding balance at the time of renegotiation, what will their new monthly payment be?

10. Janet wants to borrow $10 000 to be repaid over 10 years. From one source, money can be borrowed at $j_1 = 10\%$ and amortized by annual payments. From a second source, money can be borrowed at $j_1 = 9\frac{1}{2}\%$ if only the interest is paid annually and the principal repaid at the end of 10 years. If the second source is used, a sinking fund will be established by annual deposits that accumulate at $j_4 = 8\%$. How much can Janet save annually by using the better plan?

11. Given the following information, find the original value of the loan.

Payment #	Interest	Principal
1		10.00
2	389.00	
3		
4		13.31

12. A loan is paid off over 19 years with equal monthly payments. The total interest paid over the life of the loan is $5681.17. Using the sum-of-digits method, determine the amount of interest paid in the 163rd payment.

13. The XYZ Mortgage Company lends you $100 000 at 9% per annum convertible semi-annually. The loan is to be repaid by monthly payments at the end of each month for 20 years and the rate is guaranteed for 5 years.
    a) Find the monthly payment.
    b) Find the total amount of interest paid over the first 5 years.
    c) Split the first monthly payments into principal and interest portions
       i) based on the true amortization method, and
       ii) based on the sum-of-digits approximation method.
    d) If after 5 years of payments interest rates have increased to 11% per annum convertible semi-annually, find the new monthly payment at time of mortgage renegotiation exactly 5 years after the original loan agreement based on the true amortization method.

14. A loan of $2000 is being repaid by equal monthly payments for an unspecified length of time. Interest on the loan is $j_{12} = 15\%$.
    a) If the amount of principal in the 4th payment is $40, what amount of the 18th payment will be principal?
    b) Find the regular monthly payment.

15. On a loan at $j_{12} = 12\%$ with monthly payments, the amount of principal in the 8th payment is $62.
    a) Find the amount of principal in the 14th payment.
    b) If there are 48 equal payments in all, find the amount of the loan.

16. Given the following part of an amortization schedule, find X.

Payment #	Interest	Principal	Balance
1	50 000	180 975	819 025
2			
3			X

17. A couple purchases a home worth $150 000 by paying $30 000 down and taking out a 5-year mortgage for $120 000 at $j_2 = 10.25\%$. The mortgage will be amortized over 25 years with equal monthly payments. How much of the principal is repaid during the first year?

18. You take out an $80 000 mortgage at $j_2 = 9\%$ with a 25-year amortization period.
    a) Find the monthly payment required, rounded up to the dime.
    b) Find the reduced final payment.
    c) Find the total interest paid during the 4th year.
    d) At the end of 4 years, you pay down an additional $2500 (no penalty).
        i) How much sooner will the mortgage be paid off?
        ii) What would be the difference in total payments over the life of the mortgage?

19. You take out a car loan of $10 000 at $j_{12} = 12\%$. It is to be paid off by monthly payments for 50 months. Payments are rounded up to the next dollar and the final payment is reduced accordingly. After 15 months you decide to pay off the loan. How much is outstanding after your 15th payment if the sum-of-digits method is used?

20. A company decides to borrow $100 000 at $j_1 = 12\%$ in order to finance a new equipment purchase. One of the conditions of the loan is that the company must make annual payments into a sinking fund (the sinking fund will be used to pay off the loan at the end of 20 years). The sinking-fund investment will earn $j_1 = 6\%$.
    a) What is the amount of each sinking-fund payment if they are all to be equal?
    b) What is the total annual cost of the loan?
    c) What overall annual effective compound interest rate is the company paying to borrow the $100 000 when account is taken of the sinking-fund requirement?

21. A $50 000 loan at $j_1 = 7\%$ is to be amortized over 15 years by annual payments.
    a) Find the regular payment and the reduced final payment.
    b) The borrower accumulates the money for each annual payment by making 12 monthly deposits into a sinking fund earning $j_{12} = 6\%$. Find the size of each deposit for the first 14 years.
    c) If the sum-of-digits method is used to calculate the outstanding principal, what is the outstanding balance after 6 years?

22. A loan of $15 000 is to be paid off over 12 years by equal monthly payments at $j_{12} = 18\%$.
    a) Compare the outstanding balance at the end of 2 years using the amortization and the sum-of-digits methods.
    b) Should the borrower refinance the loan without penalty at the end of 2 years at $j_{12} = 16\%$ if the lender uses the sum-of-digits method?

CASE STUDY I

Comparison of amortization, sum-of-digits and sinking-fund methods
A $10 000 loan at $j_{12} = 12\%$ is to be paid off by monthly payments for 5 years.
Using
    a) the amortization method,
    b) the sum-of-digits method,
    c) the sinking-fund method, with $j_{365} = 6\frac{1}{2}\%$ on the sinking fund, calculate and compare
        i) the monthly expense of the loan;
        ii) the outstanding balance of the loan at the end of 2 years; and
        iii) the interest and principal payment at the end of the 1st month and at the end of 2 years.

CASE STUDY II

Increasing extra annual payments

A \$100 000 mortgage is taken out at $j_2 = 8\%$, to be amortized over 25 years by monthly payments. Payments are rounded up to the *next dime* and the final payment is reduced accordingly.

a) Find the regular monthly payment and the reduced final payment.

b) Suppose extra payments are made at the end of every year to get the mortgage paid off sooner. Find the time and amount of the last payment on the mortgage if these extra payments are:
  \$300 at the end of year 1, \$350 at the end of year 2, \$400 at the end of year 3, \$450 at the end of year 4, ... (increasing by \$50 each year)

CASE STUDY III

Mortgage amortization

A \$90 000 mortgage at $j_2 = 9\%$ is amortized over 25 years by monthly payments.

a) Find the regular monthly payment and the smaller final payment.

b) Find the total interest (total cost of financing).

c) Show the first three lines of the amortization schedule.

d) Find the outstanding balance after 2 years.

e) Find the total principal and the total interest paid in the first 2 years.

f) Suppose you paid an extra \$1000 after 2 years. How much interest would this save over the life of the mortgage?

g) How much less interest would be paid if the mortgage could be amortized over 20 years rather than 25 years?

h) In the 25-year mortgage, after 2 years, interest rates drop to $j_2 = 8.5\%$. There is a penalty of 3 months' interest on the outstanding balance for early repayment. Does it pay to refinance?

CASE STUDY IV

Accelerated mortgage payments

"Invest" that extra money back into your mortgage. (from Royal Trust)

Does it pay to accelerate mortgage payments in a low interest rate environment?

Interest rates have dropped so low, relative to what they were a few years ago, that you have to wonder if there's any merit to stepping up the mortgage payments on your home. After all, there's a temptation to invest—or spend—the difference between today's payments and the ones you made a few years ago. With the stock market so hot, you may be tempted to try to make a buck or two from stocks or mutual funds.

Forget it. Your humble, terribly dull mortgage is a far better investment.

The reason: our tax system will force you to pay tax on any earnings from your investment. If you pay down your mortgage, there's no tax on the interest you save.

"Whether the interest rate of your mortgage is 6 per cent or 12 per cent, the best investment is to pay off your mortgage," says Tom Alton, president of Bank of Montreal Mortgage Corp.

Jack Quinn, president of CIBC Mortgage Corp., agrees. "On an after-tax basis, it's pretty hard to find something that's as good as a mortgage." Someone in a 50 per cent tax bracket would have to earn at least 16 per cent just to net 8 per cent after tax on an investment.

Most lenders offer a number of strategies to save you money.

• Provided you can afford it, consider a shorter amortization period when your mortgage comes up for renewal. For instance, if you had a 11 per cent mortgage and renewed at 8 per cent, continue to pay at the old rate, instead of the reduced new one. It will shorten the amortization because the difference will be applied to the principal.

Assuming you had a 25-year \$100,000 mortgage at 11 per cent, your monthly payment would be \$962.50. If the rate drops to 8 per cent, the payment would be \$763.20, for a difference of \$199.30 a month.

What difference is another couple of hundred bucks going to make? Plenty. If you continue to make the old monthly payments, you can reduce the amortization to 14.6 years, according to Mr. Alton.

- Another strategy is to take advantage of any opportunity to make larger monthly payments. For instance, under Bank of Montreal's "10 plus 10" plan, you can increase your monthly payments by up to 10 percent. Royal Trust goes a bit further and allows borrowers to "double up" their monthly mortgage payments—that is, pay up to an additional 100 per cent of the payment any or every month of the year (see accompanying chart).
- Most institutions let you make an annual lump-sum payment of up to 10 per cent of the original principal. Canada Trust has a higher maximum: 15 per cent.
- Finally, consider an accelerated weekly mortgage. Take your monthly payment, divide it by four, and pay that amount on a weekly basis. That means that you will be painlessly paying the equivalent of 13 months in the space of a year. "If you do that you'll make larger payments, but knock about seven years of payments off a 25-year mortgage," Mr. Alton says.

HOW TO SAVE ON MORTGAGES
$100,000 mortgage, 8% interest rate
By shortening the term:

Amortization	Payment	Interest paid over life of mortgage	Interest saved versus 25-year amortization
25 years	$763	$129,098.54	--
20 years	$828	$98,927.31	$30,171.23
15 years	$948	$70,692.37	$58,406.17
10 years	$1,206	$44,794.25	$84,304.29

By doubling-up payments	Interest paid over life of mortgage	Interest saved versus no double-up payments	Time paid off sooner
No double-up payments	$129,098.54	--	
1 double-up payment a year (month 4)	$98,863.53	$30,285.06	4 years and 11 months
2 double-up payments a year (months 4, 10)	$81,925.83	$47,172.71	7 years and 11 months
4 double-up payments a year (months 3, 6, 9, 12)	$61,486.75	$67,611.69	11 years and 9 months

Data: Royal Trust

Required:
a) Verify the figures in the first 2 lines of the first table (i.e., by shortening the term).
b) Verify the figures in the second and fourth line of the second table (i.e., by doubling up payments).

Chapter Six

Bonds

Introduction and Terminology

When a corporation, municipality or government needs a large sum of money for a long period of time, they issue **bonds**, sometimes called debentures, which are sold to a number of investors.

A bond is a written contract between the issuer (borrower) and the investor (lender) that specifies:

- The **face value**, or the **denomination**, of the bond, which is stated on the front of the bond. This is usually a simple figure such as $100, $500, $1000, $5000 or $10 000.

- The **redemption date**, or **maturity date**; that is, the date on which the loan will be repaid.

- The **bond rate**, or **coupon rate**; that is, the rate at which the bond pays interest on its face value at equal time intervals until the maturity date. In most cases this rate is compounded semi-annually.

The amount of money that will be paid on the redemption date is called the **redemption value**. In most cases it is the same as the face value, and in such cases we say the bond is **redeemed at par**. Some bonds are **callable**; they contain a clause that allows the issuer to pay off the loan at a date earlier than the redemption date. Most callable bonds are called at a premium and the redemption value of a bond called before maturity is a previously specified percentage of the face value. For example, the redemption value of a $1000 bond redeemable at 103 during a certain year would be $1030 (during that year).

Bonds (as a contract) may be transferred from one investor to another. Bonds may be bought or sold on the bond market at any time. The buyer of the bond, as an investor, wants to realize a certain return on his investment, specified by the **investment** or **yield rate**. This desired rate of return will vary with the financial climate and will affect the price at which bonds are traded.

As an illustration, consider a $1000 bond redeemable at par in 5 years, paying interest at bond rate or coupon rate $j_2 = 8\%$. If the buyer paid $1000 for the bond, then he receives $40 every half-year in interest and we say his yield rate is $j_2 = 8\%$, that is, the same as the bond rate (coupon rate). If the buyer paid less than the face value, his yield rate will be higher than the bond rate $j_2 = 8\%$ because he receives $40 every half-year in interest and on the redemption date he receives the face value of $1000, i.e., more than he invested. Similarly, if the buyer paid more than the face value, his yield rate will be lower than the bond rate. In the latter case, his interest payments at $j_2 = 8\%$ will be partially offset by the loss incurred when the bond is redeemed.

In this chapter we shall use the following notation:
F = the face value or par value of the bond.
C = the redemption value of the bond.
r = the bond rate per interest period (coupon rate).
i = the yield rate per interest period (assume $i > 0$).
n = the number of interest periods until the redemption date.
P = the purchase price of the bond to yield rate i.
Fr = the bond interest payment or coupon.

The two fundamental problems relating to bonds, which will be discussed in this chapter, are

1. to determine the purchase price P of a bond to yield a given investment (yield) rate i to maturity;

2. to determine the investment rate i that a bond will yield when bought for a given price P.

OBSERVATION: In this chapter, the yield rate means the yield rate to maturity unless specified otherwise.

| Section 6.2 | **Purchase Price to Yield a Given Investment Rate to Maturity** |

We want to determine the purchase price of a bond on a bond interest date n interest periods before maturity so that it earns interest at a specified investment rate i. We shall assume that the bond rate and the yield rate have the same conversion period.

A buyer of a bond will receive two types of payments,
1. a bond interest payment, or coupon Fr, at the end of each interest period;
2. the redemption value C on the redemption date.

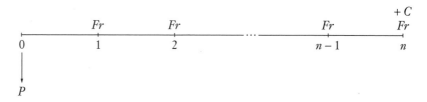

The buyer of a bond who wishes to realize an investment rate i on the investment should pay a price equal to the discounted value of the above payments at rate i. Thus,

$$P = Fr a_{\overline{n}|i} + C(1 + i)^{-n} \qquad (15)$$

It is common practice to express the price of a bond as a price per $100 face value, even when the face value is not $100.

EXAMPLE 1   A $100 bond that pays interest at $j_2 = 10\%$ is redeemable at par at the end of 5 years. Find the purchase price to yield an investor
a) 14% compounded semi-annually;
b) 8% compounded semi-annually.

Solution   The bond pays 5% of $100 semi-annually; i.e., $Fr = 100 \times .05 = \$5$, and $100 at the end of 5 years.

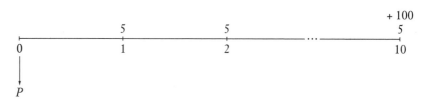

a) The purchase price P to yield $j_2 = 14\%$ is the discounted value of the above payments at $j_2 = 14\%$.

$$P = 5\ a_{\overline{10}|.07} + 100(1.07)^{-10} = 35.12 + 50.83 = \$85.95$$

The purchase price of $85.95 will yield a buyer a return of $j_2 = 14\%$ on the investment. The buyer is buying the bond for less than the redemption value; the bond is purchased at a **discount** because the yield rate is higher than the bond rate.

b) The purchase price P to yield $j_2 = 8\%$ is

$$P = 5\ a_{\overline{10}|.04} + 100(1.04)^{-10} = 40.55 + 67.56 = \$108.11$$

The purchase price of $108.11 will yield a buyer a return of $j_2 = 8\%$ on the investment. The buyer is buying the bond for more than the redemption value; the bond is purchased at a **premium** because the yield rate is lower than the bond rate.

■

EXAMPLE 2   A $5000 bond maturing at 105 on September 1, 2021, has semi-annual coupons at 7%. Find the purchase price on March 1, 2000, to guarantee a yield of $j_2 = 6.8\%$.

Solution   Each coupon is $3\frac{1}{2}\%$ of $5000 = \$175.00$ semi-annually. The bond matures on September 1, 2021, for $5000 \times 1.05 = \$5250$.

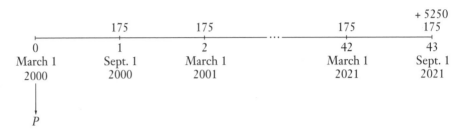

The purchaser receives 43 coupons plus the maturity value. The purchase price P on March 1, 2000, to guarantee a yield of $j_2 = 6.8\%$ is

$$P = 175.00a_{\overline{43}|.034} + 5250(1.034)^{-43}$$
$$= 3924.76 + 1246.74 = \$5171.50$$

or $\frac{5171.50}{50} = \$103.43$ per $100 unit.

In this example, the buyer wants a yield rate smaller than the coupon rate on the bond and yet, since the price is less than the redemption value, the bond has been purchased at a discount. This apparent contradiction can be explained by remembering that while the bond is redeemed for $5250, which affects the price of the bond, the coupons are determined by taking $3\frac{1}{2}\%$ of $5000, not $3\frac{1}{2}\%$ of $5250. In fact, if we determine the bond interest rate per unit of redemption, we get $\frac{175}{5250} = 3.3\%$. It is because 3.3% is less than the desired yield rate of 3.4% each half-year that results in the purchase of the bond at a discount.

■

We will study the concepts of premium and discount in more detail in Section 6.3.

The above two examples illustrate that in the bond investment market the price of a bond depends on the bond rate, the yield rate acceptable to investors, the time to maturity and the redemption value.

OBSERVATION: Note that unless it is specified otherwise, we assume that bonds are redeemed at par.

Bond Tables   Until recently, companies that bought and sold bonds used bond tables to calculate the price of a bond, or the yield rate for a price set by market forces.

Bond tables were available in large volumes with answers to 4 or 6 decimal places. The tables gave the market prices of bonds for a wide range of yield and bond interest rates and lengths of time to maturity. Almost all bond tables assumed that redemption was at par and the bond rate as well as the yield rate were compounded semi-annually.

Today, companies that buy and sell bonds use personal computers or financial calculators to calculate the information previously found in the bond tables. The formulas in the software that is used are those developed in this chapter.

An Alternative Purchase-Price Formula

We shall develop an alternative formula for the purchase price of a bond sold on a bond interest date, which is somewhat simpler than formula **(15)**. Using the identity $a_{\overline{n}|i} = \frac{1-(1+i)^{-n}}{i}$ from Chapter 3, we can get the factor $(1 + i)^{-n}$ as $(1 + i)^{-n} = 1 - ia_{\overline{n}|i}$ and eliminate it in formula **(15)**. Thus,

$$P = Fra_{\overline{n}|i} + C(1 + i)^{-n} = Fra_{\overline{n}|i} + C(1 - ia_{\overline{n}|i})$$

or

$$\boxed{P = C + (Fr - Ci)a_{\overline{n}|i}} \qquad (16)$$

Formula **(16)** is often more efficient than formula **(15)** since it requires only one calculation, $a_{\overline{n}|i}$, whereas formula **(15)** requires two: $a_{\overline{n}|i}$ and $(1 + i)^{-n}$. It also tells us immediately whether the bond is purchased at a premium or a discount and the size of the premium or discount.

EXAMPLE 3   A corporation decides to issue 15-year bonds in the amount of $10 000 000. Under the contract, interest payments will be made at the rate $j_2 = 10\%$. The bonds are priced to yield $j_2 = 8\%$ to maturity. What is the issue price of the bond? What is the price of a $1000 bond to yield $j_2 = 8\%$?

Solution   The bond issue will provide interest payments of 5% of $10 000 000 semi-annually, that is $Fr = \$500\,000$, and redemption price $10 000 000 at the end of 15 years.

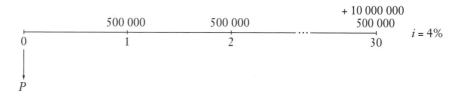

The alternative purchase-price formula will give us
$$P = 10\,000\,000 + (500\,000 - 400\,000)a_{\overline{30}|.04} \doteq \$11\,729\,203$$

The issue price of the bonds to the public is $11 729 203. The bonds will provide the investor with a yield $j_2 = 8\%$ if held to maturity. The price of a $1000 bond is
$$\frac{11\,729\,203}{10\,000} = \$1172.92$$

∎

In all examples so far we have considered the case where the bond interest payments Fr form an ordinary simple annuity, that is, where the bond interest period coincides with the period the yield rate is compounded. In cases when these periods are different, the bond interest payments form a general annuity and the interest yield rate is recalculated to coincide with the bond coupon period. The following example will illustrate the procedure.

EXAMPLE 4 Find the issue price of the bonds in Example 3 to yield a) 8% per annum compounded monthly; b) 8% per annum compounded annually.

Solution a Find rate i per half year such that

$$(1 + i)^2 = (1 + \tfrac{.08}{12})^{12}$$
$$1 + i = (1 + \tfrac{.08}{12})^6$$
$$i = (1 + \tfrac{.08}{12})^6 - 1$$
$$i = .040672622$$

Now
$$P = 10\ 000\ 000 + (500\ 000 - 406\ 726.22)a_{\overline{30}|i} \doteq \$11\ 599\ 802$$

The issue price of the bonds to the public is \$11 599 802. The purchase price of a \$1000 bond is then \$1159.98.

Solution b Find rate i per half year such that

$$(1 + i)^2 = 1.08$$
$$(1 + i) = (1.08)^{1/2}$$
$$i = (1.08)^{1/2} - 1$$
$$i = .039230485$$

Now
$$P = 10\ 000\ 000 + (500\ 000 - 392\ 304.85)a_{\overline{30}|i} \doteq \$11\ 879\ 792$$

The issue price of the bonds to the public is \$11 879 792. The purchase price of a \$1000 bond is then \$1187.98.

■

Exercise 6.2

Part A

Problems 1 to 8 use the information in the table below. Using either formula (**15**) or formula (**16**), find the purchase price of the bond.

No.	Face Value	Redemption	Bond Interest Rate	Years to Redemption	Yield Rate
1.	$ 500	at par	$j_2 = $ 9%	20	$j_2 = $ 8%
2.	$ 1 000	at par	$j_2 = $ 9%	15	$j_2 = $ 10%
3.	$ 2 000	at par	$j_2 = 6\frac{1}{2}$%	15	$j_4 = $ 7%
4.	$ 5 000	at par	$j_2 = $ 12%	20	$j_2 = $ 10%
5.	$ 1 000	at 110	$j_2 = $ 9%	18	$j_2 = $ 12%
6.	$ 2 000	at 105	$j_2 = 6\frac{1}{2}$%	20	$j_2 = $ 7%
7.	$10 000	at 110	$j_2 = 10\frac{1}{2}$%	15	$j_{12} = $ 7%
8.	$ 5 000	at 103	$j_2 = $ 11%	17	$j_4 = $ 10%

9. The XYZ Corporation needs to raise some funds to pay for new equipment. They issue $1 000 000 worth of 20-year bonds with semi-annual coupons at $j_2 = 6\%$. These bonds are redeemable at 105. At the time of issue, interest rates in the market place are $j_{12} = 5\frac{1}{2}\%$. How much money did they raise?

10. Mr. Simpson buys a $1000 bond paying bond interest at $j_2 = 6\frac{1}{2}\%$ and redeemable at par in 20 years. Mr. Simpson's desired yield rate is $j_4 = 7\%$. How much did he pay for the bond? After exactly five years he sells the bond. Interest rates have dropped and the bond is sold to yield a buyer $j_1 = 5\%$. Find the sale price.

11. A corporation issues $600 000 worth of 12-year bonds with semi-annual coupons at $j_2 = 10\%$. The bonds are priced to yield $j_2 = 9\%$. Determine the issue price per $100 unit.

12. A $1000 bond paying coupons at $j_2 = 6\%$ is redeemable at par in 20 years. Find the price to yield an investor a) $j_2 = 7\%$; b) $j_2 = 6\%$; c) $j_2 = 5\%$.

Part B

1. Show that if $Fr = Ci$, $P = C$.

2. Prove the following formula for a *par value* bond

$$P = F(1 + i)^{-n} + \frac{r}{i}[F - F(1 + i)^{-n}]$$

This purchase-price formula is known as Makeham's formula and requires the use of only $(1 + i)^{-n}$ factors.

3. An $X bond quoted as redeemable at 105 in 10 years is purchased to yield $j_2 = 10\%$. If this same bond was redeemable at par, the actual purchase price would be $113.07 less. Determine the value of X.

4. The XYZ Corporation issues a special 20-year bond issue that has no coupons. Rather, interest will accumulate on the bond at rate $j_2 = 11\%$ for the life of the bond. At the time of maturity, the total value of the loan will be paid off, including all accumulated interest. Find the price of a $1000 bond of this issue to yield $j_2 = 10\%$.

5. A $1000 bond bearing coupons at $j_2 = 6\frac{1}{2}\%$ and redeemable at par is bought to yield $j_2 = 6\%$. If the present value of the redemption value at this yield is $412, what is the purchase price? (Do not find n.)

6. Two $1000 bonds redeemable at par at the end of n years are bought to yield $j_2 = 10\%$. One bond costs $1153.72 and has semi-annual coupons at $j_2 = 12\%$. The other bond has semi-annual coupons at $j_2 = 8\%$. Find the price of the second bond.

7. A \$1000 bond with semi-annual coupons at $j_2 = 6\%$ and redeemable at the end of n years at \$1050 sells at \$980 to yield $j_2 = 7\frac{1}{2}\%$. Find the price of a \$1000 bond with semi-annual coupons at $j_2 = 5\%$ redeemable at the end of $2n$ years at \$1040 to yield $j_2 = 7\frac{1}{2}\%$. (Do not find n.)

8. A bond with a par value of \$100 000 has coupons at the rate $j_2 = 6\frac{1}{2}\%$. It will be redeemed at par when it matures a certain number of years hence. It is purchased for a price of \$96 446.90. At this price, the purchaser who holds the bond to maturity will realize a yield rate $j_2 = 7\%$. Find the number of years to maturity.

9. A \$1000 bond with annual coupons at $8\frac{1}{2}\%$ and maturing in 20 years at par is purchased to yield an annual effective rate of interest of 9% if held to maturity. The book value of the bond at any time is the discounted value of all remaining payments, using the 9% rate. Ten years later, just after a coupon payment, the bond is sold to yield the new purchaser a 10% annual effective rate of interest if held to maturity. Find the excess of the book value over the second sale price.

10. A corporation has an issue of bonds with annual coupons at $j_1 = 5\%$ maturing in five years at par that are quoted at a price that yields $j_1 = 6\%$.
    a) What is the price of a \$1000 bond?
       It is proposed to replace this issue of bonds with an issue of bonds with annual coupons at $j_1 = 5\frac{1}{2}\%$.
    b) How long must the new issue run so that the bond holders will still yield $j_1 = 6\%$? Express your answer to the nearest year.

11. A \$1000 bond bearing semi-annual coupons at $j_2 = 10\%$ is redeemable at par. What is the minimum number of whole years that the bond should run so that a person paying \$1100 for it would earn at least $j_2 = 8\%$?

12. If the coupon rate on a bond is $1\frac{1}{2}$ times the yield rate when it sells for a premium of \$10 per \$100, find the price per \$100 for a bond with the same number of coupons to run and the same yield with coupons equal to $\frac{3}{4}$ of the yield rate.

Section 6.3   Premium and Discount

A bond is just a loan agreement between the bond issuer (the borrower) and the investor (the lender). The bond issuer sets a bond interest rate (the coupon rate) and the redemption value and date, and the investor determines the amount of the loan (the purchase price of the bond).

We have seen that if $P > C$ (the purchase price of a bond exceeds its redemption value), the bond is said to have been purchased **at a premium**. In fact, from the alternative purchase price formula, the size of the premium must be

$$\boxed{\text{Premium} = P - C = (Fr - Ci)a_{\overline{m}|i}}$$

That is, a premium occurs when $Fr > Ci$. In other words, each coupon Fr exceeds the interest desired by the investor Ci, which allows the price P to exceed the redemption value C. For par value bonds (i.e., $C = F$) the bond is purchased at a premium when $r > i$.

When the bond is purchased at a premium, then only C of the original principal is returned on the redemption date. There will be a loss, equal to the premium, at the redemption date, unless part of each bond interest payment is used to amortize the premium. The remaining principal $P - C$ or the premium is returned in instalments as a part of the bond interest payments.

Each coupon payment, in addition to paying interest on the investment (at a yield rate), provides a partial return of the principal P. These payments of principal will continually reduce the value of the bond from the price on the purchase date to the redemption value on the redemption date. These adjusted values of the bond are called the **book values** of the bond and are used by many investors in reporting the asset values of bonds for financial statements. The process of gradually decreasing the value of the bond from the purchase price to the redemption value is called **amortization of a premium** or **writing down a bond**. A **bond amortization schedule** is a table that shows the division of each interest payment into the interest paid on book value (at a yield rate) and the decrease in the book value of the bond (or the book value adjustment) with the book value after each bond interest payment is paid.

EXAMPLE 1   A $1000 bond, redeemable at par on December 1, 2002, pays interest at $j_2 = 6\frac{1}{2}\%$. The bond is bought on June 1, 2000, to yield $j_2 = 5\frac{1}{2}\%$. Find the purchase price and construct the bond schedule.

Solution   The purchase price P on June 1, 2000 can be determined using the alternative purchase-price formula (**16**).

$$P = 1000 + (32.50 - 27.50)a_{\overline{5}|.0275} = 1000 + 23.06 = \$1023.06$$

The premium of $23.06 paid for the bond must be saved out of the bond interest payments in order to recover the entire principal originally invested in the bond. To construct the amortization schedule for this bond, we shall calculate how much of each coupon is used as return on the investment at the desired yield rate and how much is used to adjust the book value or the principal (i.e., amortize the premium).

At the end of the first half-year, on December 1, 2000, the investor's yield should be $1023.06 \times 0.0275 = \$28.13$. Since he actually receives $32.50, the difference of $4.37 can be regarded as part of the original principal being returned and is used to adjust (reduce) the principal, or amortize the premium. The adjusted value, or book value, after the coupon payment is $1023.06 - 4.37 = \$1018.69$. This book value can be computed independently by the alternative purchase-price formula using $F = C = 1000$, $r = 3\frac{1}{4}\%$, $i = 2\frac{3}{4}\%$ and $n = 4$. The above procedure is continued until the bond matures.

The following is a complete bond schedule with $F = C = 1000$, $r = .0325$, $i = .0275$ and $n = 5$.

Schedule for a Bond Purchased at a Premium

Date (1)	Time t (2)	Coupon $F \cdot r$ (3)	Interest on Book Value $I_t = B_{t-1} \cdot i$ (4) = (6)$_{t-1} \cdot i$	Book Value Adjustment (5) = (3) − (4)	Book Value B_t (6) = B_{t-1} − (5)
June 1, 2000	0	-	-	-	1023.06
Dec. 1, 2000	1	32.50	28.13	4.37	1018.69
June 1, 2001	2	32.50	28.01	4.49	1014.20
Dec. 1, 2001	3	32.50	27.89	4.61	1009.59
June 1, 2002	4	32.50	27.76	4.74	1004.85
Dec. 1, 2002	5	32.50	27.63	4.87	999.98*
Totals		162.50	139.42	23.08*	

* The 2¢ error is from the accumulation of round-off.

OBSERVATIONS:
1. All the book values can be reproduced using either of the purchase price formulas.
2. The sum of the book value adjustments is equal to the original amount of the premium.
3. The book value is gradually adjusted from the original purchase price to the redemption value.
4. Successive book value adjustments are in the ratio $(1 + i)$, i.e., $\frac{4.49}{4.37} \doteq \frac{4.61}{4.49} \doteq \frac{4.74}{4.61} \doteq \frac{4.87}{4.74} \doteq 1.0275$.

This provides a quick check of entries in the bond table.

∎

Similarly, if $P < C$ (the purchase price is less than the redemption value) the bond is said to have been purchased **at a discount**. From the alternative purchase-price formula, the size of the discount is

$$\boxed{\text{Discount} = C - P = (Ci - Fr)a_{\overline{n}|i}}$$

That is, a discount occurs when $Fr < Ci$. In other words, there is a "deficiency" in each bond interest payment (coupon) with respect to the desired yield and this deficiency is recovered from the discount. For a par value bond (i.e., $C = F$) the bond is purchased at a discount when $i > r$.

When the bond is purchased at a discount, the investor's return is more than just the bond interest payment. There will be a gain, equal to the discount, at the redemption date unless the book value is gradually increased. The process of gradually increasing the value of the bond from the purchase price to the redemption value is called **accumulation of a discount** or **writing up a bond**.

A **bond accumulation schedule** is a table that shows the division of the investor's interest (at a yield rate) into the bond interest payment and the increase in the book value of the bond (or the book value adjustment) with the book value after each bond interest payment is paid.

EXAMPLE 2   A \$1000 bond, redeemable at par on December 1, 2002, pays interest at $j_2 = 9\%$. The bond is bought on June 1, 2000, to yield $j_2 = 10\%$. Find the price and construct a bond schedule.

Solution   The purchase price P on June 1, 2000, can be determined by the alternative purchase-price formula (**16**)

$$P = 1000 + (45 - 50)a_{\overline{5}|.05} = 1000 - 21.65 = \$978.35$$

The discount is \$21.65.

To construct the accumulation schedule for this bond, we shall calculate the investor's interest at the end of each half-year and gradually increase the book value of the bond by the difference between investor's interest and the bond interest payment.

At the end of the first half-year, on December 1, 2000, the investor's interest should be $978.35 \times 0.05 = \$48.92$. Since the bond interest payment is only \$45, we increase the book value of the bond by $48.92 - 45.00 = \$3.92$. We say, \$3.92 is used for accumulation of a discount or -\$3.92 is the principal adjustment (book value adjustment).

The adjusted value, or book value, after the bond interest payment on December 1, 2000, is $978.35 - (-3.92) = 978.35 + 3.92 = \$982.27$. This book value can be computed independently by the alternative purchase-price formula using $F = C = 1000$, $r = 4\frac{1}{2}\%$, $i = 5\%$ and $n = 4$. The above procedure is continued until the bond matures.

The following is a complete bond schedule with $F = C = 1000$, $r = .045$, $i = .05$, and $n = 5$.

Schedule for a Bond Purchased at a Discount

Date (1)	Time t (2)	Coupon $F \cdot r$ (3)	Interest on Book Value $I_t = B_{t-1} \cdot i$ (4) = (6)$_{t-1} \cdot i$	Book Value Adjustment (5) = (3) − (4)	Book Value B_t (6) = B_{t-1} − (5)
June 1, 2000	0	-	-	-	978.35
Dec. 1, 2000	1	45.00	48.92	−3.92	982.27
June 1, 2001	2	45.00	49.11	−4.11	986.38
Dec. 1, 2001	3	45.00	49.32	−4.32	990.70
June 1, 2002	4	45.00	49.54	−4.54	995.24
Dec. 1, 2002	5	45.00	49.76	−4.76	1000.00
Totals		225.00	246.65	−21.65	

The four observations that followed Example 1, hold here as well.

■

The payments made during the term of a bond can be regarded as loan payments made by the borrower (bond issuer) to the lender (the bondholder) to repay a loan amount equal to the purchase price of the bond. The bond purchase price is calculated as the discounted value of those payments (coupons plus redemption value) at a certain yield rate (the interest rate on the loan). Thus the bond transaction can be regarded as the amortization of a loan and an amortization schedule for the bond can be constructed like the general loan amortization schedule in Section 5.1.

Also, the four observations with respect to bond schedules shown in Examples 1 and 2 follow from the general loan amortization schedule and may be summarized as follows:

OBSERVATIONS:

1. The outstanding principal (the book value of the bond) can be computed as the discounted value of the remaining payments (the purchase price of the bond) at any payment date (coupon date).

2. The sum of the principal repaid (the book value adjustments) is equal to the amount of the loan (the redemption value plus the amount of the premium or the discount).

3. The outstanding principal of the loan is gradually adjusted from the original amount to zero balance after the last payment (last coupon plus redemption value) is paid.

4. Successive principal repayments (book value adjustments) are in the ratio $1 + i$. (Last principal repaid must be reduced by redemption value to satisfy the condition that the last ratio is equal to $1 + i$.)

EXAMPLE 3   Construct the amortization schedule for the loan of a) Example 1; b) Example 2.

Solution a

Date	Payment	Interest Due (at $i = .0275$)	Principal Repaid	Outstanding Principal
June 1, 2000	-	-	-	1023.06
Dec. 1, 2000	32.50	28.13	4.37	1018.69
June 1, 2001	32.50	28.01	4.49	1014.20
Dec. 1, 2001	32.50	27.89	4.61	1009.59
June 1, 2002	32.50	27.76	4.74	1004.85
Dec. 1, 2002	1032.50	27.63	1004.87	−0.02*
Totals	1162.50	139.42	1023.08*	

* Each calculation is rounded to the nearest 1¢ and then carried forward at its rounded-off value. This results in an accumulated 2¢ error in the final balance.

Solution b

Date	Payment	Interest Due (at $i = .05$)	Principal Repaid	Outstanding Principal
June 1, 2000	-	-	-	978.35
Dec. 1, 2000	45.00	48.92	−3.92	982.27
June 1, 2001	45.00	49.11	−4.11	986.38
Dec. 1, 2001	45.00	49.32	−4.32	990.70
June 1, 2002	45.00	49.54	−4.54	995.24
Dec. 1, 2002	1045.00	49.76	995.24	0
Totals	1225.00	246.65	978.35	

∎

When $C \neq F$ we define a **modified coupon rate** $g = \frac{Fr}{C}$, so $Fr = Cg$. Then $P = C + (Cg - Ci)a_{\overline{m}|i} = C + C(g - i)a_{\overline{m}|i}$.

When a bond is purchased at a premium, $P - C = C(g - i)a_{\overline{m}|i} > 0$ if $g > i$.

When a bond is purchased at a discount, $C - P = C(i - g)a_{\overline{m}|i} > 0$ if $i > g$.

EXAMPLE 4 A 2-year $1000 bond with bond interest at $j_2 = 9\%$ is redeemable for $1050. Find the price to yield $j_4 = 10\%$ and produce a bond schedule.

Solution   First find i per half year such that

$$(1 + i)^2 = (1.025)^4$$
$$i = (1.025)^2 - 1$$
$$i = .050625$$

With $F = 1000$, $C = 1050$, $n = 4$, $r = 0.045$, $i = .050625$, and $g = \frac{1000(.045)}{1050} = .042857143$

$$P = 1050 + (45 - 53.15625)a_{\overline{4}|i} = \$1021.12$$
$$\text{or } C - P = 1050(.050625 - .042857143)a_{\overline{4}|i} = \$28.88$$

Thus, the bond is purchased at a discount of $28.88. This leads to the following bond schedule:

Time t (1)	Coupon $F \cdot r$ (2)	Interest on Book Value $B_{t-1} \cdot i$ $(3) = (5)_{t-1} \cdot i$	Book Value Adjustment $(4) = (2) - (3)$	Book Value B_t $(5) = B_{t-1} - (4)$
0	-	-	-	$1021.12
1	45.00	51.69	−6.69	1027.81
2	45.00	52.03	−7.03	1034.84
3	45.00	52.39	−7.39	1042.23
4	45.00	52.76	−7.76	1049.99*
Totals	180.00	$208.87	−28.87	

*1¢ error is from the accumulation of round-off.

∎

EXAMPLE 5 Prepare an Excel spreadsheet for the bond questions in the above examples. Show output using the data from Example 1.

Solution We summarize the entries in an Excel spreadsheet and their interpretation in the table below.

	CELL	ENTER	INTERPRETATION
Headings	A1	Time	'Time value for a coupon date'
	B1	Coupon	'Bond interest payment'
	C1	Interest on BV	'Interest on book value'
	D1	Adjustment	'Book value adjustment'
	E1	Book value	'Book value of the bond'
Line 0	A2	0	Time starts
	E2	1023.06	Purchase price at time 0
Line 1	A3	=A2+1	Time for coupon 1
	B3	32.50	Semi-annual bond interest payment
	C3	=E2*.0275	Interest on book value at time 1
	D3	=B3−C3	Book value adjustment at time 1
	E3	=E2−D3	Book value at time 1

To generate the complete schedule copy A3.E3 to A4.E7

To get totals apply Σ to B3.D7

Below is the bond schedule by an Excel spreadsheet.

	A	B	C	D	E
1	Time	Coupon	Interest on BV	Adjustment	Book Value
2	0				1,023.06
3	1	32.50	28.13	4.37	1,018.69
4	2	32.50	28.01	4.49	1,014.21
5	3	32.50	27.89	4.61	1,009.60
6	4	32.50	27.76	4.74	1,004.86
7	5	32.50	27.63	4.87	1,000.00
8		162.50	139.44	23.06	

*All output is rounded to the nearest 1¢ but carried internally to several decimals. Thus the correct final balance is produced, but some columns appear not to add up.

■

Exercise 6.3

Part A

For problems 1 to 6 in the table that follows, determine logically, before calculation, if the bond is purchased at a premium or a discount. Then find the purchase price of the bond and make out a complete bond schedule showing the amortization of the premium or the accumulation of the discount.

No.	Face Value	Redemption	Bond Interest Rate	Years to Redemption	Yield Rate
1.	$ 1 000	at par	$j_2 = 10\%$	3	$j_2 = 9\%$
2.	$ 5 000	at par	$j_2 = 6\%$	3	$j_2 = 7\%$
3.	$ 2 000	at par	$j_2 = 6\frac{1}{2}\%$	2.5	$j_2 = 5\frac{1}{2}\%$
4.	$ 1 000	at 105	$j_2 = 10\%$	2.5	$j_2 = 12\%$
5.	$ 2 000	at 103	$j_2 = 9\%$	3	$j_2 = 7\%$
6.	$10 000	at 110	$j_2 = 7\%$	2.5	$j_2 = 8\%$

7. A $1000 par value bond paying interest at $j_2 = 6\%$ has book value $1100 on March 1, 2001, at a yield rate of $j_2 = 4\frac{1}{2}\%$. Find the amount of amortization of the premium on September 1, 2001, and the new book value on that date.

Part B

1. A 20-year bond with annual coupons is bought at a premium to yield $j_1 = 8\%$. If the amount of write-down of the premium in the 3rd payment is $6, determine the amount of write-down of the premium in the 16th payment.

2. A $1000 bond, redeemable at par, with annual coupons at 10% is purchased for $1060. If the write-down in the book value is $7 at the end of the first year, what is the write-down at the end of the 4th year?

3. A bond with $80 annual coupons is purchased at a discount to yield $j_1 = 7\frac{1}{2}\%$. The write-up for the first year is $22. What was the purchase price?

4. A $1000 bond redeemable at $1050 on December 1, 2003, pays interest at $j_2 = 6\frac{1}{2}\%$. The bond is bought on June 1, 2001. Find the price and construct a bond schedule if the desired yield is a) $j_{12} = 6\%$; b) $j_1 = 5\frac{1}{2}\%$.

5. A $1000 20-year par value bond with semi-annual coupons is bought at a discount to yield $j_2 = 10\%$. If the amount of the write-up of the discount in the last entry in the schedule is $5, find the purchase price of the bond.

6. A bond with $40 semi-annual coupons is purchased at a premium to yield $j_2 = 7\%$. If the first write-down is $4.33, find the purchase price of the bond.

7. A $1000 bond pays coupons at $j_2 = 7\%$ on January 1 and July 1 and will be redeemed at par on July 1, 2004. If the bond was bought on January 1, 1996, to yield 6% per annum compounded semi-annually, find the interest due on the book value on January 1, 2000.

8. A $1000 bond providing annual coupons at $j_1 = 9\%$ is redeemable at par on November 1, 2004. The write-down in the first year was $5.63. The write-down in the eleventh year was $19.08. Determine the book value of the bond on November 1, 2000.

9. A $2000 bond with annual coupons matures at par in 5 years. The first interest coupon is $400, with subsequent coupons reduced by 25% of the previous year's coupon, each year.
a) Find the price to yield $j_1 = 10\%$.
b) Draw up the bond schedule.

10. A 10-year bond matures for $2000 and has annual coupons. The first coupon is $100, and each increases by 10%. The bond is priced to yield $j_1 = 9\%$. Find the price and draw up the bond schedule.

11. A $1000 bond with semi-annual coupons at $j_2 = 5\%$ is redeemable for $1100. If the amount for the 16th write-up is $2.50, calculate the purchase price to yield $j_2 = 6\%$.

12. You are told that a $1000 bond with semi-annual coupons at $j_2 = 8\%$ redeemable at par will be sold at $700 to an investor requiring 12% per annum compounded semi-annually.
   a) Find the price of this bond to the same investor if the above coupon rate were changed to $j_2 = 11\%$.
   b) Is the 11% bond purchased at a premium or a discount? Explain.
   c) For the 11% bond show the write-up (or write-down) entries at the first two coupon dates.

13. A $10 000 15-year bond is priced to yield $j_4 = 12\%$. It has quarterly coupons of $200 each the 1st year, $215 each the 2nd year, $230 each the 3rd year,..., $410 each the 15th year.
   a) Show that the price is $9267.05.
   b) Find the book value after 14 years.
   c) Draw up a partial bond schedule showing the first and last year's entries only.

Section 6.4   Callable Bonds

Some bonds contain a clause that allows the issuer to redeem the bond prior to the maturity date. These bonds are referred to as **callable bonds**.

   Callable bonds present a problem with respect to the calculation of the price since the term of the bond is not certain. Since the corporation issuing the bond controls when the bond is redeemed (or called), the investor must determine a price that will guarantee the desired yield regardless of the call date.

EXAMPLE 1   The XYZ Corporation issues a 20-year $1000 bond with coupons at $j_2 = 6\%$. The bond can be called, at par, at the end of 15 years. Find the purchase price that will guarantee an investor a return of
   a) $j_2 = 7\%$; b) $j_2 = 5\%$.

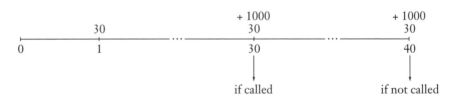

Solution a   The bond matures at the end of 20 years but may be called at the end of 15 years. Given a desired yield rate of $j_2 = 7\%$, we calculate the price of the bond for these two different dates using formula **(16)**.
   i) If it is called after 15 years
$$P = 1000 + (30 - 35)a_{\overline{30}|.035} = 1000 - 91.96 = \$908.04$$
   ii) If it matures after 20 years
$$P = 1000 + (30 - 35)a_{\overline{40}|.035} = 1000 - 106.78 = \$893.22$$

   The purchase price to guarantee a return of $j_2 = 7\%$ is the lower of these two answers, or $893.22. If the investor pays $893.22 and the bond runs the full 20 years to maturity, the investor's yield will be exactly $j_2 = 7\%$. If the investor pays $893.22 and the bond is called after 15 years, the investor's return exceeds $j_2 = 7\%$.

If the investor pays $908.04 for the bond, however, she will yield $j_2 = 7\%$ only if the bond is called after 15 years. If the bond runs to its full maturity, the rate of return will be less than $j_2 = 7\%$.

Solution b   Again we calculate the price of the bonds at the two different dates using a yield of $j_2 = 5\%$ and formula **(16)**.

i) If it is called after 15 years,

$$P = 1000 + (30 - 25)a_{\overline{30}|.025} = 1000 + 104.65 = \$1104.65.$$

ii) If it matures after 20 years,

$$P = 1000 + (30 - 25)a_{\overline{40}|.025} = 1000 + 125.51 = \$1125.51.$$

The purchase price to guarantee a return of $j_2 = 5\%$ is $1104.65. Regardless of the outcome, the investor's yield will equal or exceed $j_2 = 5\%$ at this price.

■

In the above example, we have shown, in effect, that the investor must assume that the issuer of the bond will exercise his call option to the disadvantage of the investor and must calculate the price accordingly. The example above also illustrates a useful principle *for bonds that are callable at par*. That is,

> If the yield rate is less than the coupon rate (if the bond sells at a premium), then you use the earliest possible call date in your calculation.
>
> If the yield rate is greater than the coupon rate (if the bond sells at a discount), then you use the latest possible redemption date in your calculation.

These examples can be explained logically. For a bond purchased at a premium, the earliest call date is the worst for the investor since she gets only $1000 for her bond whenever it is called. On the other hand, for a bond purchased at a discount, the earliest call date is the best for the investor, since at that early call date she will get a full $1000 for a bond that she values at something less than $1000. Thus, the most unfavourable situation is the latest possible redemption.

Unfortunately, these guidelines cannot be applied often since, when a bond is called early, it is usually done so at a premium. In that case, we are forced to calculate all possible purchase prices and pay the lowest price calculated.

EXAMPLE 2   The ABC Corporation issues a 20-year $1000 bond with bond interest at $j_2 = 6\%$. The bond is callable at the end of 10 years at $1100 or at the end of 15 years at $1050. Find the price to guarantee an investor a yield rate of $j_2 = 5\%$.

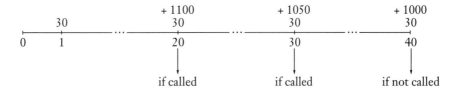

Solution   Calculate the purchase price using formula **(16)**

i) If the bond is called after 10 years,

$$P = 1100 + (30 - 27.50)a_{\overline{20}|.025} = 1100 + 38.97 = \$1138.97$$

ii) If the bond is called after 15 years,

$$P = 1050 + (30 - 26.25)a_{\overline{30}|.025} = 1050 + 78.49 = \$1128.49$$

iii) If the bond is redeemed at par after 20 years,

$$P = 1000 + (30 - 25)a_{\overline{40}|.025} = 1000 + 125.51 = \$1125.51$$

In this case, despite the fact that the desired yield rate is less than the bond coupon rate, the correct answer is found by using the latest possible redemption date. That is because of the premium value in the early call dates. ∎

There also exist debt securities called **extendible** or **retractable bonds**. These bonds are somewhat like a callable bond in that they allow the option of redeeming the bond at a time other than the stated redemption date. The difference between callable bonds and extendible/retractable bonds is that the bond owner, not the issuer, has the option.

An extendible bond allows the owner to extend the redemption date of the bond for a specified additional period. A retractable bond allows the owner to sell back the long-term bond to the issuer at a date earlier than the normal redemption date at par.

Canada Savings Bonds are an example of retractable bonds.

Because the option to extend or retract lies with the bond owner, no new mathematical analysis needs to be introduced.

Exercise 6.4

Part A

1. A $2000 bond paying interest at $j_2 = 10\%$ is redeemable at par in 20 years. It is callable at par in 15 years. Find the price to guarantee a yield rate of
a) $j_2 = 8\%$; b) $j_2 = 12\%$.

2. A $5000 bond paying interest at $j_2 = 6\%$ is redeemable at par in 20 years. It is callable at 105 in 15 years. Find the price to guarantee a yield rate of
a) $j_2 = 5\%$; b) $j_2 = 7\%$.

3. A $1000 bond with coupons at $j_2 = 9\%$ is redeemable at par in 20 years. It is callable after 10 years at 110 and after 15 years at 105. Find the price to guarantee a yield rate of a) $j_2 = 8\%$; b) $j_2 = 10\%$.

Part B

1. A $2000 bond with semi-annual coupons at $j_2 = 6\frac{1}{2}\%$ is redeemable at par in 20 years. It is callable at a 5% premium in 15 years. Find the price to guarantee a yield rate of a) $j_4 = 8\%$; b) $j_1 = 5\frac{1}{2}\%$.

2. A $1000 bond with coupons at $j_2 = 6\%$ is redeemable at par in 20 years. It also has the following call options:

Call Date	Redemption
15 years	105
16 years	104
17 years	103
18 years	102
19 years	101

Find the price to guarantee a yield rate of a) $j_1 = 5\%$; b) $j_{12} = 7\%$.

3. A $5000 callable bond pays $j_2 = 9\frac{1}{2}\%$ and matures at par in 20 years. It may be called at the end of years 10 to 15 (inclusive) for $5200. Find the price to yield at least $j_2 = 8\frac{1}{2}\%$.

4. A special callable bond with semi-annual coupons at $j_2 = 10\%$ and a face value of $1000 is sold by the issuer for a purchase price P. The redemption amount is a little unusual in that it is described as $(1200 - 20t)$ during the first 10 years and $[1000 + 20(t - 10)]$ after 10 years, where t is the number of years after issue.
   a) What is the purchase price P to yield 9% per annum compounded semi-annually assuming the bond is called at the end of 6 years?
   b) Is the bond
      i) redeemed at a premium or at a discount? Specify the amount.
      ii) purchased at a premium or at a discount? Specify the amount.
   c) Using the purchase price in a), at what other time point could the bond be called to produce the same yield rate? (Answer to the closest integral number of years n from issue.)

Price of a Bond Between Bond Interest Dates

The bond purchase-price formulas **(15)** and **(16)** were derived for bonds purchased on bond interest dates. In that case, the seller keeps the bond interest payment due on that date, and the buyer receives all the future bond interest payments. In actuality, bonds are purchased at any time and, consequently, we need a method of valuation of bonds between bond interest dates.

We will use the following notation in determining the value of a bond purchased between bond interest dates to yield the buyer interest at rate i.

$P_0 = $ the purchase price of a bond on the preceding bond interest date to yield i

$k = $ the fractional part of the interest period that has elapsed since the preceding bond interest date. $(0 < k < 1)$

$P = $ the **flat price** or total value of the bond on the actual purchase date.

The formula

$$P = P_0 (1 + i)^k$$

is the correct expression for the purchase price assuming compound interest theory.

However, in reality this formula is seldom used. Rather, a simple interest growth model is assumed for the fractional period of time k. This results in the formula

$$P = P_0(1 + ki)$$

EXAMPLE 1   A $1000 bond, redeemable at par on October 1, 2002, pays bond interest at $j_2 = 10\%$. Find the purchase price on June 16, 2000, to yield $j_2 = 9\%$ assuming a) compound interest; b) simple interest, for the fractional duration.

Solution   The preceding bond interest date is April 1, 2000. The exact time elapsed from April 1, 2000, to June 16, 2000, is 76 days. The exact time from April 1, 2000, to the next bond interest date on October 1, 2000, is 183 days. Thus, $k = \frac{76}{183}$.

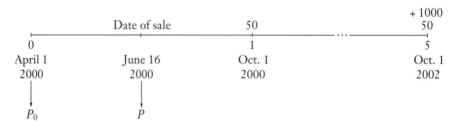

Using formula **(16)**,
$$P_0 = 1000 + (50 - 45)\, a_{\overline{5}|.045} = \$1021.95$$

a) Using the compound interest formula
$$P = P_0(1 + i)^k = 1021.95(1.045)^{\frac{76}{183}} = \$1040.80$$

b) Using the simple interest formula
$$P = P_0(1 + ki) = 1021.95\left[1 + (0.045)\left(\tfrac{76}{183}\right)\right] = \$1041.05$$

The simple interest calculation will always give a larger answer than the compound interest calculation. (For a proof of this, see Exercise 2.4, B1).

■

OBSERVATION: Unless otherwise specified, we will assume the simple interest formula for the flat price of the bond in this text.

If we were to graph the flat price of a bond, we would see a "sawtoothed" effect, as presented on the following page. As we approach each bond interest date, the flat price or actual selling price of the bond increases to give the seller the accrued value of the next bond interest payment or the **accrued bond interest**, which is equal to

$$I = k \cdot Fr$$

In our example,
$$I = \tfrac{76}{183}(50) = \$20.77$$

At each bond interest payment date, the accrued bond interest is zero and the price of the bond returns to the lower line marked Q. Q is called the **market price** or **quoted price** of the bond and is the price that is quoted in the daily paper. Q does not rise and fall as P does.

From the graph, we can see that $P = Q + I$ or

$$Q = P - I$$

In our example,

$$Q = 1041.05 - 20.77 = \$1020.28$$

If P_1 denotes the purchase price of the bond at the next bond interest date, it can also be shown that

$$Q = P_0 + k(P_1 - P_0)$$

The proof is left as an exercise (Exercise 6.5, Part B, Question 2).

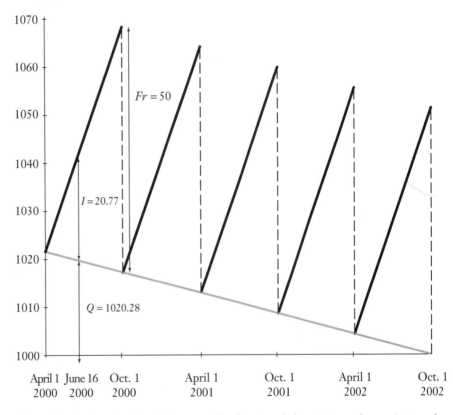

Note that under the simple interest for fractional durations or linear interpolation assumption the value of Q moves along a straight line from one coupon date to the next. The value of Q does not, however, move in a straight line from April 1, 2000, to October 1, 2002, as it may appear; Q is, in fact, a piece-wise linear function.

Exercise 6.5

Part A

Find the purchase price of the bonds in problems 1 to 6 in the table below.

No.	Face Value	Redemption	Bond Interest Rate	Yield Rate	Redemption Date	Date of Purchase
1.	$ 1 000	at par	$j_2 = 8\%$	$j_2 = 10\%$	Jan. 1, 2020	May 8, 2000
2.	$ 500	at par	$j_2 = 12\%$	$j_2 = 11\%$	Jan. 1, 2015	Oct. 3, 2000
3.	$ 2 000	at par	$j_2 = 9\%$	$j_2 = 10\%$	Nov. 1, 2011	July 20, 2000
4.	$10 000	at par	$j_2 = 6\frac{1}{2}\%$	$j_2 = 7\%$	Feb. 1, 2018	Oct. 27, 2000
5.	$ 1 000	at 105	$j_2 = 10\%$	$j_2 = 8\%$	July 1, 2011	July 30, 2000
6.	$ 2 000	at 110	$j_2 = 5\frac{1}{2}\%$	$j_2 = 6\frac{1}{2}\%$	Oct. 1, 2016	Apr. 17, 2001

7. A $1000 bond, redeemable at par on October 1, 2002, is paying bond interest at rate $j_2 = 9\%$. Find the purchase price on August 7, 2000, to yield $j_2 = 10\%$. Do a diagram similar to that found at the end of Section 6.5 for this bond.

Part B

1. Let: P_t be the value of a bond on a coupon date at time t to yield i.
   P_{t+1} be the value of a bond on the following coupon date to yield i.

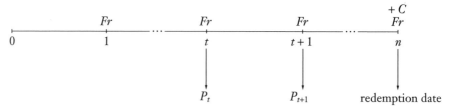

Show that $P_{t+1} = P_t(1 + i) - Fr$.

2. Using the answer to Part B, Problem 1, and given $P = P_0(1 + ki)$, prove that $Q = P_0 + k(P_1 - P_0)$.

3. A $1000 bond, redeemable at $1100 on November 7, 2010, has coupons at $j_2 = 7\%$. Find the purchase price on April 18, 2001, if the desired yield is
   a) $j_{12} = 8\%$; b) $j_1 = 6\%$.

4. An investment company is being audited and must locate the complete records of a specific bond transaction. The bond was a $500 bond with bond interest at rate $j_2 = 12\%$ redeemable at par on January 1, 2016. It was purchased for $550.89 sometime between July 1, 2001, and January 1, 2002. Find the exact date if the bond was purchased to yield $j_2 = 11\%$.

5. Repeat Part A, Problem 6, but assume compound interest for the fractional period.

6. A National Auto Company Limited $1000 bond is due at par on December 1, 2012. Interest is payable at $j_2 = 7\%$ on June 1 and December 1. The bond may be called at 104 on December 1, 2006. Find the flat (purchase) price and the market price for this bond on August 8, 2001 if the yield is to be $j_2 = 6\%$,
   a) assuming the bond is called at December 1, 2006;
   b) assuming the bond matures at par on December 1, 2012.

| *Section 6.6* | **Buying Bonds on the Market** |

In Section 6.5 we developed a method of determining the price an investor should pay for a given bond to yield a specified interest rate. In most cases, bonds are purchased on the bond market (bond exchange) where they are sold to the highest bidder. Trading of bonds is done through agents acting on behalf of the buyer and seller. The seller indicates the minimum price he or she is willing to accept (ask) and the buyer the maximum price he or she is willing to pay (bid). The agents, who work for a commission, try to get the best possible price for their client. Many newspapers publish tables of bond information, as illustrated on the next page. The columns describe the bonds as to issuer, bond interest rate and date of redemption or maturity. The yields listed are calculated assuming the bond is purchased for the price bid and held to maturity.

Bond prices will vary over time as general interest rates available vary. If investment rates of return rise, bond values will fall. If investment rates of return fall, bond values will rise. (This is illustrated in Exercise 6.6, B3.)

Since bonds are issued in different denominations, it is customary to give the **market quotation** q on the basis of a $100 bond.

EXAMPLE 1   A $1000 bond, paying interest at $j_2 = 9\frac{1}{2}\%$ is redeemable at par on August 15, 2021. This bond was sold on September 1, 2000, at a market quotation of 103.13. What did the buyer pay?

Solution   The market price was $Q = 10 \times 103.13 = \$1031.30$.

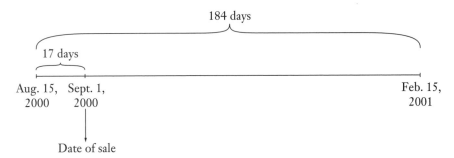

The accrued bond interest from August 15, 2000, to September 1, 2000, is
$$I = \tfrac{17}{184} \times 47.50 = \$4.39$$
and the total purchase price is
$$P = Q + I = 1031.30 + 4.39 = \$1035.69$$

∎

EXAMPLE 2   A $500 bond, paying interest at $j_2 = 8\%$, is redeemable at par on February 1, 2009. What should the market quotation be on November 15, 2000, to yield the buyer 9% per annum compounded semi-annually?

Solution

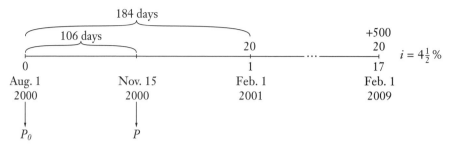

The purchase price on August 1, 2000, to yield $j_2 = 9\%$, would be

$$P_0 = 500 + (20 - 22.50)\,a_{\overline{17}|.045} = 500 - 29.27 = \$470.73$$

The purchase price on November 15, 2000, to yield $j_2 = 9\%$ would be

$$P = P_0(1 + ki) = 470.73\left(1 + \tfrac{106}{184} \times 0.045\right) = \$482.93$$

The accrued bond interest from August 1, 2000, to November 15, 2000, is

$$I = \tfrac{106}{184} \times 20 = \$11.52$$

The market price on November 15, 2000, would be

$$Q = P - I = 482.93 - 11.52 = \$471.41$$

Reducing Q to a \$100 bond we get the so-called **market quotation**

$$q = \tfrac{471.41}{5} = 94.28$$

■

CANADIAN BONDS

Provided by RBC Dominion Securities

Selected quotations, with changes since the previous day, on actively traded bond issues yesterday. Yields are calculated to full maturity. Price is the final bid-side price as of 5 pm yesterday.

Issuer	Coupon	Maturity	Price	Yield	Price $ Chg	Issuer	Coupon	Maturity	Price	Yield	Price $ Chg
GOVERNMENT OF CANADA						Ontario Hyd	7.750	Nov 03/05	104.75	6.75	-0.46
Canada	10.500	Mar 01/01	104.99	5.93	-0.07	Ontario Hyd	5.600	Jun 02/08	92.36	6.81	-0.48
Canada	7.500	Mar 01/01	101.73	5.91	-0.07	Ontario Hyd	8.250	Jun 22/26	115.69	6.95	-1.00
Canada	9.750	Jun 01/01	104.88	6.05	-0.09	Quebec	10.250	Oct 15/01	106.55	6.25	-0.15
Canada	4.500	Jun 01/01	97.98	6.02	-0.08	Quebec	5.250	Apr 01/02	97.60	6.42	-0.10
Canada	7.000	Sep 01/01	101.36	6.11	-0.10	Quebec	7.500	Dec 01/03	102.67	6.71	-0.30
Canada	5.250	Dec 01/01	98.43	6.13	-0.13	Quebec	7.750	Mar 30/06	104.45	6.86	-0.50
Canada	9.750	Dec 01/01	106.27	6.19	-0.15	Quebec	6.500	Oct 01/07	97.55	6.91	-0.50
Canada	8.500	Apr 01/02	104.57	6.26	-0.15	Quebec	11.000	Apr 01/09	127.10	6.97	-0.70
Canada	10.000	May 01/02	107.85	6.29	-0.16	**CORPORATE**					
PROVINCIAL						Rogers Cable	8.750	Jul 15/07	99.00	8.94	-0.75
Alberta	6.250	Mar 01/01	100.30	5.97	-0.06	Rogers Cant	10.500	Jun 01/06	109.50	8.54	-1.00
Alberta	6.375	Jun 01/04	99.07	6.62	-0.32	Royal Bank	5.400	Sep 03/02	97.09	6.61	-0.15
B C	9.000	Jan 09/02	104.77	6.41	-0.15	Royal Bank	5.400	Apr 07/03	96.01	6.79	-0.23
B C	7.750	Jun 16/03	103.41	6.62	-0.27	Royal Bank	6.750	Jun 04/07	98.33	7.04	-0.43
B C	5.250	Dec 01/06	91.65	6.79	-0.39	Sask Wheat	6.600	Jul 18/07	87.62	8.89	-0.37
B C	6.000	Jun 09/08	94.60	6.85	-0.48	T D Bank	5.600	Sep 05/01	98.74	6.41	-0.09
B C	9.500	Jan 09/12	120.78	6.92	-0.79	Trillium	5.690	Apr 22/03	96.60	6.86	-0.24
B C	8.500	Aug 23/13	113.72	6.93	-0.82	Trizec Hahn	7.950	Jun 01/07	93.68	9.14	-0.39
B C	6.150	Nov 19/27	89.79	6.99	-0.83	Union Gas	8.650	Nov 10/25	114.06	7.42	-0.85
B C	5.700	Jun 18/29	84.08	6.98	-0.80	Weston Geo	7.450	Feb 09/04	101.85	6.92	-0.31
B C Mun Fin	7.750	Dec 01/05	104.31	6.85	-0.51	Wstcoast Ene	6.750	Dec 15/27	91.95	7.44	-0.73
B C Mun Fin	5.500	Mar 24/08	91.19	6.92	-0.46	York Receiv	5.670	Apr 21/02	97.94	6.65	-0.13

Data provided by RBC Dominion Securities

This information can also be found at:

www.globeandmail.ca/hubs/moneymarkets.html or
www.rbcds.com/english/online/corporate/bondIndex.html

Exercise 6.6

Part A

Find the purchase price of the following $1000 bonds if bought at the given market quotation.

No.	Redemption Value	Bond Interest Rate	Market Quotation	Redemption Date	Date of Purchase
1.	par	$j_2 = 6\%$	98.50	Sept. 1, 2019	June 8, 2001
2.	par	$j_2 = 11\%$	104.25	Feb. 1, 2015	Oct. 2, 2001
3.	$1050	$j_2 = 9\%$	101.75	Oct. 1, 2016	Nov. 29, 2003
4.	$1100	$j_2 = 13\%$	112.50	Apr. 1, 2015	Jan. 12, 2000

What would be the market quotation on the following $1000 bonds?

No.	Redemption Value	Bond Interest Rate	Yield Rate	Redemption Date	Date of Purchase
5.	par	$j_2 = 11\%$	$j_2 = 8\%$	Nov. 1, 2019	Feb. 8, 2001
6.	par	$j_2 = 12\%$	$j_2 = 10\%$	Mar. 1, 2022	Aug. 19, 2003
7.	$1050	$j_2 = 5\frac{1}{2}\%$	$j_2 = 6\%$	June 1, 2015	Oct. 30, 2000
8.	$1100	$j_2 = 9\%$	$j_2 = 7\%$	Oct. 1, 2011	Nov. 2, 2003

Part B

1. A $5000 bond with semi-annual coupons at $j_2 = 9\%$ is redeemable at par on November 1, 2020.
   a) Find the price on November 1, 2000, to yield $j_4 = 10\%$.
   b) Find the book value of the bond on May 1, 2003 (just after the coupon is cashed).
   c) What should the market quotation of this bond be on August 17, 2003, if the buyer wants a yield of $j_1 = 7\%$?

2. For one of the bonds from the newspaper listing on the previous page, confirm the bid price given the other information about that bond.

3. a) A $1000 bond paying interest at $j_2 = 7\%$ is redeemable at par on September 1, 2020. Find the price on its issue date of September 1, 2000, to yield $j_2 = 9\%$.
   b) Find the book value of the bond on September 1, 2002, (just after the coupon is cashed).
   c) Find the sale price of this bond on September 1, 2002, if the buyer wants a yield of
      i) 6% compounded semi-annually;
      ii) 12% compounded semi-annually.
   d) What should the market quotation of this bond be on October 8, 2002, to yield a buyer $j_2 = 8\%$?

4. The ABC Corporation $5000 bond that pays interest at $j_2 = 7\%$ matures at par on October 1, 2011.
   a) What did the buyer pay for the bond if it was sold on July 28, 1999, at a market quotation of 89.38?
   b) What should the market quotation of this bond be on July 28, 1999, to yield a buyer $j_{12} = 6\%$?
   c) What should the market quotation of this bond be on December 13, 2001, to yield a buyer $j_1 = 8\%$?

Section 6.7 | Finding the Yield Rate

One of the fundamental problems relating to bonds is to determine the investment rate i a bond will give to the buyer when bought for a given price P. In practice, the market price is often given without stating the yield rate. The investor is interested in finding the true rate of return on his or her investment, i.e., the yield rate. Based on the yield rate, the investor can decide whether the purchase of a particular bond is an attractive investment or not, and also find out which of several bonds available is the best investment.

There are different methods available for finding the yield rate. The method most frequently used in practice was interpolation in *bond tables*, which were tables of market prices of bonds for wide ranges of bond interest rates, yield rates and terms to maturity. This method has become obsolete with the advent of personal computers and financial calculators. In this section we will calculate the yield rate in two ways.

Note that in the examples that follow, we calculate the yield rate assuming that the bond will be held to maturity unless specifically stated otherwise.

The Method of Averages

This method is simple and usually leads to fairly accurate results. It calculates an approximate value of the yield rate i as the ratio of the average interest payment over the average amount invested.

EXAMPLE 1   A $500 bond, paying interest at $j_2 = 9\frac{1}{2}\%$, redeemable at par on August 15, 2012, is quoted at 109.50 on August 15, 2000. Find the approximate value of the yield rate j_2 to maturity.

Solution   The purchase price $P = 5 \times 109.50 = \$547.50$, since the bond is sold on a bond interest date. If the buyer holds the bond until maturity, he will receive 24 bond interest payments of $23.75 each plus the redemption price $500, in total $24 \times 23.75 + 500 = \$1070$. He pays $547.50 and receives $1070. The net gain $1070 - 547.50 = \$522.50$ is realized over 24 interest periods, so that the average interest per period is

$$\frac{522.50}{24} = \$21.77$$

The average amount invested is the average of the purchase price (the original value) and the redemption value (the final value), i.e., $\frac{1}{2}(547.50 + 500) = \$523.75$. The approximate value of the yield rate is

$$i = \frac{21.77}{523.75} = 0.0416 = 4.16\% \text{ or } j_2 = 8.32\%.$$

∎

If n is the number of interest periods from the date of sale until the redemption date we can conclude that

The average interest payment $= \frac{n \times Fr + C - P}{n}$

The average amount invested $= \frac{P + C}{2}$

The approximate value of $i = \dfrac{\text{the average interest payment}}{\text{the average amount invested}}$

In most cases the answer is correct to the nearest 10th percent. If a more accurate answer is desired, the method of averages should be followed by the interpolation technique described below.

The Method of Interpolation

This method of determining the yield rate consists of finding two adjacent rates such that the market price of the bond lies between the prices determined. The standard method of interpolation between the two adjacent rates is then used to determine i or j_2. Usually the method of averages is used to get a starting value to be used in the method of interpolation.

EXAMPLE 2   Compute the yield rate j_2 in Example 1 by the method of interpolation.

Solution   By the method of averages we found that $i = 4.16\%$. Now we compute the market prices (which are equal to purchase prices) to yield $j_2 = 8\%$ and $j_2 = 9\%$.

$$Q(\text{to yield } j_2 = 8\%) = 500 + (23.75 - 20)\, a\,\overline{_{24}}_{.04} = \$557.18$$

$$Q(\text{to yield } j_2 = 9\%) = 500 + (23.75 - 22.50)\, a\,\overline{_{24}}_{.045} = \$518.12$$

Arranging the data in the interpolation table, we have

		Q	j_2	
		557.18	8%	
39.06	9.68	547.50	j_2	$\}d$
		518.12	9%	

$$\frac{d}{1\%} = \frac{9.68}{39.06}$$
$$d \doteq .25\%$$
$$\text{and } j_2 = 8.25\%$$

Check: Q (to yield $j_2 = 8.25\%) = 500 + (23.75 - 20.63)\, a\,\overline{_{24}}_{.04125} = \$546.97$

∎

The two methods described in this section apply equally well to bonds purchased between interest dates. The computations are more tedious, as is illustrated in the following example.

EXAMPLE 3   A $1000 bond paying interest at $j_2 = 11\%$ matures at par on June 1, 2010. On February 3, 2000, this bond is quoted at 95.38. What is the yield rate j_2?

Solution   First we use the method of averages to get an estimate of the yield rate assuming that the bond was quoted on the **nearest** coupon date, in this case December 1, 1999, or 21 interest periods before maturity. The market price on February 3, 2000, is $Q = \$953.80$.

The average interest payment $= \frac{21 \times 55 + 1000 - 953.80}{21} = \$57.20$.

The average amount invested $= \frac{953.80 + 1000}{2} = \$976.90$.

The approximate value of $i = \frac{57.20}{976.90} = 0.0586$ or 5.86%.

The approximate value of $j_2 = 11.71\%$.

If we want a more accurate answer, we select 2 rates, $j_2 = 11\%$ and $j_2 = 12\%$, and compute the corresponding market prices on February 3, 2000, using the method outlined in Section 6.5.

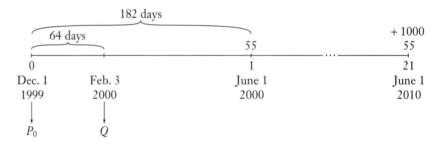

At $j_2 = 11\% : P_0 = Q = \$1000$ without calculation.

At $j_2 = 12\% : P_0 = 1000 + (55 - 60)\, a_{\overline{21}|.06} = 1000 - 58.82 = \$941.18$

$$P = 941.18\left[1 + \tfrac{64}{182}(.06)\right] = \$961.04$$

$$Q = P - I = 961.04 - \tfrac{64}{182}(55) = \$941.70$$

Arranging the data in an interpolation table, we have

Q on Feb. 3, 2000	j_2
1000.00	11%
953.80	j_2
941.70	12%

$\dfrac{d}{1\%} = \dfrac{46.20}{58.30}$

$d \doteq .79\%$

and $j_2 = 11.79\%$

■

EXAMPLE 4   A $1000 bond paying interest at $j_2 = 11\%$ matures at par on November 15, 2011. On November 15, 1996, it was purchased at 104. On November 15, 2000, it was sold at 97.50. What was the yield rate, j_2?

Solution   Using the method of averages, we find

$$\text{The average interest payment} = \frac{8 \times 55 + 975 - 1040}{8} = \$46.875$$

$$\text{The average amount invested} = \frac{975 + 1040}{2} = \$1007.50$$

$$\text{The approximate value of } i = \frac{46.875}{1007.50} = 0.0465 \text{ or } 4.65\%$$

The approximate value of $j_2 = 9.30\%$.

If we want a more accurate answer, we select two rates, $j_2 = 9\%$ and $j_2 = 10\%$, and use them in the following formula:

$$1040 = 55\, a_{\overline{8}|i} + 975(1 + i)^{-8}$$

At $j_2 = 9\%$, the right-hand side equals: $362.77 + 685.61 = \$1048.38$

At $j_2 = 10\%$, the right-hand side equals: $355.48 + 659.92 = \$1015.40$

Arranging the data in an interpolation table, we have

		Q	j_2	
		1048.38	9%	
32.98	8.38	1040.00	j_2	1%
		1015.40	10%	

$$\frac{d}{1\%} = \frac{8.38}{32.98}$$
$$d \doteq .25\%$$
$$\text{and } j_2 = 9.25\%$$

■

EXAMPLE 5   A $1000 bond paying interest at $j_2 = 12\%$ matures on June 1, 2015. On October 10, 1998, it was purchased for $1042.50 plus accrued bond interest (i.e., $q = 104.25$). On February 8, 2001, it was sold for $968.70 plus accrued bond interest (i.e., $q = 96.87$). Estimate the yield rate j_2 by the method of averages.

Solution

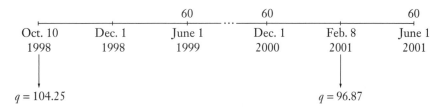

Using the method of averages, we assume that both transactions took place at the nearest respective coupon date. That is, we assume a purchase price of $1042.50 on December 1, 1998, and a sale price of $968.70 on December 1, 2000. In this period, there would be four coupons of $60 each.

The average interest payment $= \frac{4 \times 60 + 968.70 - 1042.50}{4} = \$41.55$.

The average amount invested $= \frac{968.70 + 1042.50}{2} = \$1005.60$.

The approximate value of $i = \frac{41.55}{1005.60} = 0.0413$ or 4.13%.

The approximate value of $j_2 = 8.26\%$.

The calculation of the answer using the method of interpolation is left for Part B, Problem 9 in the exercises that follow.

∎

The Yield Curve

One interesting aspect of the bond market is the yield curve. This is a graph showing the current yield-to-maturity of bonds of different maturities. The yield-to-maturity assumes the investor holds the bonds until redemption and is the yield used in previous sections. Of course, many investors may choose to sell their bond before maturity and in this case their rate of return will depend on both the purchase and the selling prices and not the yield-to-maturity at purchase.

The yield curve shows how interest rates vary for different maturities. That is, it is a visual representation of interest rates for different terms. Before looking at an example, it is important to note that each yield curve should consider bonds of similar risk.

At the time of writing, the longer-term bonds have a higher yield so that the yield curve is upward sloping (as you can see in the graph below). This is a normal or positive yield curve and represents the usual market position. However, sometimes the shorter maturities have higher yields so that the yield curve is known as downward sloping or concave.

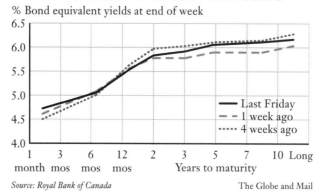

GOVERNMENT OF CANADA YIELD CURVE

Source: Royal Bank of Canada   The Globe and Mail

A normal yield curve highlights the fact that if you purchase a bond, hold on to the investment for a period of time and then sell it, it is likely that your selling yield will be lower that your purchase yield, assuming that other market forces have not changed. This provides investors with the opportunity to obtain a yield in excess of the yield-to-maturity and is known as "riding the yield curve." The next example illustrates this approach.

EXAMPLE 6

Mrs. Noori purchases a $1000 bond, paying bond interest at $j_2 = 8\%$ and redeemable at par in 10 years. The price she pays will yield $j_2 = 7\%$ if held to maturity. After holding the bond for 3 years, she sells the bond to yield the new purchaser $j_2 = 6\%$. Calculate Mrs. Noori's yield rate, j_2, over the 3-year investment period.

Solution

Using the bond pricing formula **(15)**, $Fr = 40$, $C = 1000$, $n = 20$, and $i = 0.035$

$$\text{Purchase price} = 40\, a_{\overline{20}|0.035} + 1000(1.035)^{-20} = \$1071.06$$

After 3 years, the selling price can be calculated using $Fr = 40$, $C = 1000$, $n = 14$, and $i = 0.03$

$$\text{Selling price} = 40\, a_{\overline{14}|.03} + 1000(1.03)^{-14} = \$1112.96$$

Now consider her 3-year investment. She paid $1071.06 for the bond, received a $40 coupon payment every 6 months for 3 years and then $1112.96 when she sold the bond. Therefore, her equation of value at rate i per half-year is:

$$1071.06 = 40\, a_{\overline{6}|i} + 1112.96(1 + i)^{-6}$$

We first calculate the approximate yield and then use linear interpolation to obtain a more accurate answer.

Using $Fr = 40$, $n = 6$, $C = 1112.96$ and $P = 1071.06$ and the method of averages, we calculate the approximate value of i:

$$i \doteq \frac{(6 \times 40 + 1112.96 - 1071.06)/6}{(1112.96 + 1071.06)/2} = .043 \text{ or } j_2 = 8.6\%$$

$$Q(\text{ to yield } j_2 = 8\%) = 40\, a_{\overline{6}|.04} + 1112.96(1.04)^{-6} = \$1089.27$$

$$Q(\text{ to yield } j_2 = 9\%) = 40\, a_{\overline{6}|.045} + 1112.96(1.045)^{-6} = \$1060.95$$

Arranging the data in the interpolation table, we have

	Q	j_2	
	1089.27	8%	
18.21	1071.06	j_2	d
	1060.95	9%	

$$\frac{d}{1\%} = \frac{18.21}{28.32}$$
$$d \doteq .64\%$$
and $j_2 = 8.64\%$

Hence, in this example, the investor obtains a yield in excess of 8% because she was able to sell at a lower yield. This could have been due to the shape of the yield curve or a change in market conditions.

∎

Exercise 6.7

In the questions that follow, assume that the bond will be held to maturity unless specifically stated otherwise.

Part A

In problems 1 to 4, find the yield rate, j_2, by the method of averages.

No.	Face Value	Redemption Value	Bond Interest	Years to Redemption	Purchase Price
1.	$2000	at par	$j_2 = 11\%$	12	$1940
2.	$5000	at par	$j_2 = 7\%$	10	$5340
3.	$1000	at 105	$j_2 = 12\%$	15	$1120
4.	$ 500	at 110	$j_2 = 6\%$	11	$ 450

5-8. Find the yield rate, j_2, in problems 1 to 4 by the method of interpolation.

9. A $1000 bond paying bond interest at $j_2 = 6\%$ matures at par on August 1, 2007. If this bond was quoted at 92 on August 1, 2000, what was the yield rate j_2
a) using the method of averages;
b) using the method of interpolation?

10. A $1000 bond redeemable at par at the end of 20 years and paying bond interest at $j_2 = 11\%$ is callable in 15 years at $1050. Find the yield rate j_2 if the bond is quoted now at 110 assuming
a) it is called;
b) it is not called.

11. A $1000 bond redeemable at par in 20 years pays bond interest at $j_2 = 10\%$. It is callable at the end of 10 years at 105. If it is quoted at 96, find the yield rate j_2 assuming
a) it is called;
b) it is not called.

12. A $1000 bond paying interest at $j_2 = 10\%$ matures at par on June 1, 2017. On August 17, 2001, this bond is quoted at 98.50. What is the yield rate?

13. A $1000 bond paying interest at $j_2 = 7\%$ matures at par on October 1, 2012. On April 28, 2001, this bond is quoted at 102. What is the yield rate?

14. The XYZ Corporation has a $1000 bond that pays bond interest at $j_2 = 8\%$. The bond is redeemable at par on June 1, 2016. On June 1, 2000, an investor buys this bond on the open market at 97. On June 1, 2008, he sells this bond on the open market at 101. Find the investor's yield rate j_2.

15. Ms. Holman buys a $1000 bond that pays bond interest at $j_2 = 7\%$ and is redeemable at par in 15 years. The price she pays will give her a yield of $j_2 = 8\%$ if held to maturity. After 5 years, Ms. Holman sells this bond to Mr. Dawson, who desires a yield of $j_2 = 6\%$ on his investment.
a) What price did Ms. Holman pay?
b) What price did Mr. Dawson pay?
c) What yield j_2 did Ms. Holman realize? (Use interpolation)

16. A $1000 bond paying interest at $j_2 = 11\%$ matures at par on July 15, 2019. On January 29, 2000, it was purchased for $972.50 plus accrued bond interest (i.e., 97.25). On May 7, 2003, it was sold for $1003.80 plus accrued bond interest (i.e., 100.38). Estimate the yield rate j_2 by the method of averages.

17. Bond interest, paid in the form of coupons, is taxable in full in the hands of the individual who owns the bond. However, the difference between your purchase price and your sale price (or redemption value) is either a capital gain (if sold at a profit) or a capital loss (if sold at a loss). Only two-thirds of any realized capital gain enters one's taxable income, while capital losses can be used to offset capital gains in the same year before determining one's taxable income.

On January 1, 2000, Fred buys a $1000 corporate bond paying interest at $j_2 = 8\%$ (payable June 30 and December 31) and redeemable at par in exactly 10 years. His purchase price is $980. On December 31, 2002, Fred sells the bond on a day where its market quote is 104.

By purchasing this bond, how much additional taxable income does Fred incur in: a) 2000; b) 2001?

Part B

1. The XYZ Corporation issues a $1000 bond with semi-annual coupons at $j_2 = 7\%$ redeemable at par in 20 years or callable at par in 15 years. Mr. LaBelle buys the bond to guarantee a yield rate $j_{12} = 9\%$.
 a) Find the purchase price.
 b) After 15 years, the XYZ Corporation calls the bond in and pays Mr. LaBelle his $1000. Find his overall yield stated as a rate j_{12}.

2. In our "method of averages" formula we said
$$\text{Average amount invested} = \frac{P + C}{2}$$
A slightly more accurate formula uses
$$\text{Average amount invested} = C + \frac{n + 1}{2n}(P - C)$$
Using this latter modification, try some of the above problems again and see if your approximations improve.

3. The XYZ Corporation has a bond due December 4, 2013, paying bond interest at $j_2 = 10\frac{3}{8}\%$. The price bid on December 14, 2000, is 104. What yield j_{12} does the investor desire?

4. Mr. Hunter buys a $1000 bond with semi-annual coupons at $j_2 = 6\%$. The bond is redeemable at par in 20 years. The price he pays will guarantee him a yield of $j_4 = 8\%$ if held to maturity. After 5 years, Mr. Hunter sells this bond to Miss Schnarr, who desires a yield of $j_1 = 5\frac{1}{2}\%$ on her investment.
 a) What price did Mr. Hunter pay?
 b) What price did Miss Schnarr pay?
 c) What yield, j_4, did Mr. Hunter realize?

5. The ABC Corporation issues a $1000 bond, with coupons payable at $j_2 = 7\%$ on January 1 and July 1, redeemable at par on July 1, 2011.
 a) How much would an investor pay for this bond on September 1, 2001, to yield $j_2 = 8\%$?
 b) Given the purchase price from part a), if each coupon is deposited in a bank account paying interest at $j_4 = 6\%$ and the bond is held to maturity, what is the effective annual yield rate j_1 on this investment?

6. A bond paying semi-annual coupons at $j_2 = 8\%$ matures at par in n years and is quoted at 110 to yield rate j_2. A bond paying semi-annual coupons at $j_2 = 8\frac{1}{2}\%$ matures at par in n years and is quoted at 112 on the same yield basis.
 a) Find the unknown yield rate, j_2.
 b) Determine n to the nearest half-year.

7. An issue of bonds, redeemable at par in n years, is to bear coupons at $j_2 = 9\%$. An investor offers to buy the entire issue at a premium of 15%. At the same time, she advises the issuer that if the coupon rate were raised from $j_2 = 9\%$ to $j_2 = 10\%$, she would offer to buy the entire issue at a premium of 25%. At what yield rate j_2 are these two offers equivalent?

8. In August 1991, interest rates in the bond markets were such that an investor could expect a yield of $j_2 = 10\%$. On August 1, 1991, an investor bought a bond with semi-annual coupons at $j_2 = 9\%$ which was to mature in 20 years at par. By August 1, 1992, interest rates had fallen to $j_2 = 8\%$. This same investor sold his bond on August 1, 1992, on the bond market. Find the yield j_2 on this investment.

9. For the bond question solved in Example 5 in Section 6.7, use the method of interpolation to calculate the yield rate, j_2.

10. The ABC Corporation issues a very unusual type of bond on July 1, 2001. It is issued in units where the face amount is $1000 and the redemption amount is $1500 at July 1, 2011. Semi-annual coupons are payable January 1 and July 1 and are as follows:

	2002	2003	2004	2005	2006	2007	2008	2009	2010	2011
Jan. 1	$60	64	68	72	76	80	84	88	92	96
July 1	60	64	68	72	76	80	84	88	92	96

a) What should the initial offering price per unit be if the yield to maturity is to be 10% per annum compounded semi-annually?
b) Is this a premium or a discount? Specify the amount.
c) If a purchaser on July 1, 2001, price as calculated in a), sells one unit of the bond at July 1, 2003, for $1300, what interest rate j_2 would be realized?

Section 6.8   **Other Types of Bonds**

Canada Savings Bonds

Every fall since the end of World War II, the Canadian government has borrowed the savings of individual Canadians through the issuing of Canada Savings Bonds. These bonds are available in values ranging from $100 to $10 000 and are the favourite investment of the average citizen. In fact, they account for one-quarter of the total value of all bonds—both corporate and government. Two interest options are available. Regular interest bonds pay simple interest each November by cheque or direct deposit to the customer's account. Compound interest bonds pay compound interest on maturity or redemption. Canada Savings Bonds are redeemable on any banking day at any bank for their full face value, plus any interest accrued up to that time. The bonds are guaranteed by the federal government and are registered. They are nontransferable and there is a limit on the value that can be held by any person, partnership or corporation. Canada Savings Bonds are heavily advertised every year. Any Canadian resident can buy them without any formality at any branch of any chartered bank or trust company, as well as through a stockbroker or investment dealer. They can also be purchased through payroll deductions. Since Canada Savings Bonds are redeemable at par on any banking day, purchase-price formula (**15**) or (**16**) does not apply to their valuation. For more information, see:

www.csb.gc.ca.

Indexed bonds

The relationship between inflation and interest rates was explained in Section 2.8, Example 3. The yields obtained from fixed interest investments (such as bonds) are normally in excess of the inflation rate so that a "real" yield is obtained. Yet, despite a positive "real" yield overall, investments in bonds produce a constant income during the term of the loan that diminishes in "real" value (i.e., after inflation is taken into account) and a redeemed principal that has lost a considerable amount of its "real" value. These disadvantages may be particularly relevant for retired people who invest their life savings for future income. An alternative type of bond that overcomes these difficulties is an inflation-indexed bond where the interest and the redemption value increase with a specific index (e.g., consumer price index or CPI) and therefore maintain their "real" values.

Both Canada and the United States (and many other nations) issue inflation-indexed bonds. Every six months, the face value of the bond is adjusted by an amount equal to the CPI adjustment. Bond interest, or coupons, are still payable semi-annually, but are effective on the adjusted face value of the bond. For example, assume a $1000 inflation-indexed bond with coupons at $j_2 = 3\%$. After six months, the CPI has risen 1%. At that time, the face value of the bond would rise to $1010 and the first coupon would be $15.15 (i.e., 1.5% of $1010).

By year end, the CPI is up 3% from the time the bond was issued. The face value of the bond would now be $1030 and the second coupon would be $15.45. Similar adjustments would occur over the lifetime of the bond. Note that at the time of redemption, the maturity value would be the full inflation-adjusted face value, not $1000.

In Canada, the level of outstanding inflation-indexed bonds (sometimes called Real Return Bonds) is small, but growing. These bonds are of particular interest to insurance companies and pension funds that have promises that may have to be paid in real value benefits (e.g., inflation-indexed retirement income). At the end of March 1996, there were $5.8 billion in Real Return Bonds outstanding in Canada (see **www.fin.gc.ca/** and then go to Glossary – The Government of Canada, Ministry of Finance).

As at May 9, 2000, Government of Canada Real Return Bonds due December 1, 2026 were paying bond interest at 4.25% and yielding 3.67% (see **www.globeandmail.com/hubs/rob.html**).

It should be noted that the interest rate payable on indexed bonds is considerably lower than that payable on conventional bonds as the "real" values of each interest payment and the principal are maintained throughout the term of the loan. These bonds may be valued using similar principles to those developed in earlier sections.

Serial Bonds

When companies borrow by means of an issue of bonds, they occasionally issue a series of bonds with staggered redemption dates instead of with a common redemption date. Such bonds are called **serial bonds**. Serial bonds can be thought of simply as several bonds covered under one bond contract. If the redemption date of each individual bond is known, then the valuation of any one bond can be performed by methods already described. The value of the entire issue of the bonds is just the sum of the values of the individual bonds.

EXAMPLE The directors of a company authorized the issuance of $30 000 000 of serial bonds on September 1, 2000. Interest is to be paid annually on September 1 at a nominal rate of 9%. The indenture provides the following redemption provisions:

i) $5 000 000 of the issue is to be redeemed on September 1, 2005;

ii) $10 000 000 of the issue is to be redeemed on September 1, 2010;

iii) $15 000 000 of the issue is to be redeemed on September 1, 2015.

a) Find the purchase price of the issue on September 1, 2000, which would yield $j_1 = 8\%$.

b) What would be the value of the bonds on September 1, 2007?

Solution a We consider the serial issue to be composed of 3 issues, represented by the following diagrams:

The purchase price of the entire serial issue P on September 1, 2007, is the sum of the purchase prices of the 3 individual issues $P^{(1)}$, $P^{(2)}$, $P^{(3)}$.

$$P^{(1)} = 5\ 000\ 000 + (450\ 000 - 400\ 000)\, a_{\overline{5}|.08} \qquad = \$\ 5\ 199\ 635.50$$
$$P^{(2)} = 10\ 000\ 000 + (900\ 000 - 800\ 000)\, a_{\overline{10}|.08} \qquad = \$10\ 671\ 008.14$$
$$P^{(3)} = 15\ 000\ 000 + (1\ 350\ 000 - 1\ 200\ 000)\, a_{\overline{15}|.08} \qquad = \$16\ 283\ 921.80$$
$$P = P^{(1)} + P^{(2)} + P^{(3)} \qquad\qquad\qquad = \$32\ 154\ 565.44$$

Solution b The value of the entire serial issue P on September 1, 2007, is the sum of the values of the 2nd and 3rd individual cases $P^{(2)}$ and $P^{(3)}$. (The first individual issue was already redeemed at that time.)

$$P^{(2)} = 10\ 000\ 000 + (900\ 000 - 800\ 000)\, a_{\overline{3}|.08} \qquad = \$10\ 257\ 709.70$$
$$P^{(3)} = 15\ 000\ 000 + (1\ 350\ 000 - 1\ 200\ 000)\, a_{\overline{8}|.08} \qquad = \$15\ 861\ 995.84$$
$$P = P^{(2)} + P^{(3)} \qquad\qquad\qquad\qquad = \$26\ 119\ 705.54$$

■

Strip Bonds

Investors have several reasons for choosing one investment over another. They look for high yields and safety. They also look for cash flows whose timing matches their needs. For some investors, the coupons on a bond match their cash flow needs (e.g., a corporation responsible for paying monthly retirement benefits to a worker). For other investors, the redemption value of a bond matches their cash flow needs (e.g., an insurance company that needs to satisfy the maturity value of an endowment insurance contract).

For these reasons, some investors separate the coupons from the redemption value of the bond. In some instances, an investor buys the bond contract, "strips" the coupons from the bond and sells the remainder of the asset — namely, the redemption value. This redemption-value-only bond is called a "strip" bond and would sell for a price $C(1 + i)^{-n}$. The original buyer may also then sell each coupon as a separate asset. Each coupon has value $Fr\,(1 + i)^{-n}$ where n is the number of interest conversion periods from now to the time the coupon is payable.

In rare cases, corporations issue zero-coupon bonds where the bond interest rate is 0%. Obviously, the value of this bond is the discounted value of the redemption value, i.e., $C(1 + i)^{-n}$.

EXAMPLE There exists a $1000 corporate bond with bond interest at $j_2 = 9\%$ redeemable at par in 20 years. Investor A buys the bond to yield $j_2 = 8\frac{1}{2}\%$. Investor A strips the coupons from the bond and sells the remaining "strip" bond to Investor B, who wishes a yield rate of $j_{12} = 9\%$.

Find a) the price Investor B paid for the strip bond; b) the yield rate j_2 realized by Investor A.

Solution a   Investor B will pay the discounted value of $1000 payable in 20 years at $j_{12} = 9\%$, or $i = .0075$, per month over 240 months.

$$P = 1000(1.0075)^{-240} = \$166.41$$

Solution b   First determine the price Investor A paid for the bond. With $F = C = 1000$, $r = .045$, $i = .0425$ and $n = 40$,

$$P = 1000 + (45.00 - 42.50)\,a_{\overline{40}|.0425} = \$1047.69$$

In return for the investment of $1047.69, Investor A gets 40 coupons worth $45 each over 20 years plus $166.41 payable immediately from Investor B. That is,

$$1047.69 = 166.41 + 45\,a_{\overline{40}|i}$$
$$a_{\overline{40}|i} = 19.5840$$

A starting value to solve $a_{\overline{40}|i} = 19.5840$ (see Section 3.6) is

$i = \dfrac{1-\left(\frac{19.5840}{40}\right)^2}{19.5840} = .038822092$ or $j_2 = 2i = 7.76\%$.

At $j_2 = 7\%$, $a_{\overline{40}|.035} = 21.3551$.
At $j_2 = 8\%$, $a_{\overline{40}|.04} = 19.7928$.
We obviously need to proceed to $j_2 = 9\%$.
At $j_2 = 9\%$, $a_{\overline{40}|.045} = 18.4016$.

Now we have two rates, $j_2 = 8\%$ and $j_2 = 9\%$, 1% apart, that provide upper and lower bounds for interpolation.

Arranging our data in an interpolation table, we have

$a_{\overline{40}|i}$	j_2
19.7928	8%
19.5840	j_2
18.4016	9%

$1.3912\{\ .2088\{$ ... $\}d\ \}1\%$

$\dfrac{d}{1\%} = \dfrac{.2088}{1.3919}$

$d \doteq .15\%$

and $j_2 = 8.15\%$

Exercise 6.8

Part A

1. A $10 000 serial bond is to be redeemed in instalments of $2000 at the end of each of the 21st through 25th year from the date of issue. The bonds pay interest at $j_2 = 11\%$. What is the price to yield 9% per annum compounded semi-annually?

2. A $100 000 issue of serial bonds issued July 1, 2000, and paying interest at rate $j_2 = 9\%$ is to be redeemed in instalments of $25 000 each on the first day of July in 2005, 2007, 2009 and 2011. Find the price to yield 7% per annum compounded semi-annually.

3. A corporation issues a $1000 20-year bond with bond interest at $j_2 = 10\%$. It is purchased by an investor A who wishes a yield of $j_2 = 9\tfrac{1}{2}\%$. This investor A keeps the coupons but sells the strip bond to another investor whose desired yield is $j_{12} = 10\tfrac{1}{2}\%$. Determine the overall rate j_2 to investor A.

4. A corporation issues a $1000 15-year zero coupon bond redeemable at par. Find the price paid by an investor whose desired yield is $j_{365} = 8\%$.

Part B

1. In Part B, Problem 2 of Exercise 6.2 we introduced Makeham's formula for a bond redeemable at par.

$$P = F(1 + i)^{-n} + \frac{r}{i}[F - F(1 + i)^{-n}]$$

For serial bonds with k individual issues, we have

$$P^{(k)} = F_k(1 + i)^{-n_k} + \frac{r}{i}[F_k - F_k(1 + i)^{-n_k}]$$

and $\sum_k P^{(k)} = \sum_k F_k(1 + i)^{-n_k} + \frac{r}{i}[\sum_k F_k - \sum_k F_k(1 + i)^{-n_k}]$

where $\sum_k P^{(k)}$ ... purchase price of the entire serial issue.

$\quad\quad \sum_k F_k(1 + i)^{-n_k}$ ... total present value of all redemptions

$\quad\quad \sum_k F_k$ ... total face value of k individual issues.

Try this formula on questions 1 and 2 of Part A.

2. To finance an expansion of production capacity, Minicorp Ltd. will issue on March 15, 2000, $30 000 000 of serial bonds. Bond interest at 13% per annum is payable half yearly on March 15 and September 15, and the contract provides for redemption as follows:
$10 000 000 of the issue to be redeemed March 15, 2005;
$10 000 000 of the issue to be redeemed March 15, 2010;
$10 000 000 of the issue to be redeemed March 15, 2015.

Calculate the purchase price of the issue to the public to yield $j_1 = 12\%$ on those bonds redeemable in 5 years and $j_1 = 14\%$ on the remaining bonds.

3. A $210 000 issue of serial bonds with annual coupons of 10% on the balance outstanding is to be redeemed in 20 annual instalments beginning at the end of 1 year. The amount redeemed at the end of 1 year is to be $1000; at the end of 2 years, $2000; at the end of 3 years, $3000; and so on, so that the last redemption amount is $20 000 at the end of 20 years. Find the price to be paid for this issue to yield $j_1 = 8\%$. [Note that Makeham's formula may prove to be useful.]

4. A $10 000 serial bond is to be redeemed in $1000 instalments of principal per year over the next 10 years. Coupons at the rate of 9% on the outstanding balance over the next 10 years are also to be paid annually. Calculate the purchase price to yield an investor a yield a) $j_1 = 10\%$; b) $j_2 = 8\%$.

5. A corporation issues a $1000 15-year bond with bond interest at $j_2 = 9\frac{1}{2}\%$ and redeemable at 105. It is purchased by Investor A who wishes a yield of $j_2 = 10\%$. Investor A sells the coupons to Investor B whose desired yield is $j_{12} = 10\frac{1}{2}\%$ and Investor A keeps the strip bond. Determine the overall yield rate j_2 to Investor A.

Section 6.9   # Summary and Review Exercises

- Purchase price P of a bond with face value F, redemption value C and bond rate r, on a bond interest date n interest periods prior to maturity, to yield rate i until maturity is given by

$$P = Fr\,a_{\overline{n}|i} + C(1 + i)^{-n} \quad \text{basic formula}$$

or

$$P = C + (Fr - Ci)\,a_{\overline{n}|i} \quad \text{premium/discount formula}$$

- For par value bonds with $C = F$

  premium $P - C = (Fr - Ci)\,a_{\overline{m}|i} = F(r - i)\,a_{\overline{m}|i}$

  discount $C - P = (Ci - Fr)\,a_{\overline{m}|i} = F(i - r)\,a_{\overline{m}|i}$

- For other bonds with $C \neq F$ and modified coupon rate $g = \dfrac{Fr}{C}$

  premium $P - C = (Fr - Ci)\,a_{\overline{m}|i} = C(g - i)\,a_{\overline{m}|i}$

  discount $C - P = (Ci - Fr)\,a_{\overline{m}|i} = C(i - g)\,a_{\overline{m}|i}$

- Properties of a bond schedule:
  1. All book values can be reproduced by a purchase price formula.
  2. Book value adjustments add up to the amount of discount (or the amount of premium).
  3. Book values are gradually adjusted from P to C. In the case of a discount, we accumulate the discount or write the bond up from P to C; in the case of a premium, we amortize the premium or write the bond down from P to C.
  4. Successive book value adjustments are in the ratio $(1 + i)$.

- In the bond schedule:

  Book value adjustment at time k = Coupon – Interest on book value at time $(k - 1)$.

  For bonds purchased at a discount, book value adjustments are negative and the bond is written up.

  For bonds purchased at a premium, book value adjustments are positive and the bond is written down.

- If the bond is purchased between bond interest dates, where k is the fractional part of the interest period that has elapsed since the preceding bond interest date $(0 < k < 1)$, the flat price P is given by

  $$P = P_0(1 + ki)$$

  where P_0 is the price on the preceding bond interest date.

- Market price Q is the flat price P less accrued bond interest $I = k(Fr)$

  $$Q = P - I = P_0(1 + ki) - k\,Fr$$

  Market price Q can also be calculated as the interpolated price between price P_0 on the preceding bond interest date and price P_1 on the next bond interest date

  $$Q = P_0 + k(P_1 - P_0)$$

- To estimate the yield rate by the Method of Averages we calculate

  $$i \doteq \frac{\text{average income per period}}{\text{average value invested}} = \frac{[n(Fr) + C - P]/n}{(P + C)/2}$$

  To obtain a more accurate value of i, we follow the Method of Interpolation.

Review Exercises 6.9

1. The Acme Corporation issues $10 000 000 of 20-year bonds on March 15, 2000, with semi-annual coupons at 7%. The contract requires Acme to set up a sinking fund earning interest at $j_2 = 5\%$ to redeem the bonds at maturity; the first sinking-fund deposit is to be made September 15, 2000. Find
   a) the purchase price of the bond issue to yield $j_2 = 6\%$;
   b) the necessary sinking-fund payment.

2. Refer to Problem 1. Rather than issuing 20-year bonds, assume that Acme Corporation decided to issue serial bonds on March 15, 2000, with the following redemption schedule:
   $5 000 000 on March 15, 2006;
   $3 000 000 on March 15, 2011;
   $2 000 000 on March 15, 2016.

   Compute the purchase price of this serial bond issue to the public on March 15, 2000, to yield $j_2 = 6\%$.

3. The XYZ Corporation issues a $1000 bond with coupons at $j_2 = 8\%$ and redeemable at par on August 1, 2015.
   a) How much should be paid for this bond on February 1, 2000, to yield $j_2 = 6\%$?
   b) If the purchase price was $1050, and the bond is held to maturity, determine the overall yield, j_2.

4. A $1000 bond with semi-annual coupons at $j_2 = 10\%$ payable January 1 and July 1 each year matures on July 1, 2005, for $1050.
   a) Find the price on January 1, 2000, to yield $j_2 = 10\frac{1}{2}\%$.
   b) Is the bond purchased at a premium or a discount?
   c) Calculate the entries in the bond schedule on July 1, 2000, and January 1, 2001.

5. The ABC Corporation $2000 bond, paying bond interest at $j_2 = 8\%$, matures at par on September 1, 2014.
   a) What did a buyer pay for this bond on July 20, 2001, if the market quotation for the bond was 104.75?
   b) Estimate, using linear interpolation, the yield rate for the buyer in part a).
   c) What should the market quotation for this bond be on July 20, 2001, if the desired yield is $j_2 = 7\%$?

6. The ABC Corporation issues a 20-year par value bond on February 1, 2000, with coupons at $j_2 = 8\%$ payable February 1 and August 1. Mr. Kelly buys a $1000 bond from this issue on February 1, 2000 to yield $j_2 = 9\%$. On August 1, 2004, Mr. Kelly sells this bond to Ms. Quinn, who wants a yield $j_2 = 7\%$.
   a) Find the original purchase price.
   b) Find the sale price on August 1, 2004.
   c) What yield j_2 did Mr. Kelly realize?
   d) Find the sale price if the transaction took place on September 1, 2004.

7. Ms. Machado invested $10 000 in the Ace Manufacturing Company four years ago. She was to be paid interest on the loan at $j_2 = 11\%$. The principal amount of $10 000 was to be returned after 10 years. Now, having just received the 8th interest payment in full, Ms. Machado has been informed that Ace has just been declared bankrupt. Ms. Machado has been offered, as a settlement, 25% of the present value of all monies due to her, determined at $j_1 = 13\%$. How much can she expect to receive?

8. A $1000 bond has semi-annual coupons at $j_2 = 7\%$. The bond matures after 20 years at par but can be called after 15 years at $1050.
   a) Find the price to guarantee a yield of $j_2 = 8\%$.
   b) What maturity date was assumed in answering part a)?
   c) Determine the yield j_2 realized if the bond is redeemed other than anticipated in part a). (Your answer must be larger than $j_2 = 8\%$.)

9. A bond with face value $10 000 pays $j_2 = 9\%$. An investor buys it for $10 500 and sells it 4 years later for $10 200. Find the yield rate j_2 earned by the investor over this period using a) the method of averages; b) interpolation.

10. A $1000 bond with coupons at $j_2 = 10\%$ is redeemable at par in n years. It is purchased at a premium of $300. Another $1000 bond with coupons at $j_2 = 8\%$ is also redeemable at par in n years. It is purchased at a premium of $100 on the same yield basis.
    a) Find the unknown yield rate j_2.
    b) Determine n.

11. A $1000 callable bond pays $j_2 = 10\%$ and matures at par in 20 years. It may be called at $1100 at the end of years 5–9 inclusive. Find the price to yield at least $j_2 = 9\%$.

12. A $10 000 bond has semi-annual coupons, and matures at par in 15 years. It is bought at a discount to yield $j_2 = 11\%$. The adjustment in book value at the end of 5 years is $25. Find the purchase price.

13. Find the price on Dec. 8, 2001, of a $5000 bond maturing at par on July 2, 2011, paying $j_2 = 6\%$. It is priced to yield $j_2 = 7\%$.

14. A 20-year bond with a par value of $1000, paying $j_2 = 7\%$, and a maturity value of $1100 is bought on December 20, 1998, to yield $j_2 = 6\%$ to the investor.
    a) What is the price of the bond?
    b) How much of the coupon received on June 20, 2002, can be considered as interest income by the purchaser?
    c) If the bond were sold on July 20, 2002, to yield $j_2 = 5\%$ to the purchaser, how much would the seller receive?
    d) Assume the bond is bought for the price calculated in a) and sold for the price calculated in c) on July 20, 2002. What rate j_2 has the original investor earned? Use the method of averages.

15. A $2000 bond matures at par in 10 years and pays $j_2 = 10\%$. Find the purchase price to yield to maturity a) $j_2 = 11\%$; b) $j_{12} = 9\%$.

16. A $1000 bond matures for $1050 in 2 years and pays $j_2 = 9\%$. Find the price P to yield $j_2 = 10\%$ to maturity and construct the bond schedule.

17. A $2000 bond with semi-annual coupons at $j_2 = 7\%$ is redeemable for $2100. If the amount for the 10th write-down is $6.50, calculate the purchase price to yield $j_2 = 5.8\%$.

18. For a $2000 bond redeemable at par on February 1, 2013, with bond interest at $j_2 = 8\%$, find the purchase price on May 8, 1998 to yield $j_2 = 7\%$ until maturity.

19. For a $5000 bond redeemable at 103 on May 15, 2008, paying bond interest at $j_2 = 8.5\%$, find the purchase price on March 7, 2001 at a market quotation of 91.25.

20. For a $2000 bond redeemable at par on April 18, 2010, paying bond interest at $j_2 = 7\%$, find the market quotation on June 6, 2001, that would give a yield rate to maturity $j_2 = 8.2\%$.

21. A $2000 bond maturing at par in 16 years, with bond interest at $j_2 = 9\%$, is purchased to yield $j_2 = 10\%$ to maturity. After 4 years the bond is sold to yield $j_2 = 8\%$ to maturity. Find the yield rate over the 4-year period using interpolation.

CASE STUDY I

Callable bond

A $5000 callable bond matures on September 1, 2010, at par. It is callable on September 1, 2005, 2007 or 2009 at $5250. Interest on the bond is $j_2 = 8\%$.

a) Find the price on September 1, 2001, to yield an investor $j_2 = 6\%$.

b) Draw up the bond schedule for the year September 1, 2001, to September 1, 2002.

c) If the bond were called on September 1, 2007, what yield j_2 would the investor have earned? Use the method of averages, followed by interpolation.

d) On November 1, 2003, the bond is sold to yield $j_2 = 7\%$ until maturity.
   i) Find the purchase price.
   ii) What is the market quotation?

Chapter Seven

Business Decisions, Capital Budgeting and Depreciation

Net Present Value

Most business enterprises and investors are required to decide, on a fairly regular basis, whether a particular business venture or investment is worthwhile and should proceed. In many cases, a comparison between alternative projects is also required. That is, a financial assessment of each proposal is necessary before it commences so that the appropriate decision may be made.

As introduced in Section 3.3, one method of assessing a project that involves future cash flows is to use compound interest and calculate the present value of these cash flows at a particular interest rate. The interest rate used is known as the **cost of capital** and can be considered to be the cost of borrowing money by the business or the rate of return that an investor may obtain if the money is invested with security. This process is known as finding the **net present value** for the project and may be represented by the following equation:

$$\text{Net Present Value (or NPV)} = F_0 + F_1(1 + i)^{-1} + F_2(1 + i)^{-2} + F_3(1 + i)^{-3} + ... + F_n(1 + i)^{-n}$$

where F_t is the estimated cash flow for this project at the end of the period t and may be positive, negative or zero; and i is the cost of capital per period.

If the net present value of a project is positive, then it may be considered to be profitable based on the cash-flow estimates used; while if the result is negative, then the opposite conclusion applies and the project should not proceed. In many examples the first cash flow (namely F_0) is negative, while all future cash flows during the term of the project are positive. However in other cases, additional funds may be required such that some later F_ts are negative.

EXAMPLE 1 Find the net present value of a \$100 000 investment that is estimated to return the following end-of-year cash flows, if the cost of capital is a) $j_1 = 7\%$; b) $j_1 = 14\%$.

Year End	1	2	3	4
Cash Flow	$40 000	$30 000	$30 000	$35 000

Solution a   We have $F_0 = -100\ 000$, $F_1 = 40\ 000$, $F_2 = 30\ 000$, $F_3 = 30\ 000$, $F_4 = 35\ 000$ and $j_1 = .07$,

$$\begin{aligned}\text{NPV at 7\%} &= -100\ 000 + 40\ 000(1.07)^{-1} + 30\ 000(1.07)^{-2} \\ &\quad + 30\ 000(1.07)^{-3} + 35\ 000(1.07)^{-4} \\ &= +\$14\ 777\end{aligned}$$

This result indicates that if the cost of capital is 7% then the project should proceed, for its profit in present value terms is $14 777.

Solution b   With the same values for F_t but with $j_1 = 0.14$,

$$\begin{aligned}\text{NPV at 14\%} &= -100\ 000 + 40\ 000(1.14)^{-1} + 30\ 000(1.14)^{-2} \\ &\quad + 30\ 000(1.14)^{-3} + 35\ 000(1.14)^{-4} \\ &= -\$856\end{aligned}$$

This negative net present value indicates that if the cost of capital is 14% then this project will not return as high a rate as the cost of capital and therefore should not proceed.

■

OBSERVATION: The following points should be noted.

1. Due to the uncertainty of the future, the cash-flow figures used are normally estimates and therefore the net present value is only given to the nearest dollar, or in some cases the nearest hundred or thousand dollars.

2. As the future cash flows are only estimates and may vary with changing circumstances, it is common for at least three different sets of cash flows to be used representing the expected situation as well as an optimistic and pessimistic view, which in turn provides three net present values.

As shown in previous chapters, problems involving cash flows over several periods can be presented in a spreadsheet. The following spreadsheet shows how Example 1 could be calculated at both interest rates.

	A	B	C	D	E	F
	Year End	Cash Flow	Factor at 7% p.a.	PV at 7% p.a.	Factor at 14% p.a.	PV at 14% p.a.
2	0	-100,000	1.00000	-100,000	1.00000	-100,000
3	1	+40,000	0.93458	+37,383	0.87719	+35,088
4	2	+30,000	0.87344	+26,203	0.76947	+23,084
5	3	+30,000	0.81630	+24,489	0.67497	+20,249
6	4	+35,000	0.76290	+26,702	0.59208	+20,723
7			NPV	+$14,477		-$856

In a spreadsheet, it is also possible to place a variable (say the interest rate) in a particular cell and to set up the table in terms of this cell. This means that when the variable is changed, all the relevant cells change automatically. This process enables the user to see the results at different interest rates quickly.

Most spreadsheets also have an inbuilt function to calculate the net present value. For instance, in Excel it is NPV (rate, value 1, value 2,...). However, this assumes that each value is spaced equally and that the first payment (i.e., value 1) is in one period's time. Using Example 1, the function in the spreadsheet would be:

$$= NPV(0.07,\ B2 : B6) - 100\ 000$$

Note that the $100 000 that is paid immediately, is subtracted from the function to obtain the correct answer.

One of the major advantages in calculating net present values is that they enable comparisons to be made between two alternative projects that require the same investment. However, as shown in the following example, the preferred choice may depend on the cost of capital.

■

EXAMPLE 2   The following data set out the estimated end-of-year cash flows that will be received from alternative projects A and B for an investment of $200 000. Which project should be adopted if the cost of capital is a) $j_1 = 6\%$; b) $j_1 = 8\%$?

Year	1	2	3	4
Project A	$80 000	$70 000	$60 000	$ 35 000
Project B	$30 000	$40 000	$40 000	$150 000

Solution a

$$\begin{aligned}
\text{NPV of project A at 6\%} =\ &- 200\ 000 + 80\ 000(1.06)^{-1} \\
&+ 70\ 000(1.06)^{-2} + 60\ 000(1.06)^{-3} \\
&+ 35\ 000(1.06)^{-4} \\
=\ &+ \$15\ 871
\end{aligned}$$

$$\begin{aligned}
\text{NPV of project B at 6\%} =\ &- 200\ 000 + 30\ 000(1.06)^{-1} \\
&+ 40\ 000(1.06)^{-2} + 40\ 000(1.06)^{-3} \\
&+ 150\ 000(1.06)^{-4} \\
=\ &+ \$16\ 301
\end{aligned}$$

Both net present values are positive so that both projects would be profitable and could proceed. However, with the higher NPV at 6%, project B is to be slightly preferred.

Solution b

$$\begin{aligned}
\text{NPV of project A at 8\%} =\ &- 200\ 000 + 80\ 000(1.08)^{-1} \\
&+ 70\ 000(1.08)^{-2} + 60\ 000(1.08)^{-3} \\
&+ 35\ 000(1.08)^{-4} \\
=\ &+ \$7444
\end{aligned}$$

$$\text{NPV of project B at } 8\% = -\,200\,000 + 30\,000(1.08)^{-1}$$
$$+ 40\,000(1.08)^{-2} + 40\,000(1.08)^{-3}$$
$$+ 150\,000(1.08)^{-4}$$
$$= +\$4079$$

On this occasion project A returns the higher profit, although again both projects would be profitable.

∎

OBSERVATION: This example highlights the fact that the actual interest rate used is important when using *NPV*s to make investment decisions.

EXAMPLE 3 An insurance company has $1 million to invest. Two investment options are available: the first is an investment in a 4-year 12% government bond paying half-yearly interest, redeemable at par and priced at $95; the second is the purchase of a 4-year lease on a coal mine priced at $1 million, which is likely to return the following year-end cash flows:

Year	1	2	3	4
Cash Flow	$400 000	$600 000	$500 000	−$200 000

The cash outflow in the fourth year is shown as a negative payment and is the cost of land restoration. Which option should the company follow?

Solution In this problem we will calculate the yield obtained on the government bond investment (which we assume to be secure) and then use this figure as the cost of capital to find the net present value of the mining investment. The equation of value for the bond is

$$95 = 6a_{\overline{8}|i} + 100(1 + i)^{-8}$$

which indicates that the yield will be slightly higher than 6% per half-year. For linear interpolation, the following values will be used:

$$P(\text{at } j_2 = 13\%) = 96.96$$
$$P(\text{at } j_2 = 14\%) = 94.03$$

Arranging the data in an interpolation table, we have

	P	j_2
	96.96	13
	95.00	j_2
	94.03	14

$$\frac{d}{1\%} = \frac{1.96}{2.93}$$
$$d \doteq .67\%$$
$$j_2 = 13.67\%$$

Then find i per annum equivalent to $j_2 = 13.67\%$

$$1 + i = (1 + \tfrac{.1367}{2})^2$$

$$i \doteq 14.14\%$$

We now use $i = 14.14\%$ as the cost of capital for the alternative coal mine lease.

$$\begin{aligned} \text{NPV at } 14.14\% = {} & -1\ 000\ 000 + 400\ 000(1.1414)^{-1} \\ & + 600\ 000(1.1414)^{-2} + 500\ 000(1.1414)^{-3} \\ & - 200\ 000(1.1414)^{-4} \\ = {} & +\$29\ 405 \end{aligned}$$

As the NPV is positive, it indicates that the mining investment will be more profitable than the bond investment; the investment should proceed based on these estimated cash flows.

It should be added that this example makes no allowance for the different risks involved in these two investment options. That will be analyzed in Chapter 8.

■

EXAMPLE 4   A mining company is considering whether to develop a mining property. It is estimated that an immediate expenditure of $7 000 000 will be needed to bring the property into production. Thereafter, the net cash inflow will be $1 700 000 at the end of each year for the next 10 years. An additional expenditure of $3 200 000 at the end of 11 years will have to be made to restore the property to an attractive condition. On projects of this type the company would expect to earn at least $j_1 = 20\%$. Advise whether the company should proceed.

Solution   The net present value of the project at $j_1 = 20\%$ is

$$\begin{aligned} \text{NPV at } 20\% = {} & -7\ 000\ 000 + 1\ 700\ 000a_{\overline{10}|.20} - 3\ 200\ 000(1.20)^{-11} \\ = {} & -7\ 000\ 000 + 7\ 127\ 202.55 - 430\ 681.55 \\ = {} & -\$303\ 479 \end{aligned}$$

The negative net present value of the project means that the project will not yield a 20% return per annum and therefore should be rejected. If the above calculations are repeated using a desired rate of 15%, the net present value of the project will be

$$-7\ 000\ 000 + 1\ 700\ 000a_{\overline{10}|.15} - 3\ 200\ 000(1.15)^{-11} = \$844\ 088$$

The positive net present value of the project means that the project is earning more than 15% per annum.

■

Exercise 7.1

Part A

Calculate the net present value for a $100 000 investment in problems 1 to 4 and thereby determine whether the investment should proceed.

		End-of-Year Cash Flow			
No.	Cost of Capital per Annum	Year 1	Year 2	Year 3	Year 4
1.	10%	$40 000	$30 000	$40 000	$30 000
2.	7%	$50 000	$60 000	$20 000	–
3.	12%	$20 000	$40 000	$60 000	$80 000
4.	9%	$80 000	$60 000	$20 000	–$20 000

5. A company is considering a $200 000 investment. It is expected that project A will return $50 000 at the end of each year for the next 5 years. On the other hand, it is expected that project B will return nothing for the next 2 years but $100 000 at the end of years 3, 4 and 5. Which project should be chosen if the cost of capital is $j_1 = 4\%$? Do you obtain the same answer if the cost of capital is $j_1 = 7\%$?

6. Which of the following projects should a company choose if each proposal costs $50 000 and the cost of capital is $j_1 = 10\%$?

	End-of-Year Cash Flow				
	Year 1	Year 2	Year 3	Year 4	Year 5
Project A	$20 000	$10 000	$5 000	$10 000	$20 000
Project B	$5 000	$20 000	$20 000	$20 000	$5 000

7. The Northeast Mining Company is considering the exploitation of a mining property. If the company goes ahead, the estimated cash flows are as follows:

	Cash Inflow	Cash Outflow
Now	0	$3 000 000
End of year 1	$1 000 000	$2 000 000
End of year 2	$1 000 000	0
End of year 3	$1 000 000	0
End of year 4	$1 000 000	0
End of year 5	$1 000 000	0
End of year 6	$1 000 000	0
End of year 7	$1 000 000	0
End of year 8	$1 000 000	0

The project would be financed out of working capital, on which Northeast expects to earn at least 16% per annum. Advise whether Northeast should proceed.

8. A pension fund is able to earn $j_2 = 10\%$ on its investments in government bonds. Should it proceed with an investment that costs $100 000 and returns $17 000 at the end of each half-year for the next 4 years?

9. A company is able to borrow money at $j_1 = 8\%$. A new machine costing $60 000 is now available and is estimated to produce the following savings over the next 6 years.

End of Year	1	2	3	4	5	6
Saving	$20 000	$16 000	$14 000	$12 000	$8000	$4000

Should the company borrow money to buy this machine?

Part B

1. A company is able to borrow money at $j_1 = 9\frac{1}{2}\%$. It is considering the purchase of a machine costing $110 000, which will save the company $5000 at the end of every quarter for 7 years and may be sold for $14 000 at the end of the term. Should it borrow the money to buy the machine?

2. An insurance company can invest in government bonds to obtain a yield of $j_2 = 12\%$. There are two other investments available. The first is a perpetuity paying $3000 at the end of each quarter, which may be purchased for $100 000. The second is a $50 000 investment, which will return $10 000 at the end of every year for the next 10 years. Which investment provides the company with the highest profit?

3. An investment of $100 000 produces the following year-end cash flows.

Year	1	2	3	4	5
Cash Flow	$30 000	$30 000	$20 000	$30 000	$20 000

Calculate the net present value for this project if the cost of capital is $j_1 = 2\%$, 4%, 6%, 8%, 10% or 12%. Graph your six answers showing the NPV against the cost of capital.

Internal Rate of Return

The calculation of the net present value for a proposed investment is one method used to assess whether it ought to proceed. Another method that is often used to assess the worth of future projects is the calculation of the **internal rate of return**. This is defined as the rate of interest that produces a zero net present value. In terms of the estimated cash flows, it may be expressed as the rate of interest that solves the following equation:

$$F_0 + F_1(1 + i)^{-1} + F_2(1 + i)^{-2} + F_3(1 + i)^{-3} + \ldots + F_n(1 + i)^{-n} = 0$$

where i is the internal rate of return (or IRR) or, in terms of compound interest, the yield required to solve this equation of value.

In many instances F_0 is the only cash outflow (or negative cash flow) so that the equation may be rewritten as

$$F_1(1 + i)^{-1} + F_2(1 + i)^{-2} + F_3(1 + i)^{-3} + \ldots + F_n(1 + i)^{-n} = -F_0$$

or $$F_1(1 + i)^{-1} + F_2(1 + i)^{-2} + F_3(1 + i)^{-3} + \ldots F_n(1 + i)^{-n} = A$$

where A is the initial investment.

EXAMPLE 1 Find the internal rate of return for an investment costing $10 000 that is expected to produce cash flows of $5000 at the end of year 1, $3000 at the end of year 2 and $5000 at the end of year 3.

Solution We have $F_0 = -10\ 000$, $F_1 = 5000$, $F_2 = 3000$, $F_3 = 5000$, and

$$-10\ 000 + 5000(1 + i)^{-1} + 3000(1 + i)^{-2} + 5000(1 + i)^{-3} = 0$$

or $\qquad 5000(1 + i)^{-1} + 3000(1 + i)^{-2} + 5000(1 + i)^{-3} = 10\ 000$

With trial and error, we find that:

value of left side at 14% = 10 069
value of left side at 15% = 9 904

As the cash flows are estimated we do not use linear interpolation to obtain an accurate answer. Instead, we conclude that the internal rate of return is about 14%. Quoting IRRs to the nearest % is normally sufficient.

Spreadsheet solution If B1:E1 contain the cash flow values: −10 000, 5 000, 3 000, and 5 000, respectively, then Excel function = IRR(B1:E1) would compute the values of the IRR as 14%.

∎

Internal rates of return may also be calculated directly with a preprogrammed financial calculator. However, before using this approach, students should fully understand this section.

From our definition of the internal rate of return we know that the net present value for the investment in Example 1 at about 14% will be zero. The following table and graph show the NPVs for this investment for various costs of capital.

Cost of Capital	NPV for Example 1
4%	+$2026
8%	+$1171
12%	+$ 415
14%	+$ 69
15%	−$ 96
16%	−$ 257
20%	−$ 856

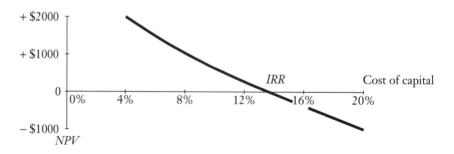

It will be noticed that if the cost of capital is less than the internal rate of return, then there is a positive NPV and hence the investment should proceed.

On the other hand, when the cost of capital is greater than the internal rate of return, the NPV is negative and the investment should not proceed. This example leads us to the following relationships, which apply in most cases.

> Let i_c be the cost of capital and i_r be the internal rate of return
>
> Where $i_c < i_r$ then $NPV > 0$ and the project will return a profit
> $\quad i_c = i_r$ then $NPV = 0$
> $\quad i_c > i_r$ then $NPV < 0$ and the project will not return the minimum required rate of return, i_c

Although the above discussion has highlighted the similarities between NPV and IRR, there are a number of disadvantages in using internal rates of return in assessing financial projects. These are:

1. Multiple solutions are possible as shown by Example 2. This problem applies when the cumulative summation of F_ts changes sign more than once during the term of the investment.

2. No reference is made to the existing interest rates in the market place. Hence it is "internal" as it only refers to the cash flows of that particular investment.

3. Should the investment need to accumulate funds for a final payment (such as land restoration) then it is implied by the IRR calculation that this positive fund will earn the internal rate of return. This is extremely unlikely.

4. The use of internal rates of return to compare alternative projects may lead to conclusions different from those found by the use of NPVs.

■

EXAMPLE 2 An investment company has been offered an investment that costs $7200 now and $27 000 in 2 years' time for an estimated return of $24 200 in 1 year's time and $10 000 in 3 years' time. Calculate the net present values for this investment at 5% intervals from 0% to 30% inclusive. Show that there exist three internal rates of return.

Solution We have $F_0 = -7200$, $F_1 = 24\ 200$, $F_2 = -27\ 000$ and $F_3 = 10\ 000$.

$$NPV = -7200 + 24\ 200(1 + i)^{-1} - 27\ 000(1 + i)^{-2} + 10\ 000(1 + i)^{-3}$$

The following table sets out the NPVs for the required rates.

Cost of Capital	Net Present Value
0%	0
5%	– $4
10%	– $1
15%	+ $3
20%	+ $4
25%	0
30%	– $9

As the internal rate of return is the interest rate that provides a zero net present value, we can immediately see that 0% and 25% are both internal rates of return. In addition, there is a third solution between 10% and 15%, actually 11.1%. The reason for these multiple solutions is that the cumulative F_rs change sign more than once. That is, $F_0 < 0$, $F_0 + F_1 > 0$, $F_0 + F_1 + F_2 < 0$ and $F_0 + F_1 + F_2 + F_3 = 0$. In all our previous examples there had only been one change of sign and this is the condition for a single IRR.

Although this particular investment is a little unusual, it highlights the possibility of multiple IRR solutions and the difficulty that may exist in the interpretation of the internal rate of return.

■

Exercise 7.2

Part A

1-4. Having calculated net present values for problems 1–4 in Part A of Exercise 7.1, calculate the internal rate of return in each case using the same data.

5. An investment of $10 000 provides an estimated cash flow of $3000 for the next 2 years and $3500 for the following 2 years. Calculate this investment's internal rate of return.

6. Find the internal rate of return for a project that costs $100 000 and is estimated to return the following year-end cash flows:

Year	1	2	3	4	5
Cash Flow	$10 000	$20 000	$30 000	$40 000	$50 000

Part B

1. An insurance company must choose between two investments, each of which costs $10 000 and produces the following cash flows at the end of each year.

Year	1	2	3	4	5
Project A	$5000	$4000	$3000	$2000	$1000
Project B	$1000	$3000	$4000	$5000	$6000

Calculate the internal rate of return for each project and their respective net present values at $j_1 = 15\%$ and $j_1 = 25\%$.

2. An investment of $100 000 generates income of $75 000 at the end of the first and second years but requires an outlay of $25 000 at the end of the third year.
a) Calculate this project's internal rate of return.
b) As the final cash flow is a negative one, a sinking fund is set up at the end of the second year so that this payment will be met. If the sinking fund earns $j_1 = 5\%$ (and not the IRR), what is the project's rate of return under this condition?

Capitalized Costs

Many business decisions involve comparing the costs of owning or operating alternative machines or other physical assets that depreciate over the long term and eventually need replacing. One method of comparison is that of capitalization. The **capitalized cost** of a particular asset is the present value of the total costs involved with the asset for an infinite term. Hence, it is defined as the original cost of the asset plus the present value of an infinite number of replacements plus the present value of the maintenance costs in perpetuity. This may be expressed as the sum of the initial cost and the value of the two perpetuities and may be written as

$$K = C + \frac{C - S}{i^*} + \frac{M}{i}$$

where K = the capitalized cost of the asset,
$\quad C$ = the original cost of the asset,
$\quad S$ = the scrap value of the asset at the end of n years,
$\quad M$ = the maintenance costs per annum (assumed at the end of each year),
$\quad i$ = the annual interest rate,
$\quad i^*$ = the equivalent interest rate convertible every n years,
$\quad n$ = the useful life of the asset.

The relationship between i and i^* may be written as

$$1 + i^* = (1 + i)^n$$
$$i^* = (1 + i)^n - 1$$

Hence, the above equation for K may be expressed as

$$\boxed{K = C + \frac{C - S}{(1 + i)^n - 1} + \frac{M}{i}} \qquad (17)$$

The concept of capitalized cost is useful in capital budgeting, when the objective is the selection of the lowest cost alternative to provide a specific level of benefits to the owner. In making comparisons between two physical assets that are employed for the same purpose but that vary with respect to original costs, scrap values, expected useful lifetimes and maintenance costs, we may compare the capitalized costs of the assets.

Some investors prefer to consider a related definition, namely the **annual cost** of each asset. As the capitalized cost is the present value of the total costs in perpetuity, the annual cost may be found as the annual interest on the capitalized cost.

$$\boxed{\textbf{Annual Cost } = Ki = Ci + \frac{C - S}{s_{\overline{n}|i}} + M}$$

> **OBSERVATION:** The annual cost may also be regarded as the sum of the annual opportunity (or interest) cost, the annual sinking fund deposit required for the asset's replacement and the annual maintenance cost.
>
> It should be noted that the above formulas make no direct allowance for inflation. However, the concepts are applicable in an inflationary period with adjustments to the interest rate.

EXAMPLE 1   A farmer can purchase a steel barn with an estimated lifetime of 20 years. It will cost $200 000 and will require $1000 a year to maintain. Alternatively, he can purchase a wooden barn with an estimated lifetime of 14 years. It will cost $160 000 and require $2000 a year to maintain. If the farmer requires a return of $j_1 = 15\%$ on his investment, what should he do?

Solution   Capitalized cost K_1 of the steel barn is

$$K_1 = 200\ 000 + \frac{200\ 000}{(1.15)^{20} - 1} + \frac{1000}{.15}$$
$$= 200\ 000 + 13\ 015.29 + 6666.67$$
$$= \$219\ 681.96$$

Capitalized cost K_2 of the wooden barn is

$$K_2 = 160\ 000 + \frac{160\ 000}{(1.15)^{14} - 1} + \frac{2000}{.15}$$
$$= 160\ 000 + 26\ 334.39 + 13\ 333.33$$
$$= \$199\ 667.72$$

The farmer should purchase the wooden barn.

■

Different assets may produce items at a different rate per unit of time. In that case, it is necessary to divide the capitalized cost by the number of items produced per unit of time. If U_1 and U_2 are the numbers of items produced per unit of time by machine 1 and machine 2 with capitalized cost K_1 and K_2 respectively, then the two machines are economically equivalent if

$$\frac{K_1}{U_1} = \frac{K_2}{U_2}$$

EXAMPLE 2   A machine costing $4000 has an estimated useful lifetime of 10 years and an estimated salvage value of $500. It produces 1600 items per year and has an annual maintenance cost of $800. How much can be spent on increasing its productivity to 2000 units per year if its lifetime, salvage value and maintenance costs remain the same? The owners want a rate of return of $j_1 = 8\%$ on their investment.

Solution   Let X be the maximum amount to be spent on increasing the productivity of the machine.

The capitalized cost K_1 of the original machine is

$$K_1 = 4000 + \frac{3500}{(1.08)^{10} - 1} + \frac{800}{.08}$$
$$= 4000 + 3020.04 + 10\ 000$$
$$= \$17\ 020.04$$

The capitalized cost K_2 after remodelling is

$$K_2 = (4000 + X) + \frac{3500 + X}{(1.08)^{10} - 1} + \frac{800}{.08}$$
$$= 17\ 020.04 + X + \frac{X}{(1.08)^{10} - 1}$$
$$= 17\ 020.04 + 1.862868609\ X$$

After remodelling, the machine will be economically equivalent to the original one if

$$\frac{K_1}{1600} = \frac{K_2}{2000}$$
$$K_2 = \frac{2000}{1600} K_1$$
$$17\ 020.04 + 1.862868609\ X = \frac{2000}{1600}(17\ 020.04)$$
$$1.862868609\ X = 4255.01$$
$$X = \$2284.12$$

Hence, if the improvements cost less than $2284, they should go ahead.

■

Exercise 7.3

1. A certain machine costs $8000; it has an expected lifetime of 10 years and scrap value of $1000. The annual maintenance cost is $1500. Find the capitalized cost of the machine if $j_1 = 12\%$.

2. A company can build a steel warehouse with an estimated lifetime of 25 years at a cost of $200 000 plus annual maintenance of $1500 a year. Alternatively, they can build a cement block warehouse with an estimated lifetime of 20 years at a cost of $180 000 plus annual maintenance of $1200 a year. If money is worth $j_1 = 7\%$, what should they do?

3. A company uses batteries costing $45 with a useful lifetime of 2 years. Another model, which cost $60, has a useful lifetime of 3 years. If money is worth $j_1 = 11\%$, which should be purchased?

4. Widget-producing machines can be purchased from Manufacturer A or Manufacturer Z. Machine A costs $18 000, will last 15 years and will have a salvage value of $2400 at that time. The cost of maintenance is $1500 a year. Machine Z costs $30 000, will last 20 years and will have a salvage value of $2000 at that time. The annual maintenance cost is $1000. If money is worth $j_1 = 7\%$, which machine should be purchased?

5. Ontario Hydro uses poles costing $200 each. These poles last 15 years and have no salvage value. How much per pole would Ontario Hydro be justified in spending on a preservative to lengthen the life of the pole to 20 years? Assume that the annual maintenance costs are equal and that money is worth $j_1 = 8\%$.

6. A town is putting a new roof on its arena. One type of roof costs $600 000 and has an expected lifetime of 20 years with no salvage value. Another type costs $700 000 but will last 30 years with no salvage value. Maintenance costs are equal. If money is worth $j_1 = 7\%$, how much can be saved each year by purchasing the more economical roof?

7. A new car can be purchased for $29 000. If it is kept for 4 years, its trade-in value will be $5200. What should the trade-in value be at the end of 3 years so that trading would be economically equivalent to keeping it 4 years, if money is worth $j_1 = 6\%$?

8. Machine 1 sells for $100 000, has an annual maintenance expense of $2000 and an estimated lifetime of 25 years with a salvage value of $5000. Machine 2 has an annual maintenance expense of $4000 and an estimated lifetime of 20 years with no salvage value. Machine 2 produces output twice as fast as Machine 1. If money is worth $j_1 = 8\%$, find the price of Machine 2 so that it would be economically equivalent to Machine 1.

9. A machine costing $40 000 has a scrap value of $5000 after 10 years. It produces 2000 units of output per year. The annual maintenance cost is $1500. How much can be spent on increasing its productivity to 3000 units per year if its period of service, scrap value, and maintenance cost remain unchanged? Assume money is worth $j_1 = 7\%$.

10. An oil furnace should be serviced annually to maintain efficient combustion.
   a) Find the capitalized cost of this servicing one year before the next servicing, assuming interest is $j_1 = 6\%$ and the cost of servicing is $100.
   b) If the first servicing is due right now, costing $100, and the cost of servicing increases by 2% per year, what is the capitalized cost at $j_1 = 6\%$?

11. A building that costs $100 000 will have a salvage value of S at the end of 5 years. The annual maintenance cost is $5000, and the total annual cost is $6380.50. The sinking fund rate and the yield rate are equal to $j_1 = 5\%$. Calculate S.

Part B

1. Egbert can purchase a car with an estimated lifetime of 7 years. It will cost $20 000 and will require $1000 a year to maintain. Alternatively, he can purchase a car with an estimated lifetime of 10 years. It will cost $25 000 and will require $750 a year to maintain. Using $j_1 = 8\%$, how much can be saved by purchasing the more economical car?

2. A company is thinking of buying a machine that will replace five skilled workers. These workers are paid $3600 each at the end of each month. The machine costs $2400 a month to maintain, will last 20 years and will have no scrap value at that time. If money is worth $j_1 = 8\%$, what is the largest amount (in thousands of dollars) that can be spent for the machine while remaining profitable?

3. Show that the capitalized cost K of an asset with zero salvage value is

$$K = \frac{1}{i}\left[\frac{C}{a_{\overline{m}|i}} + M\right]$$

4. An asset, which costs C dollars when new, has a service life of n years. At the rate of interest i per year, show that the maximum amount to be spent to extend the life of this asset by an additional m years, given $S = 0$, is

$$\frac{C\,a_{\overline{m}|i}}{s_{\overline{n}|i}} \text{ dollars}$$

5. A machine that costs $8000 when new has a salvage value of $1000 after 5 years of service. How much could be spent on increasing its output by 40% and extending its service life by 20%? Assume that the salvage value remains the same and that money is worth $j_1 = 10\%$.

6. Machine A lasts for 10 years with a salvage value of 25% of its original cost. Machine B sells for 80% of the cost of Machine A and lasts for 10 years with no salvage value. At what annual effective interest rate would a buyer be indifferent comparing the two machines? (Assume equal maintenance costs.)

7. A person buying a house under construction has the option of paying an extra $1500 to have a more efficient gas furnace installed. Gas bills for the standard furnace are expected to be $100 per month for the first year and increase by 5% each year. For the more efficient furnace, gas bills are expected to be $60 per month the first year and also have 5% annual increases. If money could be invested at $j_2 = 8\%$, how long (to the nearest month) would it take for the more efficient furnace to pay for itself?

8. An industrial plant could recover waste heat from its production process with a heat exchanger and use this heat to reduce its fuel costs. Installing the heat exchanger would cost $25 000. The system is estimated to last 15 years and have a scrap value of $100 at that time. Assume interest rates are $j_2 = 8\%$ and that inflation causes fuel and electricity costs to rise by 6% each year (i.e., an increase of 6% at the end of each year, with costs being constant during each year). Electricity costs to run the heat exchanger are $30 per month the first year. Fuel savings are $400 per month the first year.
   a) Find the equivalent constant monthly fuel savings over the 15-year period.
   b) Based on monthly costs, would the heat exchanger be cost effective?

9. Plastic trays last 8 years and cost $20. Metal trays last 24 years and cost $X. Trays are needed for 48 years, and inflation will increase the cost of the trays 5% per year. At $j_1 = 6\%$, determine X so that the buyer is indifferent to purchasing plastic or metal trays, assuming no salvage value.

10. A town pays $100 000 per year (at the end of the year, for simplicity) to operate its landfill site. In the future, this is expected to increase by 10% per year. It would cost $X, paid now, to set up a composting program, but this would save $10 000 at the end of the 1st year, with savings increasing by 10% per year (e.g., after 1 year it would cost $90 000 rather than $100 000 for landfill; after 2 years it would cost $99 000 rather than $110 000 for landfill; and so on). Based on capitalized cost, what is the maximum value of X which would make composting cost effective? Assume interest is $j_1 = 11\%$ (and ignore environmental factors!).

Section 7.4 **Depreciation**

Assets, purchased at a particular point in time, provide services over a finite future accounting period. Since the economic life of most assets (land being a notable exception) is limited, it is necessary to allocate the cost of the asset to the accounting periods in which it generates revenues. This periodic charge to income to measure the consumption of the asset is called **depreciation** and is an expense item.

An account, which usually bears the name **accumulated depreciation**, or allowance for depreciation expense, is used to record the total depreciation expense from period to period. The difference between the original cost of an asset and its accumulated depreciation is its **book value**. The book value is not necessarily the same as its market value or resale value. Rather, the book value represents the remaining amount of the original cost of the asset, which has yet to be charged as an expense.

At the end of its useful lifetime, the book value of an asset will equal its estimated **scrap** or **salvage value**. Both the economic lifetime of the asset and its salvage value are estimates made by persons familiar with the particular asset. The **depreciation base** of an asset is its original cost less its estimated scrap value and represents the total amount that should be recorded as depreciation over the useful lifetime of the asset.

There are many accounting methods for handling depreciation as an operating expense. We shall consider only a few: the straight-line method; the constant-percentage method (also called the declining-balance method), the sum-of-digits method; the physical service method; and the sinking-fund method.

In this section we will use the following notation:

C = the original cost of the asset

S = the estimated scrap, salvage or residual value of the asset at the end of its useful lifetime (this could be negative)

n = the estimated useful lifetime of the asset in years

R_k = the yearly depreciation expense or simply yearly depreciation for the kth year

B_k = the book value of the asset at the end of k years, $k \leq n$. Note that $B_0 = C$ and $B_n = S$

D_k = the accumulated depreciation expense or simply accumulated depreciation at the end of k years, $k \leq n$. Note that $D_k = R_1 + R_2 + \cdots + R_k$, $D_0 = 0$ and $D_n = C - S$

For every method of depreciation we shall set up a **depreciation schedule** showing the status of the depreciation expense at various stages during the lifetime of the asset, namely the yearly depreciation expense, the book value at the end of each year, and the accumulated depreciation to date. For all methods, the accumulated depreciation D_k plus the book value of the asset B_k must equal the original cost of the asset C. That is, $D_k + B_k = C$.

■

The Straight-Line Method

The simplest and by far most popular method for depreciating an asset is the straight-line method. This method assumes that the asset contributes its services equally to each year's operation so that the total depreciation base is evenly allocated over the lifetime of the asset and the yearly depreciation expense R_k is a constant R given by

$$\boxed{R = R_k = \frac{C - S}{n}}$$

(18)

The accumulated depreciation D_k at the end of k years is then given by

$$D_k = k\left(\frac{C-S}{n}\right) = k \cdot R$$

If we think of k as changing continuously, and not merely taking integral values, then the graph of D_k as a function of k is a straight line (see page 245). This accounts for the name given to this method.

The book value B_k of the asset at the end of k years is

$$B_k = C - k\left(\frac{C-S}{n}\right) = C - k \cdot R$$

EXAMPLE 1   Equipment costing $5000 is estimated to have a useful lifetime of 5 years and scrap value of $500. Prepare a depreciation schedule using the straight-line method.

Solution   The depreciation base of the asset is $5000 – $500 = $4500 and the yearly depreciation is

$$R_k = \frac{4500}{5} = \$900$$

The accumulated depreciation increases by $900 each year and the book value of the asset decreases by $900 each year. This is shown in the following schedule:

End of Year	Yearly Depreciation	Accumulated Depreciation	Book Value
0	0	0	5000
1	900	900	4100
2	900	1800	3200
3	900	2700	2300
4	900	3600	1400
5	900	4500	500

∎

While the straight-line method is the most popular, other depreciation methods, which charge a greater proportion of the depreciation base to the early years or conversely charge more of the expense to the later years, are used because they may match more suitably the perceived service consumption of the asset.

The Constant-Percentage Method

The constant-percentage method is used for income tax purposes in Canada for almost all capital-cost allowance calculations. The regulations in Canada state that all items in a certain class may be pooled together and depreciated at a fixed percentage of the total capital cost (book value) in that particular pool during that year. The percentage to be used is defined by the act and varies from class to class.

In general, under the constant-percentage method, each year's depreciation is a fixed percentage of the preceding book value. When this method is used, it is customary to assign a value to the **rate of depreciation** d rather than to estimate the useful lifetime n and scrap value S. The yearly depreciation for the kth year R_k is different for each year and is given by

$$R_k = B_{k-1}d$$

At the end of the 1st year:
the depreciation is Cd
the book value is $C - Cd = C(1 - d)$

At the end of the 2nd year:
the depreciation is $C(1 - d)d$
the book value is $C(1 - d) - C(1 - d)d =$
                                    $C(1 - d)(1 - d) = C(1 - d)^2$

At the end of the 3rd year:
the depreciation is $C(1 - d)^2 d$
the book value is $C(1 - d)^2 - C(1 - d)^2 d =$
                                    $C(1 - d)^2(1 - d) = C(1 - d)^3$

Continuing in this manner for k years, the book value B_k at the end of k years is given by

$$\boxed{B_k = C(1 - d)^k} \tag{19}$$

The accumulated depreciation D_k at the end of k years may then be calculated by $D_k = C - B_k$ or

$$D_k = C - C(1 - d)^k$$

EXAMPLE 2   A car costing \$26 500 depreciates 30% of its value each year. Make out a depreciation schedule for the first 3 years; find the book value at the end of 5 years and the depreciation expense for the 6th year.

Solution   A depreciation schedule is set up below.

End of Year	Yearly Depreciation	Accumulated Depreciation	Book Value
0	0	0	26 500.00
1	7950.00	7 950.00	18 550.00
2	5565.00	13 515.00	12 985.00
3	3895.50	17 410.50	9 089.50

We calculate the book value B_5 by equation **(19)**

$$B_5 = 26\ 500(1 - .3)^5$$
$$= 26\ 500(.7)^5$$
$$= 26\ 500(.16807)$$
$$= \$4453.86$$

The depreciation expense for the 6th year is

$$R_6 = dB_5$$
$$= .3(4453.86)$$
$$= \$1336.16$$

∎

If we estimate the useful lifetime n and the scrap value S of the asset, it is necessary first to compute the rate of depreciation d.

The constant-percentage method can be used only if S is positive.

EXAMPLE 3   Find the rate of depreciation and construct the depreciation schedule for the equipment in Example 1 if the constant-percentage method is used.

Solution   We have $C = 5000$, $S = 500$, $n = 5$. Noting that $S = B_5$, i.e., the scrap value is the book value at the end of 5 years, we substitute into equation **(19)** to obtain

$$B_5 = C(1 - d)^5$$
$$500 = 5000(1 - d)^5$$
$$(1 - d)^5 = .1$$
$$(1 - d) = (.1)^{1/5}$$
$$(1 - d) = .630957344$$
$$d = .369042656$$

The schedule can now be set up using the rate of depreciation $d = .369042656$.

End of Year	Yearly Depreciation	Accumulated Depreciation	Book Value
0	0	0	5000.00
1	1845.21	1845.21	3154.79
2	1164.25	3009.46	1990.54
3	734.59	3744.05	1255.95
4	463.50	4207.55	792.45
5	292.45	4500.00	500.00

∎

The Sum-of-Digits Method

This is one of the accelerated depreciation methods that accounts for larger depreciation expenses during the early years of the useful lifetime of the asset. The yearly depreciation is a specified fraction of the depreciation base $C - S$. The denominator of the fraction is the sum of digits from 1 to n. Let us denote this sum by s_n and, using the formula for the sum of an arithmetic progression (see Appendix 2), we obtain

$$s_n = 1 + 2 + 3 + \ldots + n = \frac{n(n + 1)}{2}$$

The yearly depreciation expenses are determined as follows:

1st year depreciation	$R_1 = \frac{n}{s_n}(C - S)$
2nd year depreciation	$R_2 = \frac{n-1}{s_n}(C - S)$
3rd year depreciation	$R_3 = \frac{n-2}{s_n}(C - S)$

....

| $(n-1)$st year depreciation | $R_{n-1} = \frac{2}{s_n}(C - S)$ |
| nth year depreciation | $R_n = \frac{1}{s_n}(C - S)$ |

Thus, the depreciation expense for the kth year is given by

$$R_k = \frac{n - k + 1}{s_n}(C - S)$$

(20)

It is easy to see that the sum of all yearly depreciation expenses equals the depreciation base $C - S$

$$R_1 + R_2 + R_3 + \dots + R_n = \frac{n + (n - 1) + (n - 2) + \dots + 2 + 1}{s_n}(C - S) = C - S$$

EXAMPLE 4 Construct the depreciation schedule for the equipment in Example 1 if the sum-of-digits method is used.

Solution We have $C = 5000$, $S = 500$, $n = 5$ and $C - S = 4500$. We calculate the sum of digits

$$s_5 = 1 + 2 + 3 + 4 + 5 = \frac{5(5 + 1)}{2} = 15$$

The yearly depreciation expenses are calculated below.

$$R_1 = \tfrac{5}{15}(4500) = 1500$$
$$R_2 = \tfrac{4}{15}(4500) = 1200$$
$$R_3 = \tfrac{3}{15}(4500) = 900$$
$$R_4 = \tfrac{2}{15}(4500) = 600$$
$$R_5 = \tfrac{1}{15}(4500) = 300$$

A depreciation schedule is now set up below.

End of Year	Yearly Depreciation	Accumulated Depreciation	Book Value
0	0	0	5000
1	1500	1500	3500
2	1200	2700	2300
3	900	3600	1400
4	600	4200	800
5	300	4500	500

■

The Physical-Service Method and Depletion

If assets are purchased to supply a service, the cost of using any asset should be spread in relation to the amount of service received in each period. The asset's useful lifetime is expressed in terms of service units and the depreciation expense is then expressed in dollars per service unit. The depreciation charge per period will then fluctuate with the amount of activity of the asset. The service units can be expressed in many ways. Examples include hours, units of production, mileage.

EXAMPLE 5   A machine costing $5000 is estimated to have a useful lifetime of 5 years and a scrap value of $500. It is expected to run for 20 000 hours. Determine the depreciation rate per hour of useful service and prepare a depreciation schedule assuming the actual hours worked are as follows:

Year	Production Hours
1	5000
2	4400
3	4000
4	3400
5	3200

Solution   The depreciation base of the asset is $5000 − $500 = $4500. The depreciation rate is $\frac{4500}{20\ 000} = 22\frac{1}{2}$¢ per hour.

The full depreciation schedule is as follows:

End of Year	Production Hours	Yearly Depreciation	Accumulated Depreciation	Book Value
0	0	0	0	5000
1	5000	1125	1125	3875
2	4400	990	2115	2885
3	4000	900	3015	1985
4	3400	765	3780	1220
5	3200	720	4500	500

∎

Depletion

Certain kinds of assets, such as mines, gravel pits, oil wells, timber tracts and sources of natural gas diminish in value because the natural resources they originally held become used up. This gradual loss in value due to the using up of an asset is called **depletion**. While any of the preceding methods of calculating depreciation expense could also be used in calculating depletion expenses, the normal method of calculation used is similar to the physical-service method as the following example will illustrate.

EXAMPLE 6   A mine has an original acquisition cost of $1 000 000. It is estimated to hold 50 000 grams of recoverable silver. After the mineral worth is exhausted, the land will be sold for $80 000. If, in the first year 10 000 grams are recovered, and in the second year 15 000 grams are recovered, determine the depletion deduction in years one and two.

Solution   The depletion base of the asset is $1 000 000 − 80 000 = $920 000. The depletion rate is $\frac{920\ 000}{50\ 000}$ = $18.40 per gram.
The depletion deduction in year one: $18.40 × 10 000 = $184 000.
The depletion deduction in year two: $18.40 × 15 000 = $276 000.

■

The Sinking-Fund Method

The sinking-fund method of depreciation uses a compound interest methodology to calculate the yearly depreciation expense. While it is not physically required, it is easier to understand this method if one assumes that a sinking fund is set up to accumulate to an amount equal to the depreciation base $(C - S)$ of the asset at the end of its lifetime. If the sinking fund is assumed to accumulate at rate i per year, then the equal annual sinking-fund deposit is calculated using the methods outlined in Section 5.5 and equals $\frac{C - S}{s_{\overline{n}|i}}$.

The yearly depreciation expense for the kth year R_k is equal to the sinking-fund deposit plus the interest earned by the sinking fund during the kth year.

$$R_k = \frac{C - S}{s_{\overline{n}|i}} + iD_{k-1}$$

Thus, the depreciation expense is different each year and increases over time.

The depreciation schedule in this case is the same as an ordinary sinking-fund schedule except for the added column giving the book value of the asset.

The accumulated depreciation D_k at the end of k years is equal to the accumulated value of the sinking fund.

$$\boxed{D_k = \left(\frac{C - S}{s_{\overline{n}|i}} \right) \cdot s_{\overline{k}|i}} \tag{21}$$

EXAMPLE 7   Equipment costing $5000 is estimated to have a useful lifetime of 5 years and scrap value of $500. Prepare a depreciation schedule using the sinking-fund method with $j_1 = 9\%$. (This is the same example we have used throughout this section.)

Solution   We have $C = 5000$, $S = 500$, $n = 5$, and $i = 9\%$ and we calculate the required annual sinking-fund deposit as follows:

$$\frac{C - S}{s_{\overline{n}|i}} = \frac{5000 - 500}{s_{\overline{5}|i}} = \$751.92.$$

A depreciation schedule can now be set up.

End of Year	Sinking-Fund Deposit	Interest on Fund	Yearly Depreciation	Accumulated Depreciation	Book Value
0	0	0	0	0	5000.00
1	751.92	0	751.92	751.92	4248.08
2	751.92	67.67	819.59	1571.51	3428.49
3	751.92	141.44	893.36	2464.87	2535.13
4	751.92	221.84	973.76	3438.63	1561.37
5	751.92	309.48	1061.40	4500.03	499.97*

*The 3¢ error is due to round-off.

We can check the accumulated depreciation D_3 at the end of 3 years by calculating the accumulated value of the sinking fund.

$$D_3 = 751.92 \ s_{\overline{3}|.09} = \$2464.87$$

■

Comparison of Methods

The tables and graphs below are for Examples 1, 3, 4 and 7, which all use the same data.

Straight-Line Method

Year	Yearly Depreciation	Accumulated Depreciation	Book Value
0	0	0	5000
1	900	900	4100
2	900	1800	3200
3	900	2700	2300
4	900	3600	1400
5	900	4500	500

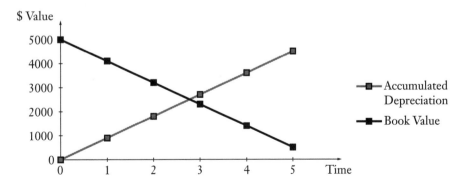

Constant-Percentage Method

Year	Yearly Depreciation	Accumulated Depreciation	Book Value
0	0	0	5000.00
1	1845.21	1845.21	3154.79
2	1164.25	3009.46	1990.54
3	734.59	3744.05	1255.95
4	463.50	4207.55	792.45
5	292.45	4500.00	500.00

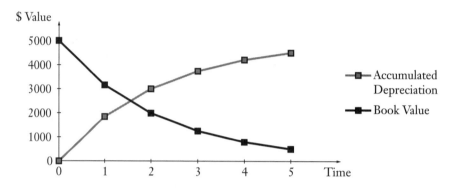

Sum-of-Digits Method

Year	Yearly Depreciation	Accumulated Depreciation	Book Value
0	0	0	5000
1	1500	1500	3500
2	1200	2700	2300
3	900	3600	1400
4	600	4200	800
5	300	4500	500

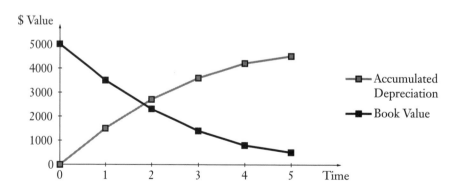

Sinking-Fund Method

Year	Yearly Depreciation	Accumulated Depreciation	Book Value
0	0	0	5000.00
1	751.92	751.92	4248.08
2	819.59	1571.51	3428.49
3	893.36	2464.87	2535.13
4	973.76	3438.63	1561.37
5	1064.40	4503.03	496.97

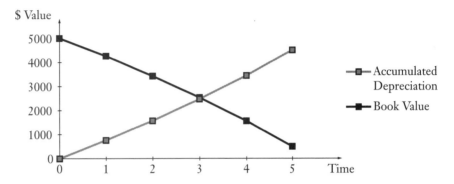

As pointed out earlier, the straight-line method is used mainly because of its simplicity. As a method of allocating the cost of an asset, however, its merits are more limited. This level allocation of the cost of the asset assumes that the services of the asset are consumed uniformly over its estimated lifetime. Further, it assumes that there is no decrease in the efficiency or quality of the service received as the asset ages. These two assumptions often may not be true.

The constant-percentage method and the sum-of-digits method can both be referred to as accelerated depreciation methods, since they allow for larger depreciation allowances in the early years and smaller ones later on. If maintenance costs of an asset increase with time, or if the use of the asset decreases with time, this may result in a better matching of revenues with expenses throughout the life of the asset.

The constant-percentage method is popular since it is the prescribed method for most assets when it comes to determine taxable income. For companies wishing to keep only one set of records, it may be easier to use the constant-percentage method throughout.

The sinking-fund method leads to a smaller depreciation allowance in the early years and a larger depreciation allowance in the later years. This pattern, however, may not be practicably applicable in many situations.

The physical-service method is set up to match the revenue and the expenses of the asset as closely as possible. That it is not possible to draw a general graph for the method should be clear after completing the exercise that follows.

Exercise 7.4

Part A

1. Equipment costing $60 000 is estimated to have a useful lifetime of 5 years and a scrap value of $8000. Prepare a depreciation schedule using
   a) the straight-line method;
   b) the constant-percentage method (you must find d);
   c) the sum-of-digits method;
   d) the sinking-fund method, using $j_1 = 6\%$.

2. A machine costing $26 000 is installed. Its expected useful lifetime is 6 years. At the end of that time it will have no scrap value. In fact, it is estimated that it will cost the company $1000 to remove the old machine (i.e., $S = -\$1000$). Prepare a depreciation schedule using
   a) the straight-line method;
   b) the sum-of-digits method;
   c) the sinking-fund method, using $j_1 = 9\%$.

3. A machine costing $50 000 has an estimated useful lifetime of 20 years and a scrap value of $2000 at that time. Find the accumulated depreciation and the book value of this asset at the end of 8 years using
   a) the straight-line method;
   b) the constant-percentage method (you must find d);
   c) the sum-of-digits method;
   d) the sinking-fund method, using $j_1 = 8\%$.

4. A machine costing $20 000 has an estimated lifetime of 15 years and zero scrap value at that time. At the end of 6 years, the machine becomes obsolete because of the development of a better machine. What is the accumulated depreciation and the book value of the asset at that time using
   a) the straight-line method;
   b) the sum-of-digits method;
   c) the sinking-fund method, using $j_{12} = 6\%$?

5. A machine costing $45 000 depreciates 20% of its remaining value each year. Make out a depreciation schedule for the first 5 years.

6. A machine that costs $40 000 will depreciate to $3000 in 12 years. Find its book value at the end of 7 years and the depreciation expense for the eighth year using the constant-percentage method.

7. Equipment worth $30 000 depreciates 10% of its remaining value each year. How long in years will it take before the equipment is worth less than $15 000?

8. A machine costing $28 000 depreciates to $4000 over 15 years. How long does it take for the machine to depreciate to less than half of its original value under the constant-percentage method?

9. A machine costing $50 000 will last 5 years, at which time it will be worth $5000. In that time it is expected to produce 90 000 units. If its production follows the pattern outlined below, do a depreciation schedule.

Year	Units Produced
1	23 000
2	16 000
3	21 000
4	17 000
5	13 000

10. An airplane costing $10 000 000 is purchased and will be used for 3 years by a company. At the end of that time it is estimated that it can be sold for $6 000 000. The accountant sets up a depreciation schedule based on miles flown and estimates that the plane will fly 450 000 miles in the next 3 years. Given the following, do a depreciation schedule.

Year	Miles Flown
1	130 000
2	210 000
3	110 000

11. A mine is purchased for $300 000 and has an expected reserve of 400 000 units of ore. After mining is complete, the land will be worth $20 000. If $100 000 units of ore are extracted in the first year, find the depletion deduction in year one.

12. Petro Canada purchases some land containing oil wells for $20 000 000. The total reserves of crude oil are estimated at 4 000 000 barrels. After the oil is gone, the land can be sold for $140 000. If Petro Canada removes 400 000 barrels in year one and 560 000 barrels in year two, determine the respective depletion deductions.

13. An $18 000 car will have a scrap value of $800 9 years from the date of purchase.
   a) Use the constant-percentage depreciation method to find the book value of the car at the end of 3 years.
   b) An identical car can be leased for 3 years at a cost of $390 paid at the beginning of each month. If money is worth $j_1 = 9\%$, determine whether it is more economical to lease or buy the car over a 3-year period. Assume the car can be sold at the end of 3 years for its book value found in (a) if the car is bought.

14. An automobile with an initial cost of $24 000 has an estimated depreciated value for tax purposes of $4033 at the end of 5 years. Our tax system uses the constant-percentage method of depreciation.
   a) Find the annual rate of depreciation allowed.
   b) Show the depreciation schedule.

Part B

1. Under the straight-line method, the book value at the end of k years is
$$B_k = C - k\left(\frac{C - S}{n}\right)$$

Because of the straight-line nature of this depreciation method, it is also possible to find the book value at the end of year k by linear interpolation between C, the original cost of the asset and S, the salvage value of the asset at the end of n years. Thus,
$$B_k = \left(1 - \frac{k}{n}\right)C + \frac{k}{n}S$$

Prove this formula is equivalent to
$$B_k = C - k\left(\frac{C - S}{n}\right)$$

and try this formula on questions 3 and 4 of Part A.

2. Use mathematical induction to prove formula **(19)**
$$B_k = C(1 - d)^k$$

3. Show that

$$d = 1 - \left(\frac{S}{C}\right)^{\frac{1}{n}}$$

where n is the estimated lifetime of the asset.

4. Given s_n is the sum-of-digits to n, prove

$$B_k = S + \frac{s_{n-k}}{s_n}(C - S)$$

5. Show that, under the sinking-fund method,
   a) the book value B_k of the asset at the end of k years is

$$B_k = C - \left(\frac{C - S}{s_{\overline{n}|i}}\right) s_{\overline{k}|i}$$

   b) the depreciation expense R_k in year k is

$$R_k = \left(\frac{C - S}{s_{\overline{n}|i}}\right)(1 + i)^{k-1}$$

6. Show that if $i = 0$, the sinking-fund method is equivalent to the straight-line method.

7. An office machine is being depreciated over 10 years to an eventual scrap value of $50, using the constant percentage method. If the book value after four years is $457.31, what was the original value of the machine?

8. A company buys two machines. Both machines are expected to last 14 years and each has a scrap value of $1050. Machine A costs $2450 and Machine B costs $$Y$. The depreciation method used for Machine A is the straight-line method, whereas the depreciation method used for Machine B is the sum-of-digits method. The present value, at $j_1 = 10\%$, of the depreciation charges made at the end of each year for Machines A and B are equal. Calculate Y.

9. Calculate Y in Question 8 if the depreciation method used for Machine B is the sinking-fund method with the interest rate on the sinking-fund at $j_1 = 6\%$.

Section 7.5   ## Summary and Review Exercises

- The net present value (or NPV) of a project over n periods with estimated cash flow F_t at the end of period t, at interest rate i per period, is given by

$$NPV = \sum_{t=0}^{n} F_t(1 + i)^{-t}$$

If NPV > 0, the project will be profitable based on the estimated cash flows.

- The rate of interest that produces a zero net present value is called the internal rate of return (or IRR) and is calculated by solving the equation

$$\sum_{t=0}^{n} F_t(1 + i)^{-t} = 0$$

- The capitalized cost K of an asset with original cost C, scrap value S, annual maintenance cost M at the end of each year, useful life of n years, is defined as the present value at annual interest rate i, of the total costs to own and operate the asset indefinitely and is given by

$$K = C + \frac{C - S}{(1 + i)^n - 1} + \frac{M}{i}$$

- The annual interest, Ki, on the capitalized cost K is called the annual cost and may be regarded as the sum of the annual interest cost Ci, the annual sinking fund deposit required for the asset's replacement $\frac{C-S}{s_{\overline{m}|i}}$, and the annual maintenance cost M.

- Notation for depreciation methods:

    C = original cost

    S = salvage value

    $C - S$ = depreciation base

    n = estimated useful lifetime in years

    R_k = depreciation for the kth year

    D_k = accumulated depreciation at the end of k years; $D_0 = 0$, $D_n = C - S$

    B_k = book value at the end of k years; $B_0 = C$, $B_n = S$

    For each depreciation method we may construct a depreciation schedule, listing k, R_k, D_k and B_k. For each $k : D_k + B_k = C$.

- Straight-line method:

    $$R_k = \frac{C - S}{n} \qquad D_k = k\frac{C - S}{n} \qquad B_k = C - k\frac{C - S}{n}$$

- Constant-percentage (Declining balance) method:
    (Assume $S > 0$, d = annual rate of depreciation)

    $$R_k = dB_{k-1} \qquad B_k = C(1 - d)^k \qquad D_k = C - C(1 - d)^k$$

    The annual rate of depreciation, d, can be found by solving $C(1 - d)^n = S$ for unknown d to obtain

    $$d = 1 - \left(\frac{S}{C}\right)^{\frac{1}{n}}$$

- Sum-of-digits method:

    Let $s_i = 1 + 2 + \ldots + i = \frac{i(i + 1)}{2}$

    $$R_k = \frac{n - k + 1}{s_n}(C - S) \qquad D_k = \frac{s_n - s_{n-k}}{s_n}(C - S) \qquad B_k = S + \frac{s_{n-k}}{s_n}(C - S)$$

- Sinking-fund method:
    (Assume that a sinking fund is set up to accumulate $C - S$ at the end of n years, by equal deposits $\frac{C - S}{s_{\overline{m}|i}}$ at the end of each year, at rate i per year).

    D_k = amount in the sinking fund = $\left(\frac{C - S}{s_{\overline{m}|i}}\right)s_{\overline{k}|i}$

    R_k = increase in the sinking fund = $\frac{C - S}{s_{\overline{m}|i}} + iD_{k-1} = \frac{C - S}{s_{\overline{m}|i}}(1 + i)^{k-1}$

    $B_k = C - \left(\frac{C - S}{s_{\overline{m}|i}}\right)s_{\overline{k}|i}$

Review Exercises 7.5

1. A pension fund can earn $j_1 = 6\%$ with investments in government securities. Determine which of the following investments the fund should accept if the initial investment required is $100 000 in each case.

	Estimated Year-End Cash Flow				
Project	Year 1	Year 2	Year 3	Year 4	Year 5
A	25 000	25 000	25 000	25 000	25 000
B	10 000	30 000	40 000	30 000	10 000
C	– 80 000	40 000	50 000	60 000	70 000
D	70 000	50 000	30 000	–	– 30 000

2. Calculate the internal rate of return for a project that costs $100 000 and returns $30 000 at the end of each of the next 4 years. Compare this rate with the internal rate of return for the following alternative projects.

	Estimated Year-End Cash Flow			
Alternative	Year 1	Year 2	Year 3	Year 4
1	60 000	–	–	60 000
2	–	60 000	60 000	–
3	–	–	–	120 000
4	120 000	–	–	–
5	140 000	– 100 000	130 000	– 50 000

3. A company is considering buying a certain machine. One machine costs $6000, will last 15 years and will have a scrap value of $800 at that time. The cost of maintenance is $500 a year. The second machine costs $10 000, will last 25 years and will have scrap value of $1000 at that time. The cost of maintenance is $800 a year. If money is worth $j_1 = 11\%$, which machine should be purchased?

4. After 10 years of service, a machine costing $20 000 has a scrap value of $2500. It produces 1000 units a year. The annual maintenance cost is $750. How much can be spent to double its output if its scrap value, lifetime and maintenance costs remain the same? Assume $j_1 = 7\%$.

5. An asset with an initial value of $10 000 has a salvage value of $1000 after 10 years. Find the difference between the depreciation expense entered in the books in the seventh year under the sinking-fund method using $j_1 = 9\%$ and under the constant-percentage method.

6. a) An asset is being depreciated over an expected lifetime of 10 years at which time it will have no salvage value ($S = 0$). If the depreciation expense in the third year is $2000, find the depreciation expense in the 8th year
   i) by the straight-line method;
   ii) by the sum-of-digits method;
   iii) by the sinking-fund method, where $j_1 = 4\%$.
  b) Find the original value of the asset C in each of the above cases.

7. An $18 000 car will have a scrap value of $2000 6 years from now. Prepare a complete depreciation schedule using
  a) the straight-line method;
  b) the constant-percentage method;
  c) the physical-use method, given the following:

Year	km Driven
1	12 500
2	14 700
3	11 800
4	16 200
5	13 800
6	11 000

8. A gravel pit is purchased for $150 000 and has expected usable gravel equal to 30 000 truck loads. After excavation is completed the land will be worth $25 000. Given the following shipments, produce a depletion schedule for the gravel pit.

Year	Truck Loads of Gravel
1	8000
2	9100
3	7300
4	5600

9. A machine sells for $100 000 and is expected to have a salvage value of $10 000 after 5 years. The maintenance expense of the machine is estimated to be $2000 per year payable at the end of each year.
a) Construct a depreciation schedule using the sinking-fund method and $j_1 = 7\%$.
b) Determine the capitalized cost of the asset.
c) What is the total annual cost of this asset assuming an investment rate at $j_1 = 6\%$?

10. An asset with an initial value of $100 000 is being depreciated by the sum-of-digits method over a 50-year period. If the depreciation charge in the 31st year is $1000, find the book value of the asset at the end of 41 years.

11. A tool is bought for $1000 and has a salvage value of $50 at the end of 8 years. Calculate the difference in book values at the end of 6 years between the sum-of-digits and declining-balance methods of depreciation.

12. Equipment with a cost of $80 000 has an estimated salvage value of $10 000 at the end of its expected lifetime of 7 years. Find the book value of the equipment at the end of 5 years, using
a) the straight-line method;
b) the constant-percentage method;
c) the sinking-fund method ($j_1 = 6\%$); and
d) the sum-of-digits method.

13. An asset with an initial value of $15 000 has a salvage value of $2000 after 10 years. Find the difference between the depreciation expenses entered in the books in the seventh year under the sinking-fund method using $j_i = 5\%$ and under the declining-balance method.

14. A machine costing $50 000 has an estimated scrap value of $5000 after 10 years. It produces 2000 units of output per year. The annual maintenance cost is $2000. How much can be spent on increasing its output by 25% and extending its service life by 30%? Assume that the salvage value and maintenance costs remain the same and that money is worth $j_i = 6\%$.

15. A machine sells for $80 000 and is expected to have a salvage value of $10 000 after 5 years. The maintenance expense of the machine is estimated to be $2000 per year payable at the end of each year.
   a) What is the capitalized cost of this asset, assuming $j_1 = 10\%$?
   b) Construct a depreciation schedule using the constant-percentage method.
   c) Construct a depreciation schedule using the sinking-fund method, with $j_1 = 6\%$.

16. A machine is purchased for $6000 and is expected to have a salvage value of $500 at the end of 10 years. Determine the annual effective rate i, if the present value of all depreciation charges under the sum-of-digits method is $3625 and under the straight-line method is $3107.50.

17. A machine is purchased for $108 000 and is expected to last 8 years. The book value of the machine at the end of the 2nd year is 44 091 by the declining-balance method. Determine the difference between the book values at the end of the 4th year using the sum-of-digits method and the declining-balance method.

CASE STUDY I

A water heater

   a) A tankless water heater costs $700 and has a projected life of 25 years. The cost of replacement will increase at an annual effective rate of 4% due to inflation. Find the capitalized cost of buying a tankless water heater if interest rates are $j_2 = 6\%$.

Disregard inflation in the prices of energy and conventional water heaters in parts b), c), d).

   b) A conventional water heater has an estimated cost of $200 with a life of 15 years. Find the capitalized cost of owning a conventional water heater at $j_2 = 6\%$.
   c) The tankless heater should save $8 per month in energy costs when compared to a conventional one. Is it cost effective to buy a tankless heater? Use $j_2 = 6\%$.
   d) For a conventional heater, what is the most that should be spent on additional tank insulation if this insulation will save $4 per month in energy costs? Again use $j_2 = 6\%$.

CASE STUDY II

Buy or lease

An $18 000 car will have a scrap value of $500, 9 years from the date of purchase.

   a) Using the constant-percentage depreciation method, find the book value of the car at the end of 4 years.
   b) If money can be invested at $j_{12} = 6\%$, what is the monthly sinking-fund deposit necessary to replace the car if it is sold for its book value in (a) after 4 years and the price of new cars has increased at an annual rate of $j_1 = 5\%$?
   c) An identical car can be leased for 4 years at a cost of $405 paid at the beginning of each month. The contract calls for the driver to pay for fuel, repairs, licenses and maintenance (but *not* for insurance). If the car is bought, there will be a cost of $800 at the time of purchase for 1 year's insurance, with insurance costs increasing by 4% at the beginning of each of the next 3 years. Still using $j_{12} = 6\%$, determine whether it is more economical to buy or lease the car over a 4-year period. Assume that it can be sold for the book value found in a) if you buy the car.

CASE STUDY III

Depreciation

The XYZ Company bought a machine on January 1, 1996 for $30 000. At that time it was decided by the accountants to depreciate the machine over 20 years using the sinking-fund method with $j_1 = 4\%$ and a salvage value of $3000. However, starting on January 1, 2002, it was decided to change to the declining-balance (constant-percentage) method to take advantage of changes in income tax laws.

a) What was the amount of the depreciation charge for 2000?

b) There are two possible ways of making the change:

I– Take the value of the machine on December 31, 2001, and depreciate it over the remaining period using the new method; or

II– Recalculate the depreciation schedule for the entire period using the new method and file amended tax returns for 1996 to 2001.

i) If method I were used, what would the depreciation charge be in 2005?

ii) If method II were used, what would be the difference between the total depreciation reported on the six income tax returns for years 1996 to 2001?

Chapter Eight

Contingent Payments

Section 8.1 ## Introduction

Interest and redemption payments in respect to corporate bonds are contingent since they depend upon the company's financial position. This situation means that two fixed interest securities with identical coupons, redemption prices and maturity dates may sell at different prices due to the market's assessment of the likelihood of each company fulfilling its future financial obligations. For instance, the bonds of a small mining company may trade at a lower price (i.e., higher yield) than similar bonds of a sound and well-established manufacturing company. In turn, these corporate bonds will sell at higher yields than government bonds as it is assumed that the government will always be able to honour its commitments. For this reason, government bonds may be known as risk-free.

In fact, there exists a rating system for companies that issue debt in the form of bonds. Companies that are rated in the highest category are deemed to be financially sound with minimal chance of default. Lower ratings imply lower levels of security (or higher levels of risk) such that an investor may wish to discount the company's bond price to take into account the possibility that each payment promised may not be received. The riskiest bonds are referred to as "junk" bonds and sell at prices well below their face value. As will be shown in Section 8.4, this additional risk may be allowed for by using a higher yield rate in the bond purchase price formula.

In the case of bankruptcy, debts in the form of bonds are paid off first, followed by preferred stocks and, finally, common stocks. Thus, bonds are the least risky of the investment vehicles.

Section 8.2 ## Probability

Benjamin Franklin once noted jokingly that "nothing is certain except death and taxes." To all other events, it is possible to assign a probability value.

We are now used to the weather reporter telling us that there is a "20% chance of rain" for example. What do phrases such as this mean?

Some probabilities are easily determined. For example, we know that the probability of tossing a head with a balanced coin is $\frac{1}{2}$ or 50%. Similarly, the probability of rolling a "2" with a fair die is 1/6. These probabilities can be determined before the event occurs and allow the following definition:

If an event can happen in h ways, and fails to happen in f ways, all of which are equally likely, the probability p of the occurrence of the event is

$$p = \frac{h}{h+f}$$

and the probability the event will fail to occur is

$$q = \frac{f}{h+f}$$

We can see that

$$p + q = \frac{h}{h+f} + \frac{f}{h+f} = \frac{h+f}{h+f} = 1$$

That is, any event either occurs or fails to occur. Furthermore when $f = 0, p = 1$ and when $h = 0, q = 1$. That is, if an event is certain to occur, the probability of its occurrence is unity. Finally, if an event occurs with probability p, the probability that the event will not occur is $q = 1 - p$. For example, if in any time period a system survives with probability, p, then the system will fail (the complement of survival) with probability, q.

Events are said to be **mutually exclusive** if the occurrence of one event precludes the occurrence of any other events. For example, in flipping a coin, the two possible outcomes are heads and tails. Since the occurrence of a head precludes any possibility of a tail (and vice-versa) these events are said to be mutually exclusive.

Given n mutually exclusive events $E_1, E_2, ..., E_n$, the probability of occurrence of at least one of the n events (i.e., E_1 or $E_2 ...$ or E_n) is the sum of the respective probabilities of the individual events. That is,

$$P(E_1 \text{ or } E_2 \text{ or } E_3 ... \text{ or } E_n) = P(E_1) + P(E_2) + ... + P(E_n)$$

Events are said to be **independent** if the occurrence of one event has no effect on the occurrence of any other event. The occurrence of a head or tail on any one toss has no effect on the occurrence of a head or tail on any other toss. Life insurance companies assume that policyholders' deaths are independent events. That is, it is assumed that the time when one policyholder dies will not affect the time when another policyholder dies. In some situations (e.g., a husband and wife) this assumption may not be completely valid, but it is made nevertheless.

The probability of the occurrence of all of a set of independent events, $E_1, E_2, ..., E_n$, is the product of the respective probabilities of the individual events. That is,

$$P(E_1 \text{ and } E_2 \text{ and } ... \text{ and } E_n) = P(E_1)P(E_2)...P(E_n)$$

For more difficult problems where the answer cannot be determined intuitively a useful device is a **probability tree**. This device will be illustrated in Example 2 that follows.

EXAMPLE 1   Find the probability that in drawing one card from a standard deck of 52 cards I get *either* an ace *or* a king.

Solution   There are 4 aces and 4 kings in a standard deck. Thus, there are 8 cards that qualify here, and 44 that don't, so the solution is

$$p = \frac{h}{h+f} = \frac{8}{8+44} = \frac{8}{52} = \frac{2}{13}$$

Equivalently, we can see that the probability of drawing an ace is $\frac{4}{52} = \frac{1}{13}$ and of drawing a king is $\frac{4}{52} = \frac{1}{13}$

$$P(\text{ace or king}) = P(\text{ace}) + P(\text{king}) = \frac{1}{13} + \frac{1}{13} = \frac{2}{13}$$

∎

EXAMPLE 2   Find the probability that in drawing two cards from a standard deck of 52 cards I get an ace *and* a king.

Solution   It is helpful here to draw what is referred to as a probability tree.

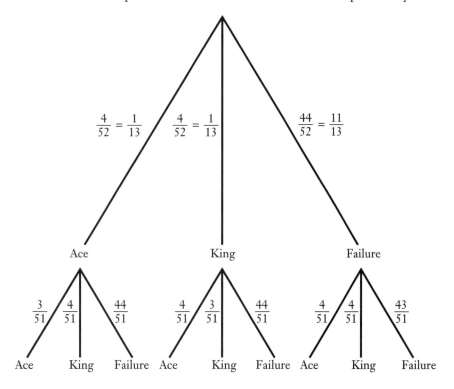

We can draw an ace first followed by a king with probability

$$P(E_1) = \frac{1}{13} \times \frac{4}{51} = \frac{4}{663}$$

OR we can draw a king first followed by an ace with probability

$$P(E_2) = \frac{1}{13} \times \frac{4}{51} = \frac{4}{663}$$

The total probability is $P(E_1 \text{ or } E_2) = P(E_1) + P(E_2) = \frac{4}{663} + \frac{4}{663} = \frac{8}{663}$

In our tree diagram, we first compute the probability of each path in the event by multiplying together the probabilities of each branch in the path. Then we add the probabilities of each path in the event together to get the total probability of the event. Example 1 fits these general rules given that it was a one-stage event.

■

EXAMPLE 3   A bag contains 5 white, 3 black and 4 red balls. Determine the following probabilities:

a) drawing a red ball;

b) drawing a white or a black ball;

c) drawing 2 white balls (no replacement);

d) drawing a black ball twice (with replacement);

e) drawing a white ball followed by a red ball (no replacement);

f) drawing a red ball followed by a black ball (with replacement).

Solution

a) Probability $= \frac{4}{12} = \frac{1}{3}$

Since there are 4 red balls and 12 balls in total.

b) Probability of white $= \frac{5}{12}$

Probability of black $= \frac{3}{12}$

Then, probability of white or black $= \frac{5}{12} + \frac{3}{12} = \frac{8}{12} = \frac{2}{3}$

c) Probability of 1st white ball $= \frac{5}{12}$

Probability of 2nd white ball $= \frac{4}{11}$

Probability of two white balls with no replacement $= \frac{5}{12} \times \frac{4}{11} = \frac{5}{33}$

d) Probability of 1st black ball $= \frac{3}{12} = \frac{1}{4}$

Probability of 2nd black ball $= \frac{1}{4}$ (replacement)

Probability of drawing black twice $= \frac{1}{4} \times \frac{1}{4} = \frac{1}{16}$

e) Probability of white 1st $= \frac{5}{12}$

Probability of red 2nd $= \frac{4}{11}$

Then, probability $= \frac{5}{12} \times \frac{4}{11} = \frac{5}{33}$

f) Using same reasoning as above,

Probability $= \frac{4}{12} \times \frac{3}{12} = \frac{1}{12}$

■

Not all events allow for a priori ("before the fact") calculation of probability. For example, if a community has 1000 birth events per year, how many of these will be multiple births (twins, triplets)?

Similarly, we could ask "What is the probability of rolling a '2' with a loaded or unfair die?"

The only method of solution here is to observe a large number of occurrences and to develop empirically, and after the fact, an estimate of the true underlying probability.

For example, we could roll our unfair die 100 times. If we count sixteen 2s, then an approximation to the true underlying probability of rolling a 2 is .16. If we roll the same die 10 000 times and observe 1715 2s, then we would refine our previous estimate and say that the "probability" of rolling a 2 is .1715.

Most real-life situations, including all insurance situations, require this type of estimated probability. It is important to note that this value for the probability is determined from past experience, and its use for prediction of future events involves the assumption that the future can be predicted by events in the past.

Also, if we say that the probability of an event occurring is $\frac{2}{3}$, this does not mean in 60 trials, there will be exactly 40 occurrences. Rather it means that as the number of trials is increased, the ratio of occurrences will approach $\frac{2}{3}$. This is known as the Law of Large Numbers. Formally, it states that: as the number of trials n increases, the probability that the average proportion of successes deviates from the true underlying number p tends toward zero.

In real life, however, we are usually forced to use the "best estimate," based on a relatively small number of observations, to predict future events.

EXAMPLE 4   According to statistics gathered in the Province of Ontario, in 1999 there were 135 000 births of which 1409 were multiple births (i.e., twins, triplets and so on). If there were 44 000 births in Alberta in 1999, estimate the number of multiple births.

Solution   We are forced to use Ontario data to predict an event in Alberta. Based on the Ontario data, the probability of having a multiple birth is

$$\frac{1409}{135\ 000} = 0.010437$$

Therefore, an estimate for the number of multiple births in Alberta is

$$44\ 000(0.010437) = 459$$

∎

EXAMPLE 5   Based on the data of a large life insurance company, we are told that the probability of surviving 10 years is 0.8, 0.7 and 0.6 for persons aged 40, 50 and 60 respectively. Determine the following probabilities:
 a) The probability of dying in the next 10 years for each group.
 b) The probability that a 40-year-old lives to age 70.
 c) The probability that a 40-year-old dies between ages 60 and 70.

Solution   In answering these questions, we are forced to assume that these probabilities will remain constant and that the mortality rates derived from past events can be used to predict future probabilities of death.

The following tree diagram will assist us.

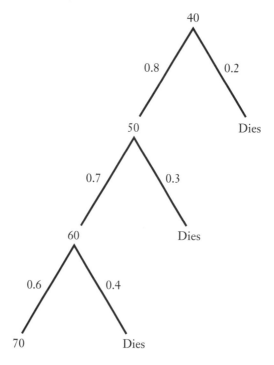

a) For each age group, the event of dying is just the complement of the event of living. Therefore,
   Prob (age 40 dies in next 10 years) = 1 − 0.8 = 0.2
   Prob (age 50 dies in next 10 years) = 1 − 0.7 = 0.3
   Prob (age 60 dies in next 10 years) = 1 − 0.6 = 0.4

b) To survive to age 70 requires survival through each of the age groups. Therefore, using the tree diagram
   Prob (40 survives to 70) = 0.8 × 0.7 × 0.6 = 0.336

c) For someone aged 40 to die between 60 and 70 requires living to age 60 and then dying in the next 10 years. Therefore, using the tree diagram
   Prob (40 dies between 60 and 70) = 0.8 × 0.7 × 0.4 = 0.224

It should be noted that current survival rates are much higher than those used in this example.

■

Exercise 8.2

Part A

1. A jar contains 4 white marbles, 2 blue marbles and 1 red marble. Find the following probabilities (for multiple drawings, assume no replacement).
   a) You draw out one marble and it is blue.
   b) You draw a blue or red marble.
   c) You draw out two marbles that are white.

    d) You draw two marbles, the first is red and the second is blue.

    e) You draw two marbles, the first is white and the second is blue.

    f) You draw out two marbles of the same colour.

2. Given a normal deck of 52 cards, determine the following probabilities:

    a) drawing the ace of spades;

    b) drawing an ace or a king;

    c) drawing the ace of spaces in two draws, with no replacement;

    d) drawing the ace of spades in two draws, given replacement of the first draw;

    e) drawing the ace of spades on the second draw, with no replacement;

    f) drawing a 2, then a 3, then a 4, if there is no replacement.

3. Given a fair die, determine the following probabilities:

    a) rolling a 2;

    b) rolling an even number;

    c) rolling a prime number;

    d) in two rolls, rolling numbers whose sum equals 9;

    e) in two rolls, rolling numbers whose sum exceeds 5.

4. Given two fair die, determine the following probabilities:

    a) rolling a total equal to 3;

    b) rolling a total equal to 9;

    c) rolling a total that exceeds 5;

    d) rolling an even number, in total;

    e) rolling a prime number, in total;

    f) in two rolls, getting an overall total equal to 23.

5. Given a fair coin, determine the following probabilities:

    a) the probability of 2 heads in three tosses;

    b) the probability of all heads in four tosses;

    c) the probability of 2 heads and then a tail;

    d) the probability of a tail and then two heads.

6. Given that the probability of having a male baby is equal to the probability of having a female baby, determine the following probabilities:

    a) the probability of having 2 boys in a family of three children;

    b) the probability of all boys in a family of four children;

    c) the probability of 2 boys and then a girl;

    d) the probability of a girl and then two boys.

7. A businessman must change planes three times in a long trip. If the probability of making any one connection successfully is 4/5, what is the probability

    a) that he completes his trip successfully?

    b) that he does not complete his trip successfully?

    (Assume the changes are independent events.)

8. The probability of team A winning a game is 3/5 (i.e., they presently have a .600 record). They are just starting a three game series. What is the probability

    a) that they win the first two, but lose the third game?

    b) that they win two of three?

    c) that they win at least one game?

    (Assume the games are independent events.)

9. The probabilities of A, B, C and D surviving a certain period are 3/4, 4/5, 5/8 and 9/10 respectively. What is the probability that

    a) all four die in the period?

    b) all four survive the period?

    c) at least one survives the period?

    d) at least one dies in the period?

    (Assume the deaths are independent events).

10. We have the following data:

Age	Probability of Surviving 10 Years
50	0.75
60	0.60
70	0.50

Determine the following probabilities:
a) someone aged 50 lives to age 80.
b) someone aged 50 dies between age 70 and 80.

11. The probability of a 40-year-old surviving to age 80 is 1/4. The probability of a 40-year-old dying between age 65 and 80 is 1/10. Find the probability of a 40-year-old surviving to age 65.

Part B

1. What is the probability that a 3-card hand, drawn at random and without replacement from an ordinary deck, consists entirely of black cards?

2. An unbiased die is thrown twice. Given that the first throw resulted in an even number, what is the probability that the sum obtained is 8?

3. A bowl contains three red chips numbered 1, 2, 3 and three blue chips numbered 1, 2, 3. What is the probability that two chips drawn at random without replacement match *either* as to colour *or* as to number?

Section 8.3   Mathematical Expectation

A term that most students are familiar with is "average." If you take six courses, your average mark is found by taking your total mark and dividing by six. If there are 34 students in your class, the class average can be found by totalling all marks and dividing by 34.

We can look at the latter problem in another manner. Let's say the mark distribution was as follows:

Mark	# of Students
40	4
50	6
60	8
70	10
80	4
90	2

Thus, the teacher could find the class average using the following formula:

$$\frac{[4(40) + 6(50) + 8(60) + 10(70) + 4(80) + 2(90)]}{34}$$

$$= 40\left(\frac{4}{34}\right) + 50\left(\frac{6}{34}\right) + 60\left(\frac{8}{34}\right) + 70\left(\frac{10}{34}\right) + 80\left(\frac{4}{34}\right) + 90\left(\frac{2}{34}\right) = 62.94$$

It should not trouble us that the class average was a mark that was not achieved by anyone and, in fact, could not have been achieved by anyone.

In statistics, the class average would be called the mean of the distribution of marks. This can also be called the expected value or mathematical expectation. Formally, we say that

If X represents possible numerical outcomes of an experiment which can take the values x_1, x_2, ... with corresponding probabilities $f(x_1)$, $f(x_2)$, ..., the **mathematical expectation** or **expected value** or **mean value** of X is defined by

$$E(X) = x_1 f(x_1) + x_2 f(x_2) + x_3 f(x_3) + ...$$

EXAMPLE 1   If, in a dice game, one wins a number of dollars equal to the number showing on the die's upward face, what is the mathematical expectation of any participant?

Solution   You can win \$1, \$2, \$3 ... up to \$6 each with probability 1/6 (assuming a fair die). Let X be the number of dollars won by a participant. Then,

$$E(X) = 1\left(\frac{1}{6}\right) + 2\left(\frac{1}{6}\right) + ... + 6\left(\frac{1}{6}\right) = \frac{21}{6} = \$3.50$$

Again, it should not trouble us that the expected value is an amount that cannot be won on any throw during the game.

If a person has to pay \$3.50 to enter this game, then:

$$E(\text{total winnings}) = E(X - 3.50) = E(X) - 3.50 = 0.$$

Such an entry fee would make this game a "fair" game, i.e., a game in which the expected winnings are zero.

■

EXAMPLE 2   Two players participate in the game described in Example 1. Player A wins (\$1, \$2 or \$3) if the face showing is 1, 2 or 3; and player B wins (\$4, \$5 or \$6) if the face showing is 4, 5 or 6. How much should each player place in the pot (i.e., wager) if the game is to be fair?

Solution   If each player in this game wagers exactly his expectation with respect to winnings, then the game will be fair. That is, in the long run, in this game, the players should expect to neither win nor lose. Thus, we find each player's expectation of winnings.

$$E(A) = \$1(1/6) + \$2(1/6) + \$3(1/6) = \$1$$
$$E(B) = \$4(1/6) + \$5(1/6) + \$6(1/6) = \$2.50$$

Therefore, if A wagers \$1 and B wagers \$2.50, this game will be fair.

■

EXAMPLE 3   Charlie pays \$1 to enter a betting game. If Charlie can get three heads in a row by flipping a fair coin, he wins \$5. Otherwise, he loses his \$1. What is his expectation?

Solution

Probability of winning = 1/8
Probability of losing = 7/8

Therefore, expectation = \$5(1/8) − \$1(7/8) = −25¢

■

EXAMPLE 4   In the Wintario Lottery, sponsored by the Province of Ontario, 5 million $1 tickets are sold each week. The cash prizes awarded are listed below. Also 250 000 people can win "Win Fall" Prizes, which entitle them to 5 tickets in the next draw. These 1 250 000 free tickets will have the same mathematical expectation as the ticket you purchase. Find the "value" of a $1 Wintario ticket.

Prize	No. Awarded	Probability
$100 000	5	$\frac{1}{1\ 000\ 000}$
$ 25 000	20	$\frac{1}{250\ 000}$
$ 5 000	45	$\frac{9}{1\ 000\ 000}$
$ 1 000	100	$\frac{1}{50\ 000}$
$ 100	1 950	$\frac{39}{100\ 000}$
$ 10	24 500	$\frac{49}{10\ 000}$

Solution   Let W be the "value" of a Wintario ticket.

$$W = \$100\ 000 \left(\frac{1}{1\ 000\ 000} \right) + \$25\ 000 \left(\frac{1}{250\ 000} \right)$$

$$+ \$5\ 000 \left(\frac{9}{1\ 000\ 000} \right) + \$1\ 000 \left(\frac{1}{50\ 000} \right) + \$100 \left(\frac{39}{100\ 000} \right)$$

$$+ \$10 \left(\frac{49}{10\ 000} \right) + W \left(\frac{1\ 250\ 000}{5\ 000\ 000} \right)$$

Solving for W we obtain $W \doteq 47\cent$.

That means that a $1 Wintario ticket has a "value" of $47\cent$.

In general, lotteries in Canada have an expected return of $41\cent$ to $49\cent$ on the dollar. This compares to an expected return of $87\cent$ on the dollar at the horse races and a $93\cent$ to $95\cent$ expected return on the dollar at Las Vegas.

∎

Exercise 8.3

Part A

1. Allan is willing to bet $1 that he can flip a coin 3 times and not get any heads. How much should Bob wager to make the game fair?

2. A box contains four $10 bills, six $5 bills and two $1 bills. You are allowed to pull two bills at random from the box. If both bills are of the same denomination, you can keep them. What is the maximum amount you can afford to pay to take part in this game without expecting to lose some money?

3. On a TV game show, a contestant has won a boat and motor valued at $13 000. He can continue to play by relinquishing his prize. If he continues to play, he can choose between curtains A or B. Behind curtain A is a gag prize worth nothing. Behind curtain B is a car valued at $35 000. What should he do to maximize his mathematical expectation?

4. Students who take driver education have probability 0.03 of having an accident in one year. Students without driver education have probability 0.08 of having an accident in one year. Statistics show that the cost of any one accident averages $6000. If car insurance were not available, should a student pay $280 for a driver education course? (Assume no one has two accidents in one year.)

5. Jane and Gilles play a dice game. Jane wins if, when the dice are thrown, the two faces showing are identical. Otherwise, Gilles wins. If Jane wagers $15, how much should Gilles wager to make the game fair?

6. For $1 a lottery offers the following prizes:

>   1 prize of $25 000
>   20 prizes of $1 000
>   168 prizes of $25

and 1701 prizes of 5 free tickets.

If 100 000 tickets are sold, find the expectation per ticket.

7. A 25-year-old man wants to buy insurance on his life for one year; it would be worth $100 000 in the event of his death. Ignoring any effect of compound interest and expenses, what is a fair price for this coverage if, on the average, 158 men out of each 100 000 die at age 25?

8. A rock-concert promoter wants to insure her next outdoor concert against the possibility of bad weather. She estimates that if it doesn't rain, 10 000 people will pay $5 each to attend the concert. If it does rain, only 5000 people will attend. Based on weather statistics for the day of the event over the past several years, it is determined that there is a 12% chance of rain. Ignoring any effect of compound interest and expenses, what is a fair price for the insurance?

Part B

1. When P plays Q, P wins, on the average, 1 out of 3 games. If P and Q agree to play until Q wins, but no more than 3 games, and if at the close of each game the loser pays the winner $1, what are Q's expected winnings?

2. A current quiz program gives the contestant six true-or-false statements and awards $5 for each correct answer. If he answers all six correctly, he gets $1000 extra. What is the value of his expectation assuming he guesses every answer?

3. Ruth wins a dollar if the coin she tosses comes up heads, and loses a dollar if it comes up tails. She tosses once and quits if she wins, but tries only once more if she loses. What is her expectation?

4. A perishable product is purchased by a retailer for $6 and is sold for $8. Based on past experience, the following probability table has been derived. If an item is not sold on the first day, the retailer absorbs a $6 loss. Find the number of units the retailer should purchase to maximize daily profits.

Daily Demand	Probability
20	0.12
21	0.30
22	0.25
23	0.18
24	0.15

5. A manager of an insurance agency is interested in trying to develop a strategy for next year. She can choose one of the following three possible courses of action:
a) stay with the present sales staff;

b) hire additional agents, all untrained;

c) hire additional agents, all trained.

The net payoffs for these actions are

Next Year's Sales	Agents		
	Same Staff	Hire Untrained	Hire Trained
No change	0	– 40 000	– 65 000
Moderate increase	25 000	40 000	70 000
Large increase	70 000	110 000	150 000

The probability predicted for next year's sales are

same as this year 0.60

moderate increase 0.25

large increase 0.15

Which course of action should she choose?

Section 8.4   Contingent Payments with Time Value

We started this chapter by pointing out that two bonds that are identical with respect to face value, coupon rate and maturity date may sell for very different prices. This is because the potential investors do not place the same probability on their actually receiving payment in full.

We are now in a position to mathematically analyze such situations, which can be referred to as contingent payments. All that is required is a combination of compound interest theory and the theory of mathematical expectation.

Suppose we receive amount S with probability, p, and amount zero with probability, $(1 - p)$. This is a special case of $E(X)$ as defined in Section 8.3. The expected value of the amount we receive is:

$$E(X) = x_1 f(x_1) + x_2 f(x_2) = Sp + 0(1 - p) = pS$$

The discounted value or the present value of the expected value, pS, to be received n periods from today, assuming money is worth i per period, is

$$pS \ (1 + i)^{-n}$$

EXAMPLE 1   Mr. Wright goes to a finance company and borrows $1000 for 1 year. If repayment were certain, the finance company would charge Mr. Wright 18% per annum. After doing a credit check on Mr. Wright, they determine that there is a 10% chance of default in which case no money will be repaid at all.

  a) How much should he repay?

  b) What rate of interest should the finance company charge Mr. Wright?

In this example, we assume that if the borrower defaults at some time, the lender gets nothing at all from the time of default. This is seldom true in reality, but is a necessary assumption for ease of analysis.

Solution a   If repayment were certain, the finance company would charge 18% and Mr. Wright would repay $1180. But, the finance company places a 90% probability of repayment. (That is, if 100 such loans are made, 90 will be repaid and 10 will not.)

Let the amount to be repaid be X. The expected value of repayment is $X(0.90)$. The present value of this is $X(0.90)(1.18)^{-1}$.

$$X(0.90)(1.18)^{-1} = \$1000$$

To find X, then,

$$X = \$1311.11$$

Solution b   If Mr. Wright does repay the loan, the rate of interest he pays is i such that

$$1000 = \$1311.11(1 + i)^{-1}$$
$$(1 + i) = 1.3111$$
$$i = 31.11\%$$

While Mr. Wright experiences a very high rate of interest, if the finance company lends $1000 to 100 people and only 90 repay the loan, they make only 18% on their total investment.

∎

EXAMPLE 2   Ms. Wolnitz is approached by an agent selling a special university scholarship award. If Ms. Wolnitz's newborn daughter survives to age 18, qualifies for and enters a Canadian university, she will be awarded a $10 000 scholarship on her 18th birthday. The cost for this plan is $2500.

Ms. Wolnitz realizes that she can make 5% per annum on her investments. She also determines from statistical tables that the probability of a newborn female surviving to age 18 is 0.9767. What probability must Ms. Wolnitz attach to her daughter entering university before the scheme is worthwhile?

Solution   Let the unknown probability be p. The scheme will be worthwhile if the discounted value of Ms. Wolnitz's expectation is at least $2500.

$$2500 = 10\ 000(1.05)^{-18}(0.9767)p$$
$$p = 0.6160$$

If the probability exceeds 61.60%, the plan would be worthwhile.

∎

EXAMPLE 3   An insurance company issues a one-year insurance policy that pays $100 000 at the end of the year if the policyholder dies during the year. If the probability that a 35-year-old female dies in the next year is 0.00086, what is a fair price for this policy, ignoring expenses and assuming $j_1 = 5\%$?

Solution   Price = $100\ 000(0.00086)(1.05)^{-1} = \$81.90$

∎

EXAMPLE 4   Mr. Rozema, who is in very poor health, is to receive $1000 at the end of each year as long as he is alive. The probability he will live n years is given in the following table.

n	Probability of Surviving n Years
1	0.80
2	0.45
3	0.00

If $j_1 = 8\%$, what value should Mr. Rozema place on these series of payments now?

Solution   The solution follows directly from the data given, and can be derived by using the following time diagram:

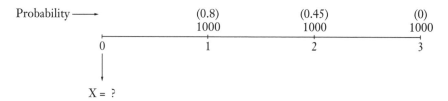

The discounted value of the expected payment is

$$X = 1000(0.8)(1.08)^{-1} + 1000(0.45)(1.08)^{-2} + 1000(0)(1.08)^{-3}$$
$$= \$1126.54$$

■

EXAMPLE 5   How much would you lend a person today if he promises to repay \$1000 at the end of each year for 10 years but there is a 10% chance of default in any year and $j_1 = 12\%$. If default occurs, no money will be received beyond the time of default.

Solution   The fact that there is a 10% chance of default in any year means that there is a 90% probability of receiving the first payment; an 81% chance, $(0.9)^2$, of receiving the second payment; a 72.9% chance, $(0.9)^3$, of receiving the third payment, and so on.

One method of solution is to sum the following geometric series (as outlined in Appendix 2). The discounted value of the expected series of payments is

$$1000(1.12)^{-1}(0.9) + 1000(1.12)^{-2}(0.9)^2 + ... + 1000(1.12)^{-10}(0.9)^{10}$$

An easier method of solution is to determine a new rate of interest j_1 that covers both the standard rate of interest and the probability of default. That is,

$$1 + j_1 = \frac{1.12}{0.9}$$
$$1 + j_1 = 1.2444$$
$$j_1 = 24.44\%$$

Now we calculate the discounted value of a simple annuity of \$1000 a year for 10 years at $j_1 = 24.44\%$

$$1000\, a_{\overline{10}|.2444} = 1000\frac{1 - (1.2444)^{-10}}{0.2444} = \$3631.64$$

■

EXAMPLE 6   A 20-year $1000 bond is offered for sale January 1, 2008. The bond interest is $j_2 = 10\%$ and the bond matures at par.
  a) Find the purchase price to yield $j_2 = 8\%$.
  b) Find the purchase price to yield $j_2 = 8\%$ if the probability of default in *any* six-month period is 5%.

Solution a

$$P = Fr\, a_{\overline{n}|i} + C(1+i)^{-n}$$
$$= 50\, a_{\overline{40}|.04} + 1000(1.04)^{-40}$$
$$= 989.64 + 208.29$$
$$= \$1197.93$$

Solution b   First we find a new interest rate, i per half year, that automatically includes the probability of default.

$$(1+i) = \frac{1.04}{.95}$$
$$1+i = 1.094736842$$
$$i = 0.094736842$$

and then calculate the purchase price

$$P = Fr\, a_{\overline{n}|i} + C(1+i)^{-n}$$
$$= 50\, a_{\overline{40}|i} + 1000(1+i)^{-40}$$
$$= 513.65 + 26.77$$
$$= \$540.42$$

While it appears that a relatively small probability of default leads to a large difference in price, a 5% default probability each half-year for 20 years gives only a 12.9% probability of full repayment of the debt. Further, we have assumed there is no value at all beyond the point of default, which is not the norm. ∎

Exercise 8.4

Part A

  1. The Friendly Finance Company could lend out their money at 18% per annum if everyone repaid their loans in full. From past experience, however, 5% of all loans are not repaid. What rate of interest j_1 should Friendly Finance Company charge for 1-year loans?

  2. Mr. Fitzpatrick wants to borrow some money. He can repay the loan with a single payment of $8000 in one year's time. The lending institution determines that there is a 10% chance that Mr. Fitzpatrick will not repay the loan. The normal lending rate at that time is $j_1 = 9\frac{1}{2}\%$. How much will they lend Mr. Fitzpatrick? If he repays the loan in full, what rate of interest was realized?

  3. Mrs. Chang wants to borrow $4000. She will repay the loan with one single payment at the end of one year's time. The lending agency estimates that there is a 5% chance of default on this loan, in which case they receive nothing at all. How much will they ask Mrs. Chang to repay if their "risk-free" rate of interest is a) $j_1 = 10\%$; b) $j_2 = 10\%$?

4. For loans repayable with single payments at the end of one year if repayment is assumed, $j_1 = 9\%$. If the XYZ Finance Company charges $j_1 = 16\%$ for the same loan, what is their expected default rate?

5. Mr. Saujani dies and leaves an estate valued at $500 000 in a bank account earning 8% per annum. He has two children: Robert, aged 8, and Tammy, aged 3. The estate will be divided among the survivors 18 years from now, when Tammy turns 21. Find the expected value of the inheritance for each child given the probability an 8-year-old boy survives for 18 years is 0.95 and the probability a 3-year-old girl survives 18 years is 0.97, and assuming independence of events. Why is your total not $500 000(1.08)^{18}$?

6. How much would you lend a person today if she promises to repay $5000 at the end of each year for 5 years and there is a 5% chance of default in any year? Assume $j_1 = 9\%$ when payment is assured.

7. An insurance company sells an annuity to a person who has the following probability of survival in the future:

Year "n"	Probability of Survival to end of Year "n"
1	0.85
2	0.65
3	0.35
4	0

If $j_1 = 5\%$ and we ignore expenses, what is a fair price for an annuity paying $2000 at the end of each year?

8. A 20-year $1000 bond is offered for sale. The bond interest is $j_2 = 7\%$ and the bond matures for $1050. Find the purchase price to yield $j_2 = 8\%$ if the probability of default in any six-month period is 1%.

9. A $1000 bond with coupons at $j_2 = 6\%$ and redeemable at par has exactly 10 years left to run. Find its price if the probability of default in any six-month period is 2% and the desired yield rate is $j_2 = 5\%$.

Part B

1. Mr. Cameron buys a $1000 20-year bond with coupons at $j_2 = 6\%$ redeemable at par. He determines his purchase price to yield $j_2 = 7\%$ and to allow for a semi-annual default probability of 2%. Exactly 5 years later Mr. Cameron sells the bond to Mrs. Yutzi, who determines her purchase price to yield $j_2 = 5\%$ but to allow for a semi-annual default probability of 3% in the remaining period. Find
   a) the original purchase price;
   b) the selling price;
   c) the yield j_2 to Mr. Cameron.

2. The XYZ finance company experiences a 90% repayment rate (a 10% default rate) on one-year loans. The ABC Bank across the street experiences a 95% repayment rate (a 5% default rate) on its one-year loans. If the ABC Bank charges $j_1 = 11\%$ on loans, what rate j_1 should the XYZ finance company charge to have the same return on loans?

3. Mrs. Langille dies and leaves an estate valued at $50 000 in a bank account earning interest at rate $j_{12} = 4\%$. She has three children: Jim, aged 7, Fred, aged 5, and Sandra, aged 4. The estate will be divided among the survivors 14 years from now, when Sandra turns 18. Find the expected value of the inheritance for each child given the following, and assuming independence.

Age Today	Probability of Survival for 14 Years
7	0.95
5	0.97
4	0.98

4. Mr. McMahon pays $880 for a $1000 bond paying bond interest at $j_2 = 9\%$ and redeemable at $1000 in 20 years. If his desired yield was $j_2 = 8\%$, what semi-annual probability of default did he expect?

5. Mr. Stockie wants to borrow some money and repay the loan with a single payment at the end of one year. Based on past experience, the bank has the following probability distribution of repayments:

Proportion of Total Debt Repaid	Probability
0	5%
50%	5%
75%	10%
90%	10%
100%	70%

If the "risk-free" rate of interest the bank uses is $j_1 = 10\%$, what rate of interest will they charge Mr. Stockie?

| Section 8.5 | **Summary and Review Exercises** |

Chapter 8 introduced the concept of risk to the evaluation of future payments.
- $E(X) = x_1 f(x_1) + x_2 f(x_2) + \ldots + x_n f(x_n)$ is the expected value of x.
- pS is the expected value of a payment, S, to be paid with probability, p.
- $pS(1 + i)^{-n}$ is the present value of pS.

The probability, p, and the present value factor $(1 + i)^{-1}$ can be combined by finding a new rate of interest, j_1, such that:

$$1 + j_1 = \frac{1 + i}{p}$$

A series of probabilistic payments can now be evaluated using basic compound interest theory at rate j_1.

Review Exercises 8.5

1. A card is drawn at random from a well-shuffled deck. What is the probability that the card will be
a) an ace or a king;
b) a red, face card (jack, queen, or king)?

2. On a particular assembly line, 8 parts in 1000 are defective. Assuming statistical independence, determine the probability that
a) out of 100 parts, none are defective;
b) two parts in a succession are defective;
c) the first two parts are not defective but the third is.

3. A coin is weighted such that the probability of a head is 2/3 and the probability of a tail is 1/3. Determine
a) the probability of 3 heads in a row;
b) the probability of no heads in two tosses;
c) the probability of a head followed by two tails.

4. Three cards are drawn from a deck of 52. If these cards are not replaced, what is the probability of a queen followed by a king followed by an ace?

5. A consumers association tests batteries to see how long certain brands will last. Compare Brand X and Brand Y below by looking at their expected lifetimes.

Time	Brand X	Brand Y
(hours)	(# having failed)	
3	5	6
4	15	18
5	30	22
6	20	35
7	20	12
8	10	7

6. A gambling game is played wherein a single card is pulled from a deck of 52. For pulling the cards 2 to 10 one wins the face value of the card (i.e., $2 to $10); for a jack, queen or king, one wins $20; and for an ace one wins $25. How much would one be expected to place in the pot to make this a fair game?

7. Perishable goods are purchased by a retailer for $10 and sold for $15 each. These goods last only one day. Given the following table determined from past experience, how many items of this good should the producer purchase to maximize his profit? What is his expected profit given that purchase?

Daily Demand	Probability
70	0.05
71	0.35
72	0.40
73	0.15
74	0.05

8. Miss O'Connor goes to the Family Trust Company to borrow some money. Family Trust charges $j_1 = 9\frac{1}{5}\%$ for customers with an established credit rating. Otherwise, they assume a 5% probability of default (in which case they get nothing). For a loan to be repaid with one payment at the end of one year, what interest rate should they charge Miss O'Connor, who has no credit rating?

9. An insurance company issues a special retirement savings policy whereby they will pay the policyholder $100 000 on his 65th birthday if he is alive at that time. If the probability that a 40-year-old male lives to age 65 is 0.810 and if $j_1 = 6\%$, what is a fair price for this policy, ignoring expenses?

10. How much would you lend a person today if he promises to repay $1000 at the end of each year for 10 years, there is a 5% chance of default in any year and $j_1 = 7\%$? (If default occurs, no payment will be made beyond the time of default.)

11. A 20-year $1000 par value bond with semi-annual coupons at $j_2 = 11\%$ is issued by a company that is not financially strong. In fact, the probability of bankruptcy is 5% in any six-month period with no value if that happens. Find the purchase price of this bond if $j_2 = 10\%$ with no default.

<table>
<tr><td>**CASE STUDY** I</td></tr>
</table>

Risky bonds

Review the Bond Quote Table on page 202. Short-term government bonds are as close to a risk-free asset as one can find. At the time of publication, their yield was approximately $j_2 = 6\%$. Corporate Bonds, on the other hand, have higher yields reflecting a positive default probability. For example, at the time of publication, a Trizec Hahn bond due June 1, 2007 was yielding $j_2 = 9.14\%$. Based solely on these two yield rates (i.e., ignoring the complications of timing of coupon payments and redemption), what probability of default, in any six-month period, do investors assume for Trizec Hahn?

Life Annuities and Life Insurance

Section 9.1 Introduction

If you review carefully the solutions to examples in Chapter 8, you will see that we have already done several questions involving life annuities and life insurance. In fact, life annuities and life insurance are nothing more than contingent payments with time value and no new theory is needed in this chapter. Rather, we will present a mortality table that could be used in Canada for such policies and then present some of the notation familiar to Canadian actuaries working with life annuities and life insurance on a day-to-day basis.

Section 9.2 Mortality Tables

Many organizations collect mortality data in Canada. For example, any time a person dies, information on that death, including age at death and cause of death, is sent to the provincial government. These "vital statistics" are compiled by Statistics Canada and analyzed every five years, just after each census, which provides us with data on the number of people alive at any age. The ratio of the number of deaths at a given age to the number of people alive at that age can be used to create mortality ratios. These ratios, in turn, can be used to create mortality tables showing the probability of death at each age, separately for Canadian males and females.

Insurance companies also collect data on age at death, but only for their policyholders. These statistics are compiled and analyzed by a committee of the Canadian Institute of Actuaries, which, from time to time, will issue mortality tables showing the probability of death at each age. These tables are done separately for females and males, and smokers and nonsmokers, and are also done separately for annuityholders versus insuranceholders.

In general, the people who buy annuities from insurance companies are in above-average health. Thus, to determine the value of an annuity, the actuary will use a mortality table based on the annuitant's experience. To determine the value of an insurance contract, the actuary will use a mortality table based on the experience of insured lives.

On page 329 at the back of this text, we have presented the Canadian Institute of Actuaries 1986–92 Mortality Table. (For information on populations' mortality tables, see: **www.statcan.ca/english/IPS/Data/84-537-XPB.htm**) This table was derived using the experience of insured lives who died in the period. This table is widely used in Canada to determine the premiums for life insurance policies. Note that by using this table the actuary is forced to assume that experience from the past will accurately predict events in the future. We have presented both the male and female tables, which are further subdivided into smokers, nonsmokers and an aggregate table.

Age	Aggregate	Smoker	Nonsmoker
15	0.26	0.27	0.25
16	0.29	0.30	0.28
17	0.31	0.33	0.30
18	0.33	0.36	0.32

Part of the female table has been reproduced above. Each column lists the probability of death at each age, multiplied by 1000. That is, for a female at age 15 who is a nonsmoker, the probability of dying in one year is 0.00025. The notation that is used for this probability at age x is q_x. That is, in this case, $q_{15}^{f,ns} = 0.00025$ where the superscript refers to a female nonsmoker.

In reality, there is no age where the probability of death is 1. In practice, however, the actuary will choose a termination age for the mortality table and (arbitrarily) set $q_x = 1$ at this age. In our table, the termination age was set at 105.

EXAMPLE 1   Using the C.I.A. 1986–92 mortality table, estimate the probability that an 18-year-old survives to age 25.

Solution   First, we must realize that the answer will differ depending upon whether the person is male or female, a smoker or nonsmoker.

Second, we only have q_x data, that is the probability of death at age x or the probability that someone who survives to age x will die before age $x + 1$. The complement of the probability of death q_x is the probability of survivorship p_x. That is,

$$\boxed{p_x = 1 - q_x}\qquad(22)$$

So at each age we can now define the probability of survivorship and these probabilities are

| PROBABILITY OF SURVIVING FROM AGE x TO AGE $x + 1$ $p_x = 1 - q_x$ | | | | | | |
| | MALE | | | FEMALE | | |
Age	Aggregate	Smoker	Nonsmoker	Aggregate	Smoker	Nonsmoker
18	0.99913	0.99904	0.99914	0.99967	0.99964	0.99968
19	0.99906	0.99895	0.99908	0.99964	0.99962	0.99966
20	0.99902	0.99888	0.99904	0.99963	0.99959	0.99965
21	0.99901	0.99885	0.99902	0.99962	0.99958	0.99965
22	0.99902	0.99884	0.99903	0.99963	0.99958	0.99966
23	0.99903	0.99884	0.99905	0.99964	0.99960	0.99967
24	0.99904	0.99884	0.99906	0.99967	0.99962	0.99970

As an aside, it is interesting to note that generally the probability of death q_x increases with age. However, in the age range above, we have mortality rates, q_x, that at first rise but then fall. This is more pronounced for males than females. The cause of the "hump" in the mortality curve is the large number of accidental deaths (plus some effects of homicides and suicides) in this age range.

We now have yearly probabilities of survival p_x from age 18 to 24. However, to determine the probability that an 18-year-old survives to age 25 requires the compound events of surviving from age 18 to age 19, then from age 19 to age 20 and so on. Notationally we say that

$$_np_x = p_x \cdot p_{x+1} \cdot p_{x+2} \cdot \ldots \cdot p_{x+n-1}$$

where $_np_x \equiv$ the probability that a person who is alive at age x will survive to age $x + n$.

Note: $_0p_x = 1$ and $_1p_x = p_x$.

So, $_7p_{18} = p_{18} \cdot p_{19} \cdot p_{20} \cdot \ldots \cdot p_{24}$ and we can derive the following probabilities

| PROBABILITY OF SURVIVING FROM AGE 18 TO AGE 25 | | | | | |
| MALE | | | FEMALE | | |
Aggregate	Smoker	Nonsmoker	Aggregate	Smoker	Nonsmoker
.99333	.99227	.99344	.99750	.99723	.99767

∎

Finally, it should be noted that

$$\boxed{_nq_x = 1 - {_np_x}} \tag{23}$$

where $_nq_x \equiv$ the probability that a person who is alive at age x dies before age $x + n$.

EXAMPLE 2   Using the data from Example 1, determine
a) the probability that a male smoker aged 19 dies before age 22.
b) the probability that a female nonsmoker aged 20 dies between ages 22 and 25.

Solution a   We want to calculate $_3q_{19} = 1 - {_3p_{19}}$ for a male smoker. Using data from Example 1 (for a male smoker) we calculate:

$$_3p_{19} = p_{19} \cdot p_{20} \cdot p_{21} = (.99895)(.99888)(.99885) = .99668$$

$$\text{and } _3q_{19} = 1 - {_3p_{19}} = 1 - .99668 = .00332$$

Solution b   We want to calculate the joint probability that a female non-smoker aged 20 lives to age 22 and then dies between age 22 and 25, that is: probability $_2p_{20} \cdot {_3q_{22}}$ for a female nonsmoker. Using data from Example 1 (for a female nonsmokers) we calculate:

$$_2p_{20} = p_{20} \cdot p_{21} = (.99965)(.99965) = .99930$$
$$_3q_{22} = 1 - {_3p_{22}} = 1 - p_{22} \cdot p_{23} \cdot p_{24} = 1 - (.99966)(.99967)(.99970) = .00097$$

$$\text{and } _2p_{20} \cdot {_3q_{22}} = (.99930)(.00097) = .00097$$

While these calculations are somewhat laborious on a pocket calculator, they are very easy to determine on a personal computer.

$\blacksquare$

Exercise 9.2

Part A

1. Using the 1986–92 C.I.A. mortality table, estimate the following probabilities:
   a) a 41-year-old male smoker survives to age 45;
   b) a 29-year-old female smoker dies before age 33;
   c) a 32-year-old male nonsmoker dies between ages 35 and 38;
   d) a 40-year-old female smoker dies between ages 42 and 45;
   e) a 50-year-old male smoker dies at age 54.

2. Using the 1986–92 C.I.A. mortality table, determine for a group of males aged 70 the age by which their population will be approximately cut in half.
   Hint: Use aggregate mortality.

3. Using the 1986–92 C.I.A. mortality table, determine how many female students out of 1000 who enter a four-year university program at age 18 will live to graduate at age 22.
   Hint: Use aggregate mortality.

Part B

1. Prove that $_tp_x q_{x+t} = {_tp_x} - {_{t+1}p_x}$
2. Using the 1986–92 C.I.A. Table, plot and compare the values of q_x on a graph.
3. Given $p_x = e^{-.01}$ for all x find
   a) $_{20}p_{25}$
   b) $_{10}q_{21}$
   c) the probability that a 30-year-old dies at age 45.
4. Repeat question 3 given $p_x = \frac{100-(x+1)}{100-x}$ or $_np_x = \frac{100-(x+n)}{100-x}$.

| Section 9.3 | # Pure Endowments |

A contract that promises to pay a person aged x today \$1 if and when the person reaches age $x + n$ is called an **n-year pure endowment** and is denoted $_nE_x$. Pure endowment contracts are not actually sold today, but are of theoretical importance.

We wish to determine a value for $_nE_x$ given a mortality table and some rate of interest.

Using the theory developed in Chapter 8, for an individual aged x today, the value of this contract is

$$_nE_x = (1 + i)^{-n} {_np_x} \qquad \text{(24)}$$

This value can be derived in another way. Define the symbol "l_x" as the number of people who survive to age x. Assume that l_x people buy this contract and each pays $\$_nE_x$ for the contract. This money can be placed in a fund earning interest at rate i for n years for a total accumulated value of $\$ \, l_x \cdot {_nE_x} \cdot (1 + i)^n$. The fund will then be divided equally among the l_{x+n} survivors so that each survivor's share is:

$$\frac{l_x \cdot {_nE_x} \cdot (1 + i)^n}{l_{x+n}}$$

But we know that each survivor is to get \$1.
Thus,

$$\frac{l_x \cdot {_nE_x} \cdot (1 + i)^n}{l_{x+n}} = 1, \text{ and}$$

$$_nE_x = (1 + i)^{-n} \cdot \frac{l_{x+n}}{l_x}.$$

But,

$$l_{x+n} = l_x \cdot {_np_x} \text{ or} \frac{l_{x+n}}{l_x} = {_np_x}.$$

So,

$$_nE_x = (1 + i)^{-n} {_np_x}$$

as before.

$_nE_x$ is sometimes called the **net single premium for an n-year pure endowment contract.** We refer to it as a net premium, since expenses have been ignored.

EXAMPLE 1 Mr. Judges works for the Acme Manufacturing Company. The company has a retirement plan, paid for by the employer, which provides each retiring employee with $50 000 at age 65. Mr. Judges is age 59 today and wants to change jobs. The Acme Manufacturing company will allow him to transfer the present value of the $50 000 to his new employer's retirement scheme. How much money will be transferred if the company uses the male aggregate mortality table and a) $j_1 = 4\%$; b) $j_1 = 9\%$?

Solution a Using formula **(24)**,

$$50\ 000_6E_{59} = 50\ 000(1 + i)^{-6}{}_6p_{59}$$
$$= 50\ 000(1.04)^{-6}(p_{59} \cdot p_{60} \cdot ... \cdot p_{64})$$
$$= 50\ 000(1.04)^{-6}(1 - q_{59})(1 - q_{60}) \cdot ... \cdot (1 - q_{64})$$
$$= \$36\ 653.37$$

Solution b This is the same as **Solution a** except $j_1 = 9\%$.

$$50\ 000_6E_{59} = 50\ 000(1.09)^{-6}{}_6p_{59}$$
$$= \$27\ 653.81$$

■

EXAMPLE 2 Find the net single premium for a 5-year pure endowment of $5000 sold to a male nonsmoker aged 55 if $j_1 = 11\%$.

Solution Using formula **(24)**, the answer is

$$5000_5E_{55} = 5000(1.11)^{-5}{}_5p_{55}$$
$$= \$2886.72$$

■

Exercise 9.3

Part A

1. Jamie Suljak, aged 15 today and a nonsmoker, wins first prize in a lottery. He can have $100 000 on his 21st birthday or $x today. What would x equal if a) $j_1 = 5\%$; b) $j_1 = 11\%$?

2. The Ayers want to place some money in a fund to provide tuition fee expenses for their daughter, now aged 12. They want the fund to pay out $6000 to their daughter on her 18th birthday if she is still alive. How much must be placed in the fund today if a) $j_1 = 4\%$; b) $j_1 = 8\%$? Assume aggregate mortality.

3. Find the net single premium for a 5-year pure endowment of $10 000 sold to a female smoker aged 45 if a) $j_1 = 6\%$; b) $j_1 = 10\%$.

4. Find the present value of $5000 due in 5 years' time at $j_1 = 4\%$ if
   a) the payment is certain to be made;
   b) the payment is contingent upon a 35-year-old male smoker surviving to age 40.

5. How large a pure endowment, payable at age 65, can a female nonsmoker aged 60 buy with $1000 cash if $j_1 = 7\%$?

Part B

1. Given $l_x = l_0(1 - \frac{x}{105})$, find the net single premium for a pure endowment of $1000 due in 20 years purchased by a 21-year-old male if $j_1 = 11\%$.

2. Show that
   a) $_mE_x \cdot {_nE_{x+m}} = {_{m+n}E_x}$
   b) $_nE_x = {_1E_x} \cdot {_1E_{x+1}} \cdot {_1E_{x+2}} \cdot \ldots \cdot {_1E_{x+n-1}}$

Section 9.4   # Life Annuities

A **life annuity** is a level series of payments made to a policyholder aged x, called an **annuitant**, as long as the annuitant is alive. If not otherwise specified, a life annuity has its first payment at the end of the first year, i.e., an **ordinary life annuity**.

Life annuities can be paid more frequently than annually, but we will present formulas only for annual payment schemes. We will continue to use the C.I.A. 1986–92 mortality table despite the fact that, strictly speaking, it is used for insurance policies and not annuities. The formulas and the methodology are not affected.

To determine the discounted value of a $1 ordinary life annuity issued to someone aged x, denoted a_x, we can return to the methodology of Chapter 8 and see that

$$a_x = (1 + i)^{-1}p_x + (1 + i)^{-2}{_2p_x} + (1 + i)^{-3}{_3p_x} + \cdots$$

or

$$\boxed{a_x = \sum_{t=1}^{\infty}(1 + i)^{-t}{_tp_x}} \tag{25}$$

The upper bound on the summation is given as infinity, but in reality we need proceed no further than age 105, since our mortality table ends at age 105. That is,

$$a_x = \sum_{t=1}^{105-x}(1 + i)^{-t}{_tp_x}$$

Another way of looking at life annuities is to perceive the payments made as a series of pure endowments. That is,

$$a_x = \sum_{t=1}^{\infty}{_tE_x} = \sum_{t=1}^{105-x}{_tE_x}$$

EXAMPLE 1   Find the cost of a life annuity issued to a male nonsmoker aged 95 paying $1000 a year (first payment at age 96) if $j_1 = 6\%$.

Solution   Using formula (25), the cost of the annuity is

$$1000a_{95} = 1000 \sum_{t=1}^{10}(1 + i)^{-t}{}_tp_{95} = \$1913.65$$

It is interesting to note that the cost to a female nonsmoker for the same annuity would be $2242.87. The extra cost is due to the fact that women, as a group, live significantly longer than do men.

This cost is referred to as a **net single premium**. It is a net premium instead of a gross premium since we have ignored expenses.

■

EXAMPLE 2   Find the net single premium for a life annuity paying $1000 a year (first payment at the end of the first year) to a 60-year-old given $j_1 = 7\%$ and the annuitant is
  a) a male nonsmoker;
  b) a female nonsmoker;
  c) a male smoker;
  d) a female smoker.

Solution   We want $1000a_{60}$ throughout where

$$a_{60} = \sum_{t=1}^{105-x}(1 + i)^{-t}{}_tp_{60}$$

We must do the calculation four times using the four different mortality assumptions. This can be done easily on a personal computer once the values of $1000q_x$ have been entered as seen in the Excel spreadsheet that follows.

The answers are as follows:
  a) $10 116.91
  b) $10 914.28
  c) $8369.55
  d) $9967.53

As one would expect, nonsmokers will pay more than smokers since they are expected (on average) to live longer. For the same reason, females will pay more than males.

	A	B	C	D	E	F	G	H
1								
2	The annuity for a male smoker							
3								
4		Cell	Enter			Cell	Enter	
5		D	1–C			G	(1+A$10)^(–E)	
6		F10	D10/1000			H	SUMPRODUCT(F:G)	
7	Fi(11~75)	F(I–1)*Di/1000						
8								
9	Interest	Age	1000qx	1000px	t	tpx	(1+I)^–t	1000Annuity
10	0.07	60	20.14	979.86	1	0.97986	0.9346	8369.55
11		61	22.29	977.71	2	0.95802	0.8734	
12		62	24.63	975.37	3	0.93442	0.8163	
13		63	27.17	972.83	4	0.90903	0.7629	
14		64	29.91	970.09	5	0.88185	0.7130	
15		65	32.88	967.12	6	0.85285	0.6663	
16		66	36.09	963.91	7	0.82207	0.6227	
17		67	39.54	960.46	8	0.78957	0.5820	
18		68	43.25	956.75	9	0.75542	0.5439	
19		69	47.23	952.77	10	0.71974	0.5083	
20		70	51.49	948.51	11	0.68268	0.4751	
21		71	56.04	943.96	12	0.64442	0.4440	
22		72	60.91	939.09	13	0.60517	0.4150	
23		73	66.08	933.92	14	0.56518	0.3878	
24		74	71.59	928.41	15	0.52472	0.3624	
25		75	77.44	922.56	16	0.48409	0.3387	
26		76	83.65	916.35	17	0.44359	0.3166	
27		77	90.22	909.78	18	0.40357	0.2959	
28		78	97.16	902.84	19	0.36436	0.2765	
29		79	104.50	895.50	20	0.32628	0.2584	
30		80	112.23	887.77	21	0.28967	0.2415	
31		81	120.39	879.61	22	0.25479	0.2257	
32		82	128.97	871.03	23	0.22193	0.2109	
33		83	138.01	861.99	24	0.19130	0.1971	
34		84	147.51	852.49	25	0.16308	0.1842	
35		85	157.51	842.49	26	0.13740	0.1722	
36		86	168.02	831.98	27	0.11431	0.1609	
37		87	179.07	820.93	28	0.09384	0.1504	
38		88	190.71	809.29	29	0.07595	0.1406	
39		89	202.98	797.02	30	0.06053	0.1314	
40		90	215.92	784.08	31	0.04746	0.1228	
41		91	229.59	770.41	32	0.03656	0.1147	
42		92	244.06	755.94	33	0.02764	0.1072	
43		93	259.40	740.60	34	0.02047	0.1002	
44		94	275.70	724.30	35	0.01483	0.0937	
45		95	293.06	706.94	36	0.01048	0.0875	
46		96	311.58	688.42	37	0.00722	0.0818	
47		97	331.38	668.62	38	0.00482	0.0765	
48		98	352.57	647.43	39	0.00312	0.0715	
49		99	375.28	624.72	40	0.00195	0.0668	
50		100	401.79	598.21	41	0.00117	0.0624	
51		101	443.01	556.99	42	0.00065	0.0583	
52		102	511.27	488.23	43	0.00032	0.0545	
53		103	620.86	379.14	44	0.00012	0.0509	
54		104	783.19	216.81	45	0.00003	0.0476	
55		105	1000.00	0.00	46	0.00000	0.0445	

∎

In Chapter 4 we looked at other simple annuities; for example, annuities due and deferred annuities. **Life annuities due** are used a great deal by life insurance companies.

A $1 life annuity due has the first payment now. Therefore, the discounted value of a $1 life annuity due, denoted $\ddot{a}_x$, is given by

$$\ddot{a}_x = \sum_{t=0}^{\infty}(1 + i)^{-t}{}_tp_x \tag{26}$$

Note: $\ddot{a}_x = a_x + 1$. The proof is left to the student as an exercise.

EXAMPLE 3   Find the cost of a life annuity due issued to a female nonsmoker aged 95 that pays $1000 a year if $j_1 = 6\%$.

Solution   We want

$$1000\ddot{a}_{95} = 1000 \sum_{t=0}^{10}(1 + i)^{-t}{}_tp_{95} = \$3242.87$$

Note: The answer is simply $1000 larger than the answer given for the female in Example 1.

■

EXAMPLE 4   Determine the cost to a male nonsmoker aged 85 of a life annuity of $1000 a year where payments start at age 96 if $j_1 = 6\%$.

Solution   Let P denote the cost of the annuity as shown.

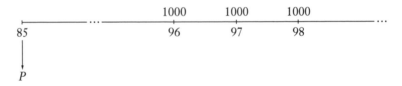

$$P = 1000({}_{11}E_{85} + {}_{12}E_{85} + {}_{13}E_{85} + \cdots)$$
$$= 1000((1 + i)^{-11}{}_{11}p_{85} + (1 + i)^{-12}{}_{12}p_{85} + (1 + i)^{-13}{}_{13}p_{85} + \cdots)$$

However, it can be shown that this is equivalent to

$$P = 1000(1 + i)^{-10}{}_{10}p_{85}((1 + i)^{-1} \cdot p_{95} + (1 + i)^{-2}{}_2p_{95} + (1 + i)^{-3}{}_3p_{95} + \cdots)$$
$$= 1000(1 + i)^{-10}{}_{10}p_{85} \cdot a_{95} \quad (\text{or } 1000{}_{10}E_{85}a_{95})$$

From Example 1, $1000a_{95} = \$1913.65$

and we calculate $_{10}E_{85} = (1 + i)^{-10}{}_{10}p_{85} = .0847274$

so that $P = 1913.65(.0847274) = \$162.14.$ ∎

Example 4 illustrates what is known as a **deferred life annuity**. The notation used is $_n|a_x$ for the discounted value of a \$1 ordinary deferred life annuity and $_n|\ddot{a}_x$ for the discounted value of a \$1 deferred life annuity due where n is the period of deferment.

In general

$$_n|a_x = (1 + i)^{-n}{}_np_x \cdot a_{x+n} = {}_nE_x \cdot a_{x+n}$$

$$_n|\ddot{a}_x = (1 + i)^{-n}{}_np_x \cdot \ddot{a}_{x+n} = {}_nE_x \cdot \ddot{a}_{x+n}$$

EXAMPLE 5   Determine the net single premium P for a life annuity of \$500 a year payable for a maximum of 5 years, issued to a female smoker aged 40 if $j_1 = 10\%$.

Solution   The following time diagram illustrates the payments.

	500	500	500	500	500
40	41	42	43	44	45

$\downarrow$
P

From the diagram, we see that our payments start at age 41 and end at age 45. These payments will be valued at age 40. Thus, we need evaluate only five terms and

$P = 500[(1.10)^{-1}p_{40} + (1.10)^{-2}{}_2p_{40} + (1.10)^{-3}{}_3p_{40} + (1.10)^{-4}{}_4p_{40} + (1.10)^{-5}{}_5p_{40}]$
$= \$1881.82$ ∎

Example 5 illustrates what is known as a **temporary life annuity**. The notation for the discounted value of a \$1 temporary life annuity is $a_{x:\overline{n}|}$ and $\ddot{a}_{x:\overline{n}|}$ for the discounted value of a \$1 temporary life annuity due where n is the length of the payment period.

In general,

$$a_{x:\overline{n}|} = \sum_{t=1}^{n}(1 + i)^{-t}{}_tp_x$$

$$\ddot{a}_{x:\overline{n}|} = \sum_{t=0}^{n-1}(1 + i)^{-t}{}_tp_x$$

Exercise 9.4

1. Prove that $\ddot{a}_x = a_x + 1$.

2. In Example 5 we showed that for a female smoker aged 40 the net single premium for a 5-year temporary annuity of $500 at $j_1 = 10\%$ a year was $1881.82. From Chapter 3 we know that the cost of a 5-year annuity certain of $500 a year at $j_1 = 10\%$ is $500a_{\overline{5}|.10} = \$1895.39$. Explain why a temporary life annuity costs less than an annuity certain of the same duration.

3. Find the net single premium for a life annuity issued to a female nonsmoker aged 96 paying $600 a year (first payment at age 97) if $j_1 = 8\%$.

4. Find the cost of a life annuity due of $2000 a year payable to a male smoker aged 98 if $j_1 = 4\%$.

5. Determine the net single premium for a female nonsmoker aged 75 of an annuity of $6000 a year where the first payment will be at age 97 if $j_1 = 7\%$.

6. Determine the cost of a temporary life annuity of $1500 a year payable for 5 years issued to a male smoker aged 25 if $j_1 = 9\%$.

7. Determine the net single premium for a 5-year temporary life annuity due issued to a female aged 40 that pays $900 per annum if $j_1 = 4\%$. Use aggregate mortality.

8. Mrs. Anderson, a nonsmoker aged 37, inherits $10 000 and uses this money to buy a 5-year temporary life annuity at $j_1 = 8\%$. What annual payments will she receive from the annuity if
a) the first payment is now;
b) the first payment is at age 38?

9. Robbie McBean, aged 12, wins $50 000 in a lottery. The money is left to accumulate at $j_1 = 9\%$ and will be paid to him in the form of a 5-year temporary life annuity, first payment at age 18 if he is then alive. Find the size of each payment. Assume aggregate mortality.

10. A 56-year-old male earning $42 000 a year who is a nonsmoker is left paralyzed as the result of an industrial accident. He sues the company for the present value of his lost income through to age 65. Find the value of the lawsuit, if successful, if we assume his salary is paid at the end of each year, and if rates of investment returns are expected to exceed salary increases in the future by the rate of $j_1 = 2\%$.

11. A 25-year-old female nonsmoker takes out a life insurance policy on which premiums are $300 a year (at the beginning of each year) for 5 years. Find the present value of these payments if $j_1 = 4\%$.

12. A male and a female worker, both nonsmokers and both working for the same period of employment with the same company at the same past salary, both retire at age 65. The employer has built up a fund of $100 000 for each of them to buy them a 10-year temporary life annuity (first payment due at age 65) at $j_1 = 6\%$. What annual income will each worker receive?

13. A 5-year-old girl inherits $100 000. The money is set aside in a fund to provide 5 annual payments to her, first payment at age 21, as long as she is alive. Find the size of each payment she will receive assuming $j_1 = 11\%$ throughout and aggregate mortality.

1. Use a computer to determine the value of a life annuity of $1000 per year issued to a male nonsmoker aged 65 if $j_1 = 9\frac{1}{2}\%$.

2. Show that $\ddot{a}_x = \ddot{a}_{x:\overline{n}|} + {}_{n|}\ddot{a}_x = \ddot{a}_{x:\overline{n}|} + {}_nE_x \cdot \ddot{a}_{x+n}$.

3. Mr. Evanoff, a nonsmoker aged 65, buys a retirement annuity with the first payment made at age 76 and with annual payments of $4500 each. The first 20 payments are guaranteed. After that, payments are made contingent upon Mr. Evanoff being alive. Find the value of this annuity if $j_1 = 8\%$.

4. Show that
   a) $\ddot{a}_x = 1 + (1 + i)^{-1} \cdot p_x \cdot \ddot{a}_{x+1}$
   b) $a_x = (1 + i)^{-1}p_x + (1 + i)^{-2} \cdot {}_2p_x \cdot \ddot{a}_{x+2}$

5. Show that
   a) $\ddot{a}_{x:\overline{n}|} = 1 + a_{x:\overline{n-1}|}$
   b) $\ddot{a}_{x:\overline{n}|} = a_{x:\overline{n}|} + 1 - {}_nE_x$

6. Given $l_x = l_0(1 - \frac{x}{105})$, calculate $\ddot{a}_{20:\overline{5}|}$ if interest is 9% per annum.

7. Find the net single premium for a 10-year temporary life annuity issued to a male smoker aged 65 if $j_1 = 8\%$ in the first 5 years and $j_1 = 6\%$ in the next 5 years.

Section 9.5   Life Insurance

People buy life insurance to protect their dependants from the financial consequences of their untimely death.

The principal types of life insurance are:

a) **Whole life insurance**, wherein the insurance company pays the face value of the policy to the beneficiary upon the death of the insured, whenever that occurs.

b) **n-year term insurance**, wherein the insurance company pays the face value of the policy to the beneficiary upon the death of the insured, but only if the insured dies within the n-year period defined.

c) **n-year endowment insurance**, wherein the insurance company pays the face value of the policy to the beneficiary upon the death of the insured if the insured dies within the n-year period or pays the face value of the policy to the insured at the end of the n-year period if the policyholder is still alive. That is, an n-year endowment insurance policy combines the benefits of an n-year term insurance policy and an n-year pure endowment.

In Example 3 of Section 8.4, we solved a problem involving a one-year term insurance policy. We will now return to a similar problem and formalize the methodology used to determine the cost of certain insurance policies. While in reality death benefits are paid as soon as proof of death is given to the insurance company, we will be assuming that all death benefits are paid at the end of the year of death. This assumption is necessary to allow for ease of calculation.

EXAMPLE 1 Find the net single premium for a $10 000, one-year term insurance policy issued to a male smoker aged 31 if $j_1 = 11\%$.

Solution As in Example 3 of Section 8.4, the premium P is

$$P = 10\ 000(1.11)^{-1} \cdot q_{31}$$

Using q_{31} from the 1986–92 C.I.A. Table, we get $P = \$14.14$.

From here, we can build up the formulae for the other insurance contracts commonly offered.

EXAMPLE 2 Find the net single premium for a $5000, 5-year term insurance issued to a female nonsmoker aged 27 if $j_1 = 12\%$.

Solution The data presented in the 1986–92 C.I.A. Table provide us with the probability of dying in any one year. We are assuming that the death benefit is paid at the end of the year of death. Hence, the net single premium P can be determined as follows:

$$P = 5000[(1.12)^{-1}q_{27} + (1.12)^{-2} \cdot p_{27} \cdot q_{28} + (1.12)^{-3} \cdot {}_2p_{27} \cdot q_{29}$$
$$+ (1.12)^{-4}{}_3p_{27} \cdot q_{30} + (1.12)^{-5} \cdot {}_4p_{27} \cdot q_{31}]$$
$$= \$7.57$$

∎

In general, the symbol for the **net single premium for a $1, n-year term insurance policy** issued to a person aged x is $A^1_{x:\overline{n}|}$ and

$$A^1_{x:\overline{n}|} = \sum_{t=0}^{n-1}(1 + i)^{-(t+1)}{}_tp_xq_{x+t} \tag{27}$$

The "1" over the x signifies that the event that causes a payment to be made is the death of a person aged x at issue before the passage of n years of time.

In Example 1, the 1-year term insurance policy could be denoted $10\,000A^1_{x:\overline{1}|}$.

∎

EXAMPLE 3 Find the net single premium for a whole life policy for $50\,000 issued to a male nonsmoker aged 96 if $j_1 = 8\%$.

Solution As stated before, this policy provides coverage for the whole of life and will pay the benefit at the end of the year of death whenever death occurs. This is just term insurance that proceeds to the end of the table. That is,

$$P = 50\,000 \sum_{t=0}^{\infty}(1 + i)^{-(t+1)}{}_tp_xq_{x+t}$$

$$= 50\,000 \sum_{t=0}^{9}(1 + i)^{-(t+1)}{}_tp_{96}\, q_{96+t}$$

$$= \$40\,087.03$$

∎

In general, the notation for the **net single premium for a $1 whole life insurance policy** issued to a person aged x is A_x and

$$A_x = \sum_{t=0}^{\infty}(1 + i)^{-(t+1)}{}_tp_xq_{x+t} \tag{28}$$

EXAMPLE 4   Find the net single premium for a $100 000, whole life insurance policy issued to a person aged 40 with $j_1 = 7\%$ given
  a) a male nonsmoker;
  b) a female nonsmoker;
  c) a male smoker;
  d) a female smoker.

Solution   We want 100 000 A_{40} throughout, where

$$A_{40} = \sum_{t=0}^{65}(1 + i)^{-(t+1)}\,{}_tp_{40}\,q_{40+t}$$

We must do the calculation four times using the four different mortality assumptions. This can be done easily on a personal computer once the values of 1000 q_x have been entered as seen in the Excel spreadsheet that follows.

The answers are as follows:
  a) $9253.52
  b) $7205.09
  c) $14 644.50
  d) $10 648.75

	A	B	C	D	E	F	G	H	I
2	The insurance for a male smoker								
3									
4		Cell	Enter			Cell	Enter		
5		D	1-C			G	(1+A$10)^(-E)		
6		F10	D10/1000			H	D*F/1000		
7		Fi(11~75)	F(i-1)*Di/1000			I10	sumproct(G11:G76,H10:H75)		
8									
9	Interest	Age	1000qx	1000px	t	tpx	(1+I)^-t	tpxq(x+t)	100,000 Insurance
10	0.07	40	2.24	997.76	0	1	1.0000	0.0022	14644.50
11		41	2.46	997.54	1	0.99776	0.9346	0.0025	
12		42	2.73	997.27	2	0.99531	0.8734	0.0027	
13		43	3.03	996.97	3	0.99259	0.8163	0.0030	
14		44	3.38	996.62	4	0.98958	0.7629	0.0033	
15		45	3.78	996.22	5	0.98624	0.7130	0.0037	
16		46	4.24	995.76	6	0.98251	0.6663	0.0042	
17		47	4.75	995.25	7	0.97834	0.6227	0.0046	
18		48	5.34	994.66	8	0.97370	0.5820	0.0052	
19		49	5.99	994.01	9	0.96850	0.5439	0.0058	
20		50	6.72	993.28	10	0.96269	0.5083	0.0065	
21		51	7.54	992.46	11	0.95622	0.4751	0.0072	
22		52	8.46	991.54	12	0.94902	0.4440	0.0080	
23		53	9.47	990.53	13	0.94099	0.4150	0.0089	
24		54	10.60	989.40	14	0.93208	0.3878	0.0099	
25		55	11.84	988.16	15	0.92220	0.3624	0.0109	
26		56	13.21	986.79	16	0.91128	0.3387	0.0120	
27		57	14.71	985.29	17	0.89924	0.3166	0.0132	
28		58	16.36	983.64	18	0.88601	0.2959	0.0145	
29		59	18.17	981.83	19	0.87152	0.2765	0.0158	
30		60	20.14	979.86	20	0.85568	0.2584	0.0172	
31		61	22.29	977.71	21	0.83845	0.2415	0.0187	
32		62	24.63	975.37	22	0.81976	0.2257	0.0202	
33		63	27.17	972.83	23	0.79957	0.2109	0.0217	
34		64	29.91	970.09	24	0.77784	0.1971	0.0233	
35		65	32.88	967.12	25	0.75458	0.1842	0.0248	

36		66	36.09	963.91	26	0.72977	0.1722	0.0263	
37		67	39.54	960.46	27	0.70343	0.1609	0.0278	
38		68	43.25	956.75	28	0.67562	0.1504	0.0292	
39		69	47.23	952.77	29	0.64640	0.1406	0.0305	
40		70	51.49	948.51	30	0.61587	0.1314	0.0317	
41		71	56.04	943.96	31	0.58416	0.1228	0.0327	
42		72	60.91	939.09	32	0.55142	0.1147	0.0336	
43		73	66.08	933.92	33	0.51783	0.1072	0.0342	
44		74	71.59	928.41	34	0.48361	0.1002	0.0346	
45		75	77.44	922.56	35	0.44899	0.0937	0.0348	
46		76	83.65	916.35	36	0.41422	0.0875	0.0346	
47		77	90.22	909.78	37	0.37957	0.0818	0.0342	
48		78	97.16	902.84	38	0.34533	0.0765	0.0336	
49		79	104.50	895.50	39	0.31178	0.0715	0.0326	
50		80	112.23	887.77	40	0.27919	0.0668	0.0313	
51		81	120.39	879.61	41	0.24786	0.0624	0.0298	
52		82	128.97	871.03	42	0.21802	0.0583	0.0281	
53		83	138.01	861.99	43	0.18990	0.0545	0.0262	
54		84	147.51	852.49	44	0.16369	0.0509	0.0241	
55		85	157.51	842.49	45	0.13955	0.0476	0.0220	
56		86	168.02	831.98	46	0.11757	0.0445	0.0198	
57		87	179.07	820.93	47	0.09781	0.0416	0.0175	
58		88	190.71	809.29	48	0.08030	0.0389	0.0153	
59	0.07	89	202.98	797.02	49	0.06498	0.0363	0.0132	14644.5
60		90	215.92	784.08	50	0.05179	0.0339	0.0112	
61		91	229.59	770.41	51	0.04061	0.0317	0.0093	
62		92	244.06	755.94	52	0.03129	0.0297	0.0076	
63		93	259.40	740.60	53	0.02365	0.0277	0.0061	
64		94	275.70	724.30	54	0.01752	0.0259	0.0048	
65		95	293.06	706.94	55	0.01269	0.0242	0.0037	
66		96	311.58	688.42	56	0.00897	0.0226	0.0028	
67		97	331.38	668.42	57	0.00617	0.0211	0.0020	
68		98	352.57	647.43	58	0.00413	0.0198	0.0015	
69		99	375.28	624.72	59	0.00267	0.0185	0.0010	
70		100	401.79	598.21	60	0.00167	0.0173	0.0007	
71		101	443.01	556.99	61	0.00100	0.0161	0.0004	
72		102	511.77	488.23	62	0.00056	0.0151	0.0003	
73		103	620.86	379.14	63	0.00027	0.0141	0.0002	
74		104	783.19	216.81	64	0.00010	0.0132	0.0001	
75		105	1000.00	0.00	65	0.00002	0.0123	0.0000	
76					66	0.00000	0.0115	0.0000	

As one would expect, nonsmokers pay less than smokers since they are expected (on average) to live longer. For the same reason, females will pay less than males.

■

EXAMPLE 5   Mary Boyce decides to buy a $4000 10-year endowment insurance policy for her niece on her eleventh birthday. Find the net single premium for this policy if $j_1 = 10\%$. Assume aggregate mortality.

Solution   This policy provides term insurance of $4000 for 10 years and, if the policyholder is then alive, it pays out a pure endowment of $4000. That is, the net single premium

$$P = 4000(A^1_{11:\overline{10}|} + {}_{10}E_{11})$$
$$= 4000(\sum_{t=0}^{9}(1 + i)^{-(t+1)}{}_tp_{11}q_{11+t} + (1 + i)^{-10}{}_{10}p_{11})$$
$$= \$1543.85$$

Out of this cost, $5.54 is for the term insurance and $1538.31 is for the pure endowment.

■

In general, the notation for the **net single premium for a $1 n-year endowment insurance policy** issued to a person aged x is $A_{x:\overline{n}|}$ and

$$A_{x:\overline{n}|} = A^1_{x:\overline{n}|} + {}_nE_x \qquad (29)$$

Exercise 9.5

Part A

1. Find the net single premium for a $30 000 1-year term insurance policy issued to a female nonsmoker aged 29 if $j_1 = 12\%$.
2. Find the net cost of a $50 000 5-year term insurance policy issued to a male smoker aged 36 if $j_1 = 10\%$.
3. Find the net cost of a whole life policy for $40 000 issued to a female nonsmoker aged 97 if $j_1 = 8\%$.
4. Determine the net single premium for a $10 000 5-year endowment insurance policy issued to a female smoker aged 60 if $j_1 = 9\%$.
5. Compare the cost of a $100 000 whole life policy issued to a male and to a female smoker, both aged 97, if $j_1 = 7\%$.
6. How much whole life insurance can a female nonsmoker aged 96 buy with $8000 if $j_1 = 10\%$?

Part B

1. Use a computer to determine the net single premium for a $50 000 whole life insurance policy issued to a female smoker aged 35 if $j_1 = 9\%$.
2. Prove that $A_x = (1 + i)^{-1}(q_x + p_x \cdot A_{x+1})$
3. Given $l_x = l_0(1 - \frac{x}{105})$, calculate $A^1_{21:\overline{5}|}$ if $j_1 = 10\%$.
4. Calculate $A^1_{35:\overline{10}|}$ for a 35-year-old male nonsmoker if $j_1 = 10\%$ for 5 years and then 8% for the last 5 years.
5. Prove that $A_x = (1 - i)^{-1}\ddot{a}_x - a_x$

Section 9.6   **Annual Premium Policies**

While it is possible to buy either an annuity or an insurance contract by paying a single premium, this is seldom done. Looking at the cost of some of the policies in Section 9.5 provides one obvious reason. Instead, individuals usually pay a periodic premium to the insurance company to pay for the contract benefits that they desire. While premiums may be paid in a variety of patterns (e.g., weekly, monthly), we will assume that all premiums are paid annually, and at the beginning of each year.

To determine the fair value of the net annual premium P for an annuity or an insurance contract, we must use the theory of mathematical expectation. In particular, we can say that a fair price will be determined if

Mathematical Expectation of Premium Income	$=$	Mathematical Expectation of Benefits to be Paid

These values are now easily determined. The mathematical expectation of the premium income will always be of the form $P \cdot \ddot{a}_x$ since premiums are paid at the beginning of each year, but only if the policyholder is alive. The mathematical expectation of the benefits to be paid is just the particular net single premium for the policy in question. These values have been derived in Sections 8.3, 8.4 and 8.5 depending on the contract. Thus, no new theory is required, as the following examples will illustrate.

EXAMPLE 1 Mr. Raithby, a nonsmoker aged 66, decides to set up a fund for chronic care. He will pay an annual premium P at the beginning of each year for 10 years, if alive. The fund will then earn interest for 20 years and there will be no withdrawals. He can then make annual withdrawals of $15 000, starting with the first withdrawal at age 97 if he is alive. If $j_1 = 8\%$, find P.

Solution We want the mathematical expectation of premium income to equal the mathematical expectation of benefits to be paid as evaluated at age 66.

$$P\ddot{a}_{66:\overline{10|}} = \$15\ 000_{30}E_{66}a_{96}$$

$$P = \frac{(15\ 000)((1 + i)^{-30}{}_{30}p_{66})(\sum_{t=1}^{\infty}(1 + i)^{-t}{}_{t}p_{96})}{\sum_{t=0}^{9}(1 + i)^{-t}{}_{t}p_{66}}$$

$$= \frac{(15\ 000)(0.00446)(1.67650)}{6.72306}$$

$$= \$16.68$$

∎

EXAMPLE 2 Find the net annual premium for a $100 000 whole life policy issued to a male nonsmoker aged 96 if $j_1 = 8\%$.

Solution We have

$$P\ddot{a}_{96} = 100\ 000A_{96}$$

$$P = \frac{100\ 000(\sum_{t=0}^{\infty}(1 + i)^{-(t+1)}{}_{t}p_{96}\ q_{96+t})}{\sum_{t=0}^{\infty}(1 + i)^{-t}{}_{t}p_{96}}$$

$$= \frac{80\ 174.06}{2.67650}$$

$$= \$29\ 954.81$$

∎

In general, the notation for the **net annual premium for a $1 whole life insurance policy** issued to a person aged x is P_x and

$$P_x = \frac{A_x}{\ddot{a}_x} = \frac{\sum\limits_{t=0}^{\infty}(1+i)^{-(t+1)}{}_tp_x \cdot q_{x+t}}{\sum\limits_{t=0}^{\infty}(1+i)^{-t}{}_tp_x} \qquad (30)$$

EXAMPLE 3   Find the net annual premium for a $100 000 whole life insurance policy issued to an applicant aged 35 if $j_1 = 7\frac{1}{2}\%$ and the applicant is

a) a male nonsmoker;
b) a female nonsmoker;
c) a male smoker;
d) a female smoker.

Solution   In each case, we want $100\ 000P_{35} = \frac{100\ 000A_{35}}{\ddot{a}_{35}}$

where:
$$A_{35} = \sum_{t=0}^{\infty}(1+i)^{-(t+1)}{}_tp_{35}\ q_{35+t}$$

and
$$\ddot{a}_{35} = \sum_{t=0}^{\infty}(1+i)^{-t}{}_tp_{35}.$$

We must do the calculation four times using the four different mortality assumptions. This can be done on a personal computer and the answers are

a) $451.19
b) $334.21
c) $762.73
d) $527.41

Note: the size of the annual premium is smaller for nonsmokers than smokers and smaller for females than males, as expected.   ■

EXAMPLE 4   Find the net annual premium for an $80 000 20-pay whole life insurance policy issued to a female smoker aged 29 if $j_1 = 6\%$.

Solution   Premiums are payable for 20 years, but the insurance coverage is whole life.

We have
$$P\ddot{a}_{29:\overline{20|}} = 80\ 000\ A_{29}$$
or
$$P = \frac{80\ 000\ A_{29}}{\ddot{a}_{29:\overline{20|}}} = \$524.17$$

■

EXAMPLE 5   Find the net annual premium for a $100 000 5-year term insurance policy issued to a male smoker aged 41 if $j_1 = 9\%$.

Solution   We have

$$P\ddot{a}_{41:\overline{5}|} = 100\ 000\ A^1_{41:\overline{5}|}$$

$$P = \frac{100\ 000\ A^1_{41:\overline{5}|}}{\ddot{a}_{41:\overline{5}|}}$$

$$= \frac{100\ 000\left(\sum\limits_{t=0}^{4}(1+i)^{-(t+1)}\,_tp_{41}q_{41+t}\right)}{\sum\limits_{t=0}^{4}(1+i)^{-t}\,_tp_{41}}$$

$$= \frac{1167.98}{4.21859}$$

$$= \$276.87$$

∎

In general, the notation for the **net annual premium for a \$1 n-year term insurance policy** issued to a person aged x is $P^1_{x:\overline{n}|}$ and

$$P^1_{x:\overline{n}|} = \frac{A^1_{x:\overline{n}|}}{\ddot{a}_{x:\overline{n}|}} = \frac{\sum\limits_{t=0}^{n-1}(1+i)^{-(t+1)}\,_tp_x\,q_{x+t}}{\sum\limits_{t=0}^{n-1}(1+i)^{-t}\,_tp_x} \qquad (31)$$

EXAMPLE 6   Find the net annual premium for a $100 000 5-year endowment insurance policy issued to a male smoker aged 41 if $j_1 = 9\%$.

Solution   We have

$$P\ddot{a}_{41:\overline{5}|} = 100\ 000\ A_{41:\overline{5}|}$$

$$P = \frac{100\ 000\ (A^1_{41:\overline{5}|} + {}_5E_{41})}{\ddot{a}_{41:\overline{5}|}}$$

which is the answer to Example 5 plus $\dfrac{100\ 000\ {}_5E_{41}}{\ddot{a}_{41:\overline{5}|}}$.

Solving, $P = 276.87 + 15\ 170.86 = \$15\ 447.73$

∎

In general, the notation for the **net annual premium for a \$1 n-year endowment insurance policy** to a person aged x is $P_{x:\overline{n}|}$ and

$$P_{x:\overline{n}|} = \frac{A_{x:\overline{n}|}}{\ddot{a}_{x:\overline{n}|}} = \frac{A^1_{x:\overline{n}|} + {}_nE_x}{\ddot{a}_{x:\overline{n}|}} \qquad (32)$$

EXAMPLE 7   Find the net annual premium for a $10 000 20-year endowment insurance policy purchased for a male aged 1 if premiums are to be paid for only 15 years. Assume aggregate mortality and $j_1 = 8\%$.

Solution   We have

$$P\ddot{a}_{1:\overline{15|}} = 10\ 000\ A_{1:\overline{20|}}$$

or

$$P = \frac{10\ 000\ A_{1:\overline{20|}}}{\ddot{a}_{1:\overline{15|}}}$$

$$= \$234.05$$

∎

In general, the notation for the **net annual premium for a $1 *m*-pay, *n*-year endowment insurance policy** issued to a person aged x is $_mP_{x:\overline{n|}}$ and

$$\boxed{_mP_{x:\overline{n|}} = \frac{A_{x:\overline{n|}}}{\ddot{a}_{x:\overline{m|}}}} \tag{33}$$

Exercise 9.6

Part A

1. Find the annual premium starting at age 32 and paid to age 59 inclusive that is required for a female nonsmoker who wishes to retire on $10 000 a year, first payment at age 60, if $j_1 = 4\%$.

2. Find the net annual premium for a $40 000, 5-year term insurance policy issued to a female smoker aged 26 if $j_1 = 10\%$.

3. Find the net annual premium for a $100 000 whole life policy issued to a female nonsmoker aged 97 if $j_1 = 9\%$.

4. Determine the net annual premium for a $15 000, 5-year endowment insurance policy issued to a male smoker aged 45 if $j_1 = 8\%$.

5. Find an expression for the net annual premium for a $50 000 whole life policy issued to a female aged 31 if
   a) premiums are payable for 20 years;
   b) premiums are payable to age 65 (i.e., to age 64 inclusive).

6. How much 5-year term insurance can a male smoker aged 34 purchase with a premium of $180 a year if $j_1 = 6\%$?

Part B

1. Use a computer to determine the net annual premium for a $150 000 whole life policy issued to a female non-smoker aged 35 if $j_1 = 9\%$.

2. Given $l_x = l_0(\frac{105-x}{105})$. Find $_5P_{30:\overline{20|}}$ if $j_1 = 7\%$.

| Section 9.7 | **Summary and Review Exercises** |

Chapter 9 has introduced a number of variables and formulas used by actuaries in the pricing of insurance including:

- $q_x \equiv$ the probability that an individual aged x dies before age $x + 1$
- $p_x = 1 - q_x \equiv$ the probability that an individual aged x survives to age $x + 1$

The following apply to persons aged x at issue of the policy.

- $_nE_x = (1 + i)^{-n}{}_np_x \equiv$ the discounted value of an n-year pure endowment

- $\ddot{a}_x = \sum_{t=0}^{\infty}(1 + i)^{-t}{}_tp_x \equiv$ the discounted value of a \$1 life annuity due

- $A^1_{x:\overline{n}|} = \sum_{t=0}^{n-1}(1 + i)^{-(t+1)}{}_tp_x\, q_{x+t} \equiv$ the net single premium for a \$1 n-year term insurance policy

- $A_x = \sum_{t=0}^{\infty}(1 + i)^{-(t+1)}{}_tp_x\, q_{x+t} \equiv$ the net single premium for a \$1 whole life insurance policy

- $A_{x:\overline{n}|} = A^1_{x:\overline{n}|} + {}_nE_x \equiv$ the net single premium for a \$1 n-year endowment insurance policy

- $P_x = \dfrac{A_x}{\ddot{a}_x} \equiv$ the net annual premium for a \$1 whole life insurance policy

- $P^1_{x:\overline{n}|} = \dfrac{A^1_{x:\overline{n}|}}{\ddot{a}_{x:\overline{n}|}} \equiv$ the net annual premium for a \$1 n-year term insurance policy

- $P_{x:\overline{n}|} = \dfrac{A_{x:\overline{n}|}}{\ddot{a}_{x:\overline{n}|}} \equiv$ the net annual premium for a \$1 n-year endowment insurance policy

- $_mP_{x:\overline{n}|} = \dfrac{A_{x:\overline{n}|}}{\ddot{a}_{x:\overline{m}|}} \equiv$ the net annual premium for an m-pay, \$1 n-year endowment insurance policy

Review Exercises 9.7

1. An insurance policy issued to a male smoker aged 20 can be purchased by paying 5 annual premiums (first one due right now) of \$100 each. What net single premium would buy the same policy if $j_1 = 7\%$?

2. Mr. Anderson, a smoker, wants to buy a \$25 000 10-year endowment insurance policy at age 32. Find the net annual premium for this policy if $j_1 = 5\%$. What would the annual premium be if he used a 5-pay basis? What is the net single premium for this policy?

3. The XYZ Insurance company sells a Life-Paid-Up-at-65 policy that offers whole life insurance coverage with premiums payable to age 65 (i.e., to age 64 inclusive). Find an expression for the net annual premium to a person aged 38 for this coverage for \$10 000.

4. To what age does a 21-year-old male nonsmoker have a 50% chance of living (based on the C.I.A. Table)?

5. If $l_x = l_0(\frac{100-x}{100})$, find a) p_{70}; b) $_3p_{40}$; c) $_5q_{20}$.

6. Complete the following table:

x	l_x	p_x	q_x
102	128		
103	48		$\frac{2}{3}$
104		$\frac{1}{4}$	
105			1

7. Determine the present value to a female smoker aged 28 of a deferred life annuity having 3 payments of $1000 each, the first payment being at age 40, if $j_1 = 11\%$.

8. Determine the cost of a life annuity issued to a male nonsmoker aged 98 for $500 a year if $j_1 = 10\%$.

9. Find an expression for the present value at age 30 of a whole life annuity of $1 per year with the first payment at age 30; at age 31; at age 65.

10. Find an expression for the net annual premium for a whole life insurance of $1 issued to a 35-year-old with annual premiums payable
a) for life;
b) for 20 years;
c) payable to age 60 (i.e., to age 59 inclusive);
d) the net single premium.

Exponents and Logarithms

Section A1.1 Exponents

The product $a \cdot a \cdot a \cdot a$ may be abbreviated to a^4, which is known as the fourth power of a. The symbol a is called the **base**, the number 4, which indicates the number of times the base a is to appear as a factor, is the **exponent**. For a first power we have $a^1 = a$, a second power is called a *square*, a third power is a *cube*.

EXAMPLE 1
 a) $243 = 3 \cdot 3 \cdot 3 \cdot 3 \cdot 3 = 3^5$
 b) $(1 + i)^3 = (1 + i)(1 + i)(1 + i)$
 c) $625 = 5 \cdot 5 \cdot 5 \cdot 5 = 5^4$

∎

Section A1.2 Laws of Exponents

If m and n are positive integers and $a \neq 0$, $b \neq 0$, we have

1. $a^m \cdot a^n = a^{m+n}$

2. $a^m \div a^n = a^{m-n}$

3. $(a^m)^n = a^{mn}$

4. $(ab)^n = a^n b^n$

5. $(\frac{a}{b})^n = \frac{a^n}{b^n}$

We illustrate law (1) below:

$$a^m \cdot a^n = (a \cdot a \ldots \text{to } m \text{ factors})(a \cdot a \ldots \text{to } n \text{ factors})$$
$$= (a \cdot a \ldots \text{to } m + n \text{ factors}) = a^{m+n}$$

EXAMPLE 2

a) $2^3 2^6 = 2^{3+6} = 2^9$

b) $\frac{x^5}{x^2} = x^{5-2} = x^3$

c) $(a^3)^2 = a^6$

d) $(2x^3)^4 = 2^4 x^{12} = 16x^{12}$

e) $(\frac{a^2}{b})^3 = \frac{(a^2)^3}{b^3} = \frac{a^6}{b^3}$

■

Section A1.3   Zero, Negative and Fractional Exponents

We extend the notion of an exponent to include zero, negative integers and common fractions by the following definitions:

$$a^0 = 1, \ a \neq 0$$

$$a^{-n} = \frac{1}{a^n}, \ n \text{ positive integer}$$

$$a^{\frac{m}{n}} = \sqrt[n]{a^m}, \ m \text{ and } n \text{ positive integers.}$$

It can be shown that the laws of exponents (1) to (5) hold when m and n are rational numbers (i.e., zero, positive and negative integers, and common fractions).

EXAMPLE 3

a) $1 = \frac{3^4}{3^4} = 3^{4-4} = 3^0$

b) $4^{-2} = \frac{1}{4^2} = \frac{1}{16}$

c) $(27)^{1/3} + (25)^{1/2} = \sqrt[3]{27} + \sqrt{25} = 3 + 5 = 8$

d) $(9)^{3/2} = (9^{1/2})^3 = (\sqrt{9})^3 = 3^3 = 27$

e) $\frac{a^3}{a^{-2}} = a^3 \, a^2 = a^5$

f) $\sqrt{\frac{x^{-2}}{y^6}} = (\frac{x^{-2}}{y^6})^{1/2} = \frac{x^{-1}}{y^3} = x^{-1}y^{-3}$ or $\frac{1}{xy^3}$

■

EXAMPLE 4   Using a pocket calculator, calculate

a) $\sqrt[5]{3} = 3^{1/5} = 1.2457309$

b) $\sqrt[3]{\frac{6034 \times .4185}{1.507}} = (\frac{6034 \times .4185}{1.507})^{1/3} = 11.87761303$

c) $15\ 000(1.068)^{-3} = 12\ 313.3862$

d) $\frac{3}{\sqrt[4]{608}} = \frac{3}{(608)^{1/4}} = 3(608)^{-1/4} = .604150807$

■

Section A1.4 | # Exponential Equations — Unknown Base

Applying the laws of exponents and definitions of the rational exponents, we may solve exponential equations with the unknown in the base using a pocket calculator.

EXAMPLE 5   Solve the following equations:

a) $100(1 + i)^8 = 200$

$$(1 + i)^8 = 2$$
$$1 + i = \sqrt[8]{2}$$
$$1 + i = 2^{1/8}$$
$$i = 2^{1/8} - 1$$
$$i = .090507733$$

b) $8800(1 - d)^{10} = 1500$

$$(1 - d)^{10} = \tfrac{1500}{8800}$$
$$1 - d = \sqrt[10]{\tfrac{1500}{8800}}$$
$$1 - d = \left(\tfrac{1500}{8800}\right)^{1/10}$$
$$d = 1 - \left(\tfrac{1500}{8800}\right)^{1/10}$$
$$d = .162160447$$

c) $(1 + i)^{12} = (1.055)^4$

$$1 + i = \sqrt[12]{(1.055)^4}$$
$$1 + i = (1.055)^{4/12}$$
$$i = (1.055)^{1/3} - 1$$
$$i = .01800713$$

∎

Section A1.5 | # Logarithms

Let N be a positive number and let b be a positive number other than 1. Then the **logarithm, base b**, of the number N is the exponent L on base b such that $b^L = N$. The statement that L is the logarithm, base b, of N is written briefly as $L = \log_b N$. For example,

$$\log_2 16 = 4 \text{ since } 2^4 = 16 \text{ and}$$
$$\log_5 125 = 3 \text{ since } 5^3 = 125$$

Logarithms, base 10, are called **common logarithms** and are used as aids in computation. We shall write simply $\log N$ instead of $\log_{10} N$ for the common logarithm of a number N. Hereafter, the word logarithm will refer to a common logarithm. By definition,

$$\log 1000 = 3 \qquad \text{since } 10^3 = 1000$$
$$\log 100 = 2 \qquad \text{since } 10^2 = 100$$

$$\log 10 = 1 \qquad \text{since } 10^1 = 10$$
$$\log 1 = 0 \qquad \text{since } 10^0 = 1$$
$$\log 0.1 = -1 \qquad \text{since } 10^{-1} = 0.1$$
$$\log 0.01 = -2 \qquad \text{since } 10^{-2} = 0.01$$
$$\log 0.001 = -3 \qquad \text{since } 10^{-3} = 0.001, \text{ etc.}$$

It should be remembered that $\log N$ is defined for positive numbers N only but $\log N$ may be any real number that is positive, negative, or zero.

Section A1.6 Basic Properties of Logarithms

Since logarithms are exponents, the rules for exponents will be used to prove the rules for logarithms. Let $A = 10^a$ and $B = 10^b$ so that $\log A = a$ and $\log B = b$. Since

$$A \cdot B = 10^a \cdot 10^b = 10^{a+b}$$
$$\frac{A}{B} = \frac{10^a}{10^b} = 10^{a-b}$$
$$A^k = (10^a)^k = 10^{ka}, \text{ a real number,}$$

it follows that

$$\log A \cdot B = a + b = \log A + \log B$$
$$\log \frac{A}{B} = a - b = \log A - \log B$$
$$\log A^k = ka = k \log A$$

We have shown the three fundamental laws of logarithms.

1. The logarithm of the product of two positive numbers is the sum of the logarithms of the numbers.

$$\log A \cdot B = \log A + \log B$$

2. The logarithm of the quotient of two positive numbers is the logarithm of the numerator minus the logarithm of the denominator.

$$\log \frac{A}{B} = \log A - \log B$$

3. The logarithm of the kth power of a positive number is k times the logarithm of the number.

$$\log A^k = k \log A$$

EXAMPLE 6 Given $\log 2 = 0.301029996$ and $\log 3 = 0.477121255$, then

a) $\log 6 = \log(2 \times 3) = \log 2 + \log 3 = 0.301029996 + 0.477121255$
$$= 0.778151251$$

b) $\log 1.5 = \log \frac{3}{2} = \log 3 - \log 2 = 0.477121255 - 0.301029996$
$$= 0.176091259$$

c) $\log 8 = \log 2^3 = 3 \log 2 = 3(0.301029996) = 0.903089988$

d) $\log 200 = \log 2 \times 10^2 = \log 2 + \log 10^2 = 0.301029996 + 2$
$$= 2.301029996$$

e) $\log 0.003 = \log 3 \times 10^{-3} = \log 3 + \log 10^{-3} = 0.477121255 + (-3)$
   $= -2.522878745$

f) $\log \sqrt[3]{2} = \log 2^{\frac{1}{3}} = \frac{1}{3}\log 2 = \frac{1}{3}(0.301029996) = 0.100343332$

∎

Section A1.7 Characteristic and Mantissa, Antilogarithms

Every positive number can be written as a so-called **basic number** (a number between 1 and 10) multiplied by an integral power of 10. For example,

$$5836 = 5.836 \times 10^3$$
$$0.0032 = 3.2 \times 10^{-3}$$

Taking logarithms of these numbers we obtain

$$\log 5836 = \log 5.836 + 3$$
$$\log 0.0032 = \log 3.2 - 3$$

We notice that the logarithm of a basic number is always between 0 and 1 (since log 1 = 0 and log 10 =1) and the logarithm of an integral power of 10 is, by definition, an integer. Thus, the logarithm of a positive number may be thought of as consisting of two parts:

(a) an integral part, called the **characteristic**. The characteristic is the logarithm of the integral power of 10 and is solely determined by the position of the decimal point in the number. The characteristic may be any whole number (integer) that is positive, negative or zero.

(b) a decimal part, called the **mantissa**. The mantissa is the logarithm of the basic number and is determined by the sequence of digits in the number without regard to the position of the decimal point. The mantissa is always a positive decimal.

Standard textbooks in the mathematics of finance often contain a table of mantissas (rounded off to six decimal places) of all numbers of four or fewer digits, i.e., all numbers from 1.000 to 9.999. In this textbook we deleted the table of mantissas and assumed that each student has a pocket calculator with a built-in logarithmic function.

When calculators with a built-in logarithmic function are used, then the logarithm of a number N, such that $0 < N < 1$, will be shown on the display of the calculator as a single negative number. For $0 < N < 1$ the negative characteristic is combined with the positive mantissa into a single negative number, representing log N. In this case, the decimal part of the displayed negative number does not represent the mantissa. For example, given

$$\log 2 = 0.301029996, \text{ then } \log 0.002 = 0.301029996 - 3$$

Combining the negative characteristic with the positive mantissas, i.e., subtracting 3 from 0.301029996, we obtain log 0.002 = -2.698970004. The calculator will display log 0.002 as -2.698970004. Note that 0.698970004 is not the mantissa of 2.

So far we have discussed the problem of finding the logarithm L of a given positive number N. Another problem when using logarithms in computation is: given the logarithm L of a number N, find the number N. The number that corresponds to a given logarithm is called the **antilogarithm**. We write $N =$ antilog L when log $N = L$. For example,

antilog $1.301029996 = 20$ (since $\log 20 = 1.301029996$)
antilog $0.84509804 - 1 = 0.7$ (since $\log 0.7 = 0.84509804 - 1$)

To calculate antilog L on a pocket calculator we use the inverse logarithmic function (INV LOG or 10^x). For details consult the owner's manual of your calculator.

Computing with Logarithms

The importance of logarithms in computations has diminished with the development of electronic calculators. Students now can quickly perform operations such as long multiplication or division; taking powers or roots; and using pocket calculators without the aid of logarithms.

However in some cases, such as solving exponential equations for an unknown exponent (see section A1.9 of this appendix), logarithms must still be used.

In the following examples we shall illustrate the use of logarithms in situations where they provide an alternative method of solution. We assume the use of calculators with a built-in logarithmic function so that logarithms or antilogarithms can be found quickly without using logarithmic tables. We also show direct solutions without the aid of logarithms.

EXAMPLE 7 The following equation comes from a compound interest problem. Determine i, which represents the interest rate per interest conversion period, if

$$800(1 + i)^{20} = 5000$$

Using logarithms, we have

$$\log 800 + 20\ \log(1 + i) = \log 5000$$
$$20\ \log(1 + i) = \log 5000 - \log 800$$
$$20\ \log(1 + i) = 3.698970004 - 2.903089987$$
$$20\ \log(1 + i) = .7895880017$$
$$\log(1 + i) = 0.039794001$$
$$(1 + i) = 1.095958226$$
$$i = .095958226$$
$$i \doteq 9.60\%$$

Direct solution without the aid of logarithms.

$$800\,(1+i)^{20} = 5000$$

$$(1+i)^{20} = \tfrac{5000}{800}$$

$$1+i = \left(\tfrac{5000}{800}\right)^{1/20}$$

$$i = \left(\tfrac{5000}{800}\right)^{1/20} - 1$$

$$i = 0.095958226$$

$$i \doteq 9.60\%$$

■

EXAMPLE 8   The following equation comes from a depreciation problem. Determine d, the annual compounded rate of depreciation, if

$$83\ 000(1-d)^{10} = 10\ 500$$

Using logarithms, we have

$$\log 83\ 000 + 10 \log (1-d) = \log 10\ 500$$

$$10 \log (1-d) = \log 10\ 500 - \log 83\ 000$$

$$10 \log (1-d) = 4.021189299 - 4.919078092$$

$$10 \log (1-d) = -.897888793$$

$$\log (1-d) = -.089788879$$

$$1-d = .813225748$$

$$d = .186774252$$

$$d \doteq 18.68\%$$

Direct solution without the aid of logarithms.

$$83\ 000(1-d)^{10} = 10\ 500$$

$$(1-d)^{10} = \tfrac{10\ 500}{83\ 000}$$

$$1-d = \left(\tfrac{10\ 500}{83\ 000}\right)^{1/10}$$

$$d = 1 - \left(\tfrac{10\ 500}{83\ 000}\right)^{1/10}$$

$$d = .186774252$$

$$d \doteq 18.68\%$$

■

Exponential Equations — Unknown Exponent

In solving exponential equations for an unknown exponent there is no direct solution available and logarithms provide an efficient method of solution.

EXAMPLE 9 The following equation comes from a compound interest problem. Determine n, which represents the number of interest conversion periods, if

$$250(1.015)^n = 750$$
$$(1.015)^n = 3$$
$$n \log 1.015 = \log 3$$
$$n = \frac{\log 3}{\log 1.015}$$
$$n = 73.78876232$$ ∎

EXAMPLE 10 The following equation comes from an annuity problem. Determine n, which represents the number of payments, if

$$\frac{1 - (1.01)^{-n}}{.01} = 5$$
$$1 - (1.01)^{-n} = .05$$
$$(1.01)^{-n} = .95$$
$$-n \log 1.01 = \log .95$$
$$n = -\frac{\log .95}{\log 1.01}$$
$$n = 5.154933554$$

This result means that there will be 5 full payments and a reduced 6th payment.

∎

Exercises — Appendix 1

1. Simplify.

a) $a^3 \cdot a^6$

b) $a \cdot a^2 \cdot a^3$

c) $\frac{a^8}{a^4}$

d) $(a^2)^3$

e) $\frac{(a^3)^2 a}{a^5}$

f) $(a^2 b^3)^2$

g) $\left(\frac{aa^2}{b^2 b}\right)^3$

h) $(1.05)^3 (1.05)^{12} (1.05)^4$

2. Simplify.

a) $a^{1/2} a^{1/3}$

b) $a^{1/2} \div a^{1/3}$

c) $\frac{aa^{-2}}{a^3}$

d) $a^{(2/3)^{3/2}} \cdot a^{-1}$

e) $25^{-1/2}$

f) $\left(\frac{a^3}{b^2}\right)^{-1/4}$

3. Simplify using rational exponents

a) $\sqrt{a}\sqrt[3]{a}$

b) $\dfrac{a \cdot \sqrt[3]{a}}{\sqrt{a^3}}$

c) $\left(\dfrac{\sqrt{a^4}\sqrt[3]{b^2}}{ab^3}\right)^{-3}$

4. Using a pocket calculator, calculate

a) $\sqrt[3]{.0468}$

b) $\sqrt[15]{\dfrac{24.60}{396}}$

c) $\dfrac{37(23.3)^2}{\sqrt[3]{111.3}}$

d) $\dfrac{(1.065)^{15} - 1}{.065}$

e) $375(1.03)^{-2/3}$

f) $\sqrt[4]{\dfrac{21.2}{(.082)^2}}$

g) $\sqrt{3}\sqrt[3]{5}\sqrt[4]{7}$

h) $\dfrac{1 - (1.11)^{-13}}{.11}$

5. Solve the following exponential equations
  i) directly without the aid of logarithms;
  ii) using logarithms.

a) $3500(1 + i)^8 = 5000$

b) $823.21(1 + i)^{60} = 15\ 000$

c) $17\ 800(1 - d)^{20} = 500$

d) $8000(1 - d)^{11} = 800$

e) $1000(1 + i)^{-20} = 35$

f) $(1 + i)^{-10} = .9490$

g) $(1 + i)^{1/4} = 1.0113$

h) $(1 + i)^{20} - 1 = 80$

i) $(1 + i)^4 = (1.01)^{12}$

j) $(1 + i)^{12} = (1.05)^2$

6. Solve the following exponential equations

a) $50\,(1.035)^n = 200$

b) $500 = 20\,(2.06)^x - 150$

c) $808\,(1.092)^{-n} = 90$

d) $(1.0463)^{-n} = .3826$

e) $(1.02)^n - 1 = .5314$

f) $3^x = 5(2^x)$

g) $126\,(.75)^x = 30$

h) $1 + 2^x = 81$

i) $\dfrac{3^x + 1}{2} = 21$

j) $\dfrac{2^x - 1}{3} = 12$

7. Solve the following equations for n and check your answer by substitution.

a) $\dfrac{(1.083)^n - 1}{.083}$

b) $\dfrac{(1.11)^n - 1}{.11} = 11$

c) $\dfrac{(1.005)^n - 1}{.005} = 10$

d) $\dfrac{1 - (1.087)^{-n}}{.087} = 4.5$

e) $\dfrac{1 - (1.0975)^{-n}}{.0975} = 6$

f) $\dfrac{1 - (1.025)^{-n}}{.025} = 3$

Progressions

Arithmetic Progressions

Any sequence of numbers with the property that any two consecutive numbers in the sequence are separated by a common difference is called an **arithmetic progression**. Thus,

3, 5, 7, 9,...is an arithmetic progression with common difference 2.

20, 17, 14, 11,...is an arithmetic progression with common difference –3.

Let us look at the first few terms of an arithmetic progression whose first term is t_1 and whose common difference is d.

1st term: t_1
2nd term: $t_1 + d$
3rd term: $t_1 + 2d$
4th term: $t_1 + 3d$, etc.

We notice that the coefficient of d is always one less than the number of the term. Thus, the 40th term would be $t_1 + 39d$.

In general, the nth term on an arithmetic progression, denoted by t_n, with first term t_1 and common difference d is given by

$$t_n = t_1 + (n - 1)d$$

Given any three of the four parameters, we can solve for the fourth.

EXAMPLE 1   In an arithmetic progression
 a) Given $t_1 = 3$, $t_n = 9$, $n = 7$, find d and t_{10}.
 b) Given $t_4 = 12$, $t_8 = -4$, find t_1 and d.

Solution
 a) We have $t_7 = t_1 + 6d = 3 + 6d = 9$ and calculate $d = 1$ and
   $$t_{10} = 3 + 9 = 12.$$
 b) We have $t_4 = t_1 + 3d = 12$
   $$t_8 = t_1 + 7d = -4$$

and solve these equations for t_1 and d. Subtracting $t_8 - t_4$ we get $4d = -16$ and $d = -4$. Substituting for d in the first equation we obtain $t_1 = 12 - 3d = 12 - 3(-4) = 24$.

■

Section A2.2 | The Sum of an Arithmetic Progression

Let S_n denote the sum of the first n terms of an arithmetic progression. We may write

$$S_n = t_1 + (t_1 + d) + (t_1 + 2d) + \ldots + (t_n - 2d) + (t_n - d) + t_n$$

and

$$S_n = t_n + (t_n - d) + (t_n - 2d) + \ldots + (t_1 + 2d) + (t_1 + d) + t_1$$

Adding the above expressions for S_n, term by term, we obtain

$$2S_n = (t_1 + t_n) + (t_1 + t_n) + (t_1 + t_n) + \ldots$$
$$+ (t_1 + t_n) + (t_1 + t_n) + (t_1 + t_n)$$
$$= n(t_1 + t_n)$$

and

$$S_n = \frac{n}{2}(t_1 + t_n)$$

Thus,

$$3 + 5 + 7 + 9 + 11 + 13 = \tfrac{6}{2}(3 + 13) = 48$$

and

$$20 + 17 + 14 + 11 + 8 + 5 + 2 + (-1) + (-4) = \tfrac{9}{2}(20 + (-4)) = 72$$

It is easy to see that the sum of the first n integers is

$$S_n = \frac{n}{2}(1 + n)$$

Thus,

$$1 + 2 + \ldots + 100 = \tfrac{100}{2}(1 + 100) = 50(101) = 5050$$

EXAMPLE 2 A woman borrows $1500, agreeing to pay $100 at the end of each month to reduce the outstanding principal and agreeing to pay the interest due on any unpaid balance at rate 12% per annum (i.e., 1% per month). Find the sum of all interest payments.

Solution Interest payments are calculated as 1% of the progression 1500, 1400,..., 100. The sum of all interest payments is the sum of 15 terms of an arithmetic progression 15, 14, 13,..., 1

$$S_{15} = \tfrac{15}{2}(15 + 1) = \$120$$

■

Section A2.3 **Geometric Progression**

A sequence of numbers with the property that the ratio of any 2 consecutive numbers in the sequence is constant is called a **geometric progression**. Thus,

2, 4, 8, 16, 32,...is a geometric progression with common ratio 2.

3, 3x, 3x^2, 3x^3, 3x^4,... is a geometric progression with common ratio x.

Let us look at the first few terms of a geometric progression whose first term is t_1 and whose common ratio is r.

1st term: t_1
2nd term: $t_1 r$
3rd term: $t_1 r^2$
4th term: $t_1 r^3$, etc.

We notice that the exponent of r is always one less than the number of the term. Thus, the 20th term would be $t_1 r^{19}$.

In general, the nth term of a geometric progression, denoted by t_n, with the first term t_1 and common ratio r is given by

$$t_n = t_1 r^{n-1}$$

Given any three of the four parameters, we can solve for the fourth.

EXAMPLE 3 In a geometric progression if $r = 10$ and the eighth term is 2000, find the first and the fifth term.

Solution We have $t_8 = t_1(10)^7 = 2000$. Thus, $t_1 = 2000(10)^{-7} = .0002$ and $t_5 = t_1(10)^4 = .0002(10^4) = 2$.

■

Section A2.4 **The Sum of a Geometric Progression**

Let S_n denote the sum of the first n terms of a geometric progression. Thus,

$$S_n = t_1 + t_2 + \ldots + t_n$$
$$= t_1 + t_1 r + t_1 r^2 + \ldots + t_1 r^{n-1}$$

Also,

$$r \cdot S_n = t_1 r + t_1 r^2 + \ldots + t_1 r^{n-1} + t_1 r^n$$

Subtracting the above expressions for S_n and rS_n, term by term, we obtain

$$(1 - r)S_n = t_1 - t_1 r^n$$

and

$$S_n = t_1 \frac{1 - r^n}{1 - r} \quad \text{or} \quad S_n = t_1 \frac{r^n - 1}{r - 1}$$

Thus,

$$2 + 4 + 8 + 16 + 32 + 64 = 2\frac{(1 - 2^6)}{1 - 2}$$
$$= \frac{2(2^6 - 1)}{2 - 1}$$
$$= 2(2^6 - 1)$$
$$= 2(63)$$
$$= 126$$

and

$$3 + 3x + 3x^2 + \ldots + 3x^n (n + 1 \text{ terms}) = 3\left(\frac{1 - x^{n+1}}{1 - x}\right)$$

$$= 3\left(\frac{x^{n+1} - 1}{x - 1}\right)$$

EXAMPLE 4   A person starts a chain letter and gives it to 4 friends (step one) who pass it to 4 friends each (step two), etc. If there are no duplications, how many people will have been contacted after 10 steps?

Solution   We have a geometric progression

$$4, 16, 64, \ldots$$

with $t_1 = 4$, $r = 4$, $n = 10$; we calculate

$$S_n = 4\frac{(4^{10}) - 1}{4 - 1} = 1\ 398\ 100$$

After 10 steps, 1 398 100 people will have been contacted.

Section A2.5 # Applications of Geometric Progressions to Annuities

In Chapter 3 we used geometric progressions to derive

$$s_{\overline{n}|i} = \frac{(1 + i)^n - 1}{i} \quad \text{and} \quad a_{\overline{n}|i} = \frac{1 - (1 + i)^{-n}}{i}$$

In Chapter 4 we used geometric progressions to solve annuity problems where payments vary.

In the following examples we illustrate how geometric progressions can also be used in the ordinary general annuity problems of Chapter 4.

EXAMPLE 5   Find the accumulated and discounted value of $1000 paid at the end of each year for 10 years if the interest rate is 10% compounded quarterly.

Solution   From first principles, using geometric progressions, we calculate the accumulated value S

$$S = 1000[1 + (1.025)^4 + (1.025)^8 + \ldots 10 \text{ terms}]$$

$$= 1000\frac{(1.025)^{40} - 1}{(1.025)^4 - 1} = \$16\ 231.74$$

and the discounted value A

$$A = 1000[(1.025)^{-4} + (1.025)^{-8} + (1.025)^{-12} + \ldots 10 \text{ terms}]$$

$$= 1000(1.025)^{-4}\frac{1 - (1.025)^{-40}}{1 - (1.025)^{-4}} = \$6045.20$$

Section A2.6	**Infinite Geometric Progressions**

Consider the geometric progression

$$1, \frac{1}{2}, \frac{1}{4}, \frac{1}{8}, \frac{1}{16}, \ldots$$

whose first term is $t_1 = 1$ and whose ratio is $r = \frac{1}{2}$.

The sum of the first n terms is

$$S_n = \frac{1 - (\frac{1}{2})^n}{1 - \frac{1}{2}} = \frac{1}{1 - \frac{1}{2}} - \frac{(\frac{1}{2})^n}{1 - \frac{1}{2}} = 2 - (\frac{1}{2})^{n-1}$$

We note that for any n the difference $2 - S_n = (\frac{1}{2})^{n-1}$ remains positive and becomes smaller and smaller as n increases. We say, as n increases without bound (as n becomes infinite), the sum S_n of the first n terms approaches 2 as a limit. We write

$$\lim_{n \to \infty} S_n = 2$$

For the general geometric progression

$$t_1, t_1 r, t_1 r^2, \ldots$$

we can write the sum of the first n terms

$$S_n = t_1 \frac{1 - r^n}{1 - r} = \frac{t_1}{1 - r} - \frac{t_1 r^n}{1 - r}$$

When $-1 < r < 1$, $\lim_{n \to \infty} S_n = \frac{t_1}{1 - r}$ and we call $S = \dfrac{t_1}{1 - r}$ the **sum of the infinite geometric progression**.

EXAMPLE 6   Find the sum of the infinite geometric progression

$$100, 100(1.01)^{-1}, 100(1.01)^{-2}, 100(1.01)^{-3}, \ldots$$

Solution   We have $t_1 = 100$, $r = (1.01)^{-1} < 1$ and calculate

$$S = \frac{100}{1 - (1.01)^{-1}} = \frac{100}{1 - \frac{1}{1.01}} = \frac{101}{.01} = 10\ 100$$

∎

Section A2.7	**Exercises — Appendix 2**

1. Determine which of the following progressions are arithmetic and which are geometric, then write the next term of each and find the term and the sum as indicated.

   a) $1, -\frac{1}{2}, \frac{1}{4}, \ldots$ Find the 8th term and the sum to 10 terms.

   b) $-1, 2, 5, \ldots$ Find the 15th term and the sum to 12 terms.

   c) $19, 31, 43, \ldots$ Find the 9th term and the sum to 10 terms.

   d) $40, \frac{120}{7}, \frac{360}{49}, \ldots$ Find the 7th term and the sum to 12 terms.

   e) $\frac{1}{3}, \frac{1}{12}, -\frac{1}{6}, \ldots$ Find the 8th term and the sum to 10 terms.

   f) $9.2, 8, 6.8, \ldots$ Find the 10th term and the sum to 15 terms.

2. Find the sum of
   a) the first 300 positive integers;
   b) the first 100 positive even numbers;
   c) all odd integers from 15 to 219 inclusive;
   d) all even integers from 18 to 280 inclusive.

3. Find the 10th term and the sum of the first 10 terms of the progressions
   a) $2, 4, 6, \ldots$
   b) $625, 125, 25, \ldots$
   c) $1, 1.08, (1.08)^2, \ldots$
   d) $(1.05)^{-1}, (1.05)^{-2}, (1.05)^{-3}, \ldots$

4. In an arithmetic progression
   a) Given $t_1 = 2$, $d = 3$, $n = 10$, find t_n and S_n.
   b) Given $t_n = -11$, $d = -4$, $n = 7$, find t_1 and S_n.
   c) Given $t_3 = 18$, $t_6 = 42$, find t_1 and S_6.
   d) Given $t_1 = 7$, $t_n = 77$, $S_n = 420$, find n and d.
   e) Given $t_1 = 13$, $d = -3$, $S_n = 20$, find n and t_n.

5. In a geometric progression
   a) Given $t_1 = 5$, $r = 2$, $n = 12$, find t_n and S_n.
   b) Given $t_1 = 12$, $r = \frac{1}{2}$, $t_n = \frac{3}{8}$, find n and S_n.
   c) Given $t_5 = \frac{1}{20}$, $r = \frac{1}{4}$, find t_1 and S_5.
   d) Given $t_2 = \frac{7}{4}$, $t_5 = 14$, find t_{10} and S_{10}.
   e) Given $t_1 = 1.03$, $r = 1.03$, find t_{15} and S_{15}.

6. A man borrows $5000, agreeing to reduce the principal by $200 at the end of each month and to pay 15% interest per annum (i.e., $1\frac{1}{4}$% per month) on all unpaid balances. Find the sum of all interest payments.

7. In buying a house a couple agrees to pay $3000 at the end of the 1st year, $3500 at the end of the 2nd year, $4000 at the end of the 3rd year, and so on. How much do they pay for the house if they make 20 payments in all?

8. In drilling a well the cost is $5.50 for the first 10 cm, and for each succeeding 10 cm the cost is $1 more than that for the preceding 10 cm. How deep a well may be drilled for $1000?

9. The value of a certain machine at the end of each year is 80% as much as its value at the beginning of the year. If the machine originally costs $10 000, find its value at the end of 10 years.

10. Each stroke of a vacuum pump extracts 5% of the air in a tank. What decimal fraction of the original air remains after 40 strokes?

11. A rubber ball is dropped from a height of 50 metres to the ground. If it always bounces back one-half of the height from which it falls, find
    a) how far it will rise on the 8th bounce;
    b) the total distance it has travelled as it hits the ground the 10th time;
    c) the total distance it has travelled before coming to rest.

12. If a person were offered a job paying 1 cent the first day, 2 cents the second day, 4 cents the third day, and so on, each day's salary being double that for the preceding day, how much would she receive
    a) on the 30th day;
    b) for a 30-day working period;
    c) for the each of the three consecutive 10-day working periods?

13. Find the accumulated and the discounted value from first principles, using geometric progressions,
    a) of $100 paid at the end of each month for 10 years if the interest rate is 12% compounded semi-annually.
    b) of $500 paid at the end of each half year for 5 years if the interest rate is 12% compounded monthly.

14. Find the sum of the infinite geometric progressions
    a) $1, -\frac{1}{3}, -\frac{1}{27}, \ldots$
    b) $3, 0.3, 0.03, 0.003, \ldots$
    c) $1, .8, (.8)^2, (.8)^3, \ldots$
    d) $(1+i)^{-1}, (1+i)^{-2}, (1+i)^{-3}, (1+i)^{-4}, \ldots$

Linear Interpolation

The first use of linear interpolation in the text is Example 2 of Section 2.5. In this appendix, we expand upon the solution from Section 2.5 to show, in general, how linear interpolation should be applied.

EXAMPLE 1 How long will it take $500 to accumulate to $850 at $j_{12} = 12\%$ if the practical method of accumulation is in effect?

Solution As shown in Section 2.5, this reduces to asking: Find n given $(1.01)^n = 1.7$. Solving this equation by logarithms we calculated $n = 53.3277$. Using a calculator, we find:

$$(1.01)^{53} = 1.6945$$
$$(1.01)^n = 1.7000$$
$$(1.01)^{54} = 1.7114$$

This can be shown graphically as follows:

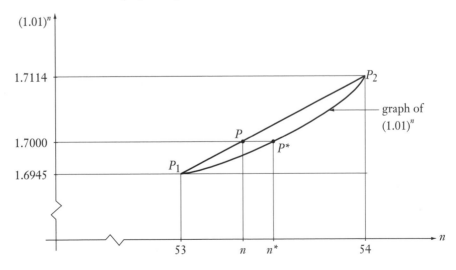

The true value n^* corresponds to the point P^* where the curve gives the value $(1.01)^n = 1.7$. Using logarithms we can find this exact value of n^*. Otherwise, we are forced to make some approximating assumption.

Under the method of linear interpolation, we assume the graph is linear between the points P_1 and P_2, as indicated by the straight line. The method of interpolation will determine the value n corresponding to the point P on the straight line. Clearly n is an approximate value of n^*.

To find the value n we use the fact that the three points $P_1 = (53, 1.6945)$, $P = (n, 1.7000)$ and $P_2 = (54, 1.7114)$ lie on a straight line, i.e.,

$$\frac{n - 53}{54 - 53} = \frac{1.7000 - 1.6945}{1.7114 - 1.6945}$$

Solving for n we obtain

$$\frac{n - 53}{1} = \frac{.0055}{.0169}$$
$$n = 53.32544379$$
$$n \doteq 53.33$$

This linear approximation, called **linear interpolation**, may be illustrated schematically by the following diagram:

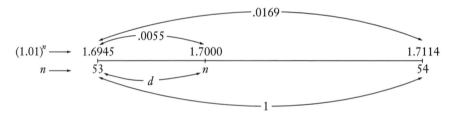

The value n lies the same distance between 53 and 54 as the value 1.7000 between 1.6945 and 1.7114.
Thus,

$$\frac{d}{1} = \frac{.0055}{.0169}$$
$$d = .32544379$$
$$\text{and } n \doteq 53.33$$

The data needed for linear interpolation is usually given in tabular form, called an **interpolation table**, shown below.

	$(1.01)^n$	n
	1.6945	53
	1.7000	n
	1.7114	54

$.0169\{ .0055\{ \quad \}d \quad \}1$

$$\frac{d}{1} = \frac{.0055}{.0169}$$
$$d \doteq .33$$
and $n = 53 + d = 53.33$

In Section 3.6 we used linear interpolation to find the interest rate (see Examples 1 and 2 of Section 3.6).

EXAMPLE 2   Find the interest rate j_4 at which deposits of $250 at the end of every 3 months will accumulate to $5000 in 4 years.

Solution   As shown in Section 3.6, Example 1, this reduces to solving the equation below for unknown rate i.

$$s_{\overline{16}|i} = \frac{(1+i)^{16} - 1}{i} = 20$$

Using a calculator we find

For $j_4 = 11\%$ $(i = .0275)$ $s_{\overline{16}|i} = 19.7640$
For $j_4 = 12\%$ $(i = .03)$ $s_{\overline{16}|i} = 20.1569$

We want to use a linear approximation to find the value of $j_4 = 4i$ such that $s_{\overline{16}|i} = 20$. This linear approximation, called linear interpolation, may be illustrated schematically by the following diagram:

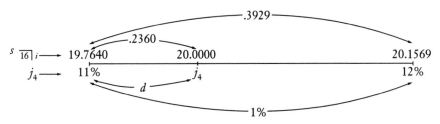

The value j_4 lies the same distance between 11% and 12% as the value 20.0000 does between 19.7640 and 20.1569. Thus,

$$\frac{d}{1\%} = \frac{.2360}{.3929}$$
$$d \doteq .60\%$$
$$\text{and } j_4 = 11.60\%.$$

Usually we arrange our data in an interpolation table, as below.

	$s_{\overline{16}|i}$	j_4
19.7640	11%
20.0000	j_4
20.1569	12%

.3929 { .2360 { 19.7640 | 11% } d } 1%   with braces .2360 spanning 19.7640/20.0000, .3929 spanning all, d spanning 11%/j_4, 1% spanning all

$$\frac{d}{1\%} = \frac{.2360}{.3929}$$
$$d \doteq .60$$
$$\text{and } j_4 = 11.60\%$$

■

In Section 6.8, we used linear interpolation to find the yield rate of a bond (see Examples 2 and 3 of Section 6.8).

EXAMPLE 3   A $500 bond paying interest at $j_2 = 9\frac{1}{2}\%$, redeemable at par on August 15, 2013 is quoted at 109.50 on August 15, 2001. Find the yield rate, j_2, to maturity by the method of linear interpolation.

Solution   As shown in Section 6.8, Example 2, this results in linear interpolation between two market prices.

Q (to yield $j_2 = 8\%$) = $557.18 and
Q (to yield $j_2 = 9\%$) = $518.12

The linear interpolation may be illustrated schematically by the following diagram:

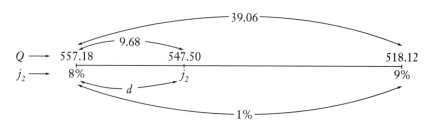

The value j_2 lies the same distance between 8% and 9% as the value 547.50 does between 557.18 and 518.12. Thus,

$$\frac{d}{1\%} = \frac{9.68}{39.06} \text{ or } d \doteq .25\% \text{ and } j_2 = 8.25\%.$$

■

Section A3.1   Exercises — Appendix 3

In the following exercises, illustrate linear interpolation schematically on a diagram and also set up the interpolation table.

1. Solve the following equations for n using linear interpolation.

a) $(1.05)^n = 2$

b) $(1.025)^n = 3.8$

c) $800(1.03)^n = 1100$

d) $(1.045)^{-n} = .5$

e) $(1.0125)^{-n} = \frac{1}{4}$

f) $1000(1.0225)^{-n} = 700$

2. Solve the following equations for i using linear interpolation.

a) $\frac{(1+i)^{12}-1}{i} = 15$

b) $\frac{(1+i)^{100}-1}{i} = 200$

c) $\frac{1-(1+i)^{-20}}{i} = 10$

d) $100 + 90\frac{1-(1+i)^{-6}}{i} = 600$

3. Find the interest rate j_{12} at which deposits of $100 at the end of each month will accumulate to $4000 in 3 years.

4. An insurance company will pay $100 000 to a beneficiary or monthly payments of $1150 for 15 years. What rate j_{12} is the insurance company using?

5. A $1000 bond paying interest at $j_2 = 10\frac{1}{4}\%$ matures at par on August 1, 2020. If this bond is quoted at 105.50 on August 1, 2003, find the yield rate j_2 using the method of interpolation.

6. A $1000 bond that pays interest at $j_2 = 11\%$ matures at par on November 1, 2009. If this bond is sold on May 1, 2004, at a market quotation 78.25, find the yield rate j_2 that the buyer will realize using the method of linear interpolation.

Glossary of Terms

Accumulated Depreciation   The total depreciation allocated to accounting periods at any point in the life of an asset.

Accumulated Value   The total of the amount of money originally invested plus the interest earned.

Accumulated Value (of an Annuity)   The equivalent dated value of the periodic payments at the end of the term of the annuity.

Accumulation of Bond Discount   The periodic increase in the book value of a bond bought at a price below redemption so that the book value will equal the redemption value at maturity.

Amortization Schedule   A schedule showing in detail how a debt is repaid by the amortization method.

Amortize   To repay a debt by means of equal periodic payments of interest and principal.

Amortizing a Bond Premium   The periodic reduction of the book value of a bond bought at a price above redemption, so that the book value will equal the redemption value at maturity.

Annual Cost (of an Asset)   The annual interest on the capitalized cost of an asset including operating expenses and repair costs.

Annuity   A series of payments made or received at regular intervals of time.

Annuity Certain   A series of payments involving a fixed number of such payments.

Annuity Due   A series of payments in which the payments are made at the beginning of each interval of time.

Bond   A certificate of indebtedness agreeing to reimburse the purchaser and to pay periodic interest payments, called coupons.

Bond Interest Rate   The interest rate specified in a bond, upon which the actual periodic interest payments are based. Also called the coupon rate.

Book Value (of an Asset)   The difference between the original cost of an asset and its accumulated depreciation.

Book Value (of a Bond)   The value of a bond, at any particular time, according to its purchaser's accounting records.

Call Price   A lump sum of money paid to the purchaser to redeem a bond earlier than its maturity date.

Capitalized Cost (of an Asset)   The present value of the total costs involved with an asset for an infinite term.

Compound Interest   Interest that is earned upon previously earned interest.

Contingent Payment   A payment that will be made only if some predesignated condition is met, such as the recipient being alive.

Coupon   A detachable portion of a bond that represents the bond interest payment.

Deferred Annuity   A series of payments that has its first payment postponed for one or more periods.

Demand Loan   A loan for which repayment in full or in part may be required at any time.

Depreciation   A periodic charge to income to measure the consumption of an asset.

Depreciation Schedule   A chart showing the details of allocating the consumption of an asset to the various accounting periods.

Discount (Bond Purchased at)   The amount by which the purchase price of a bond is less than its redemption value.

Discounted Value   The amount of money that must be invested on the evaluation date in order to accumulate a specified amount at a later date.

Discounted Value (of an Annuity)   The equivalent dated value of the periodic payments at the beginning of the term of the annuity.

Discounting   Finding the discounted value at a specified rate of interest or discount of an amount due in the future.

Effective Interest Rate   An annual interest rate that produces the same accumulated values as the nominal rate compounded more frequently than annually.

Endowment Insurance   Insurance that provides a benefit either if death occurs during a specified number of years or at the end of that time if the person is then alive.

Equation of Values   The equation obtained when comparing the dated values of the original payments at a focal date to the dated values of the replacement payments at the same focal date.

Equivalent Rates   Two rates of interest or discount that produce the same accumulated value in the same period of time.

Equivalent Single Payment   One payment that can replace several other payments because it equals the dated value of the other payments.

Face Value (of a Bond)   The amount that is stated on a bond, upon which the interest payments and redemption price are based.

Face Value (of a Promissory Note)   The amount of money specified on the promissory note.

Flat Price (of a Bond)   The total purchase price of a bond, including any accrued interest.

Focal Date   A specific point in time selected to compare the dated values of sets of payments. Also called the valuation date.

Grace Period   The period of time (usually 3 days) following the date a payment of a short-term promissory note is due; during that time payment may be made without the loss of any rights.

Indexed Bond   A bond where the coupon and redemption value increase with a specific index such as the consumer price index (CPI)

Interest Conversion Period   The period of time between interest compoundings.

Internal Rate of Return   The rate of interest for which the net present value of a capital investment project is equal to zero.

Life Annuity   A series of payments, each of which is made only if a designated person is then alive, with the payments continuing for that person's entire lifetime. Each such payment is made at the *end* of an interval of time.

Life Annuity Due   A series of payments, each of which is made only if a designated person is then alive, with the payments continuing for that person's entire lifetime. Each such payment is made at the *beginning* of an interval of time.

Maturity Date (of a Note)   The legal due date, which includes 3 days of grace, for a short-term promissory note in Canada.

Maturity Value (of a Note)   The face value plus interest that must be paid on the legal due date.

Mortality Rate   The probability of dying within one year after attaining a specified age.

Net Premium   An amount necessary to provide insurance benefits, calculated by using the assumed rate of interest and the tabular mortality rate.

Nominal Interest Rate   An annual interest rate that is quoted with the understanding that interest is compounded more than once a year.

Noninterest-Bearing Promissory Note   A note that does not require the payment of interest (the maturity value of such a note is the same as its face value).

Ordinary Annuity   A series of payments in which the payments are made at the end of each interval of time. Also called an immediate annuity.

Periodic Cost (of a Debt)   The sum of interest paid and the sinking-fund deposit when a debt is retired by the sinking-fund method.

Perpetuity   An annuity for which the payments continue forever.

Premium (Bond Purchased at)   The amount by which the purchase price of a bond exceeds its redemption value.

Probability   The likelihood of an event occurring.

Proceeds (of a Promissory Note)   The amount of money for which a promissory note is sold.

Promissory Note   A written promise to pay a specified amount of money after a specified period of time with or without interest, as specified.

Pure Endowment   An amount that is paid at the end of the endowment period only if a designated person is then alive.

Quoted Price (of a Bond)   The price of a bond without accrued bond interest.

Redemption Value (of a Bond)   A lump sum of money paid to the bondholder to redeem a bond on or after the day of maturity.

Salvage Value (of an Asset)   The value of an asset at the end of its useful life. Also called the scrap or the residual value.

Serial Bond   A bond whose principal is repaid by a series of payments rather than by a single redemption.

Sinking Fund   A fund that is being accumulated by periodic payments for the purpose of attaining a certain amount by a certain date.

Sinking Fund Deposit   A regular periodic payment into a sinking fund.

Strip Bond   A bond that has its coupons removed and sold separately.

Temporary Life Annuity   A series of payments, each of which is made only if a designated person is then alive, with the number of such payments limited to a specified number. Each such payment is made at the *end* of an interval of time.

Temporary Life Annuity Due A series of payments, each of which is made only if a designated person is then alive, with the number of such payments limited to a specified number. Each such payment is made at the *beginning* of an interval of time.

Term Insurance Insurance that provides a benefit only if death occurs during a specified period.

Whole Life Insurance Insurance that provides a benefit whenever death occurs.

Yield Rate The interest rate that an investor will actually realize on an investment.

CANADIAN INSTITUTE OF ACTUARIES
1986–1992 MORTALITY TABLE
$1000q_x$

ATTAINED	MALE			FEMALE		
AGE	AGGREGATE	SMOKER	NONSMOKER	AGGREGATE	SMOKER	NONSMOKER
0	0.77	–	–	0.69	–	–
1	0.47	–	–	0.42	–	–
2	0.34	–	–	0.27	–	–
3	0.25	–	–	0.19	–	–
4	0.20	–	–	0.15	–	–
5	0.17	–	–	0.14	–	–
6	0.16	–	–	0.14	–	–
7	0.15	–	–	0.13	–	–
8	0.15	–	–	0.11	–	–
9	0.15	–	–	0.10	–	–
10	0.15	–	–	0.10	–	–
11	0.18	–	–	0.10	–	–
12	0.23	–	–	0.12	–	–
13	0.30	–	–	0.16	–	–
14	0.40	–	–	0.21	–	–
15	0.52	0.56	0.52	0.26	0.27	0.25
16	0.65	0.70	0.64	0.29	0.30	0.28
17	0.77	0.84	0.76	0.31	0.33	0.30
18	0.87	0.96	0.86	0.33	0.36	0.32
19	0.94	1.05	0.92	0.36	0.38	0.34
20	0.98	1.12	0.96	0.37	0.41	0.35
21	0.99	1.15	0.98	0.38	0.42	0.35
22	0.98	1.16	0.97	0.37	0.42	0.34
23	0.97	1.16	0.95	0.36	0.40	0.33
24	0.96	1.16	0.94	0.33	0.38	0.30
25	0.96	1.19	0.94	0.33	0.38	0.29
26	0.98	1.23	0.96	0.35	0.42	0.31
27	1.01	1.29	0.98	0.40	0.48	0.35
28	1.03	1.35	1.00	0.46	0.57	0.40
29	1.05	1.41	1.02	0.52	0.64	0.44
30	1.09	1.48	1.05	0.56	0.70	0.47
31	1.13	1.57	1.09	0.58	0.75	0.48
32	1.19	1.68	1.14	0.60	0.79	0.49
33	1.24	1.78	1.18	0.63	0.84	0.51
34	1.26	1.85	1.19	0.68	0.92	0.53

(continued)

| ATTAINED | MALE | | | FEMALE | | |
AGE	AGGREGATE	SMOKER	NONSMOKER	AGGREGATE	SMOKER	NONSMOKER
35	1.26	1.88	1.18	0.72	1.00	0.56
36	1.25	1.90	1.16	0.77	1.09	0.59
37	1.24	1.93	1.15	0.84	1.20	0.63
38	1.26	1.99	1.15	0.92	1.34	0.69
39	1.31	2.09	1.18	1.04	1.54	0.76
40	1.37	2.24	1.22	1.17	1.76	0.85
41	1.49	2.46	1.31	1.30	1.98	0.93
42	1.62	2.73	1.40	1.41	2.17	1.00
43	1.78	3.03	1.52	1.53	2.39	1.07
44	1.96	3.38	1.64	1.67	2.64	1.16
45	2.16	3.78	1.78	1.80	2.87	1.23
46	2.39	4.24	1.94	1.94	3.13	1.32
47	2.66	4.75	2.12	2.09	3.40	1.42
48	2.95	5.34	2.31	2.25	3.71	1.52
49	3.28	5.99	2.53	2.44	4.04	1.64
50	3.65	6.72	2.77	2.63	4.40	1.78
51	4.06	7.54	3.04	2.85	4.79	1.93
52	4.52	8.46	3.34	3.10	5.22	2.10
53	5.03	9.47	3.67	3.36	5.69	2.29
54	5.60	10.60	4.03	3.66	6.21	2.51
55	6.23	11.84	4.45	3.99	6.77	2.75
56	6.92	13.21	4.91	4.35	7.38	3.03
57	7.69	14.71	5.42	4.75	8.06	3.34
58	8.55	16.36	6.00	5.19	8.80	3.69
59	9.49	18.17	6.66	5.69	9.60	4.09
60	10.52	20.14	7.40	6.24	10.49	4.55
61	11.66	22.29	8.24	6.85	11.45	5.06
62	12.92	24.63	9.19	7.52	12.50	5.64
63	14.30	27.17	10.26	8.28	13.65	6.29
64	15.82	29.91	11.47	9.12	14.91	7.03
65	17.49	32.88	12.84	10.05	16.28	7.86
66	19.32	36.09	14.39	11.09	17.78	8.81
67	21.34	39.54	16.14	12.25	19.41	9.87
68	23.54	43.25	18.10	13.54	21.19	11.07
69	25.96	47.23	20.31	14.98	23.12	12.42

(continued)

ATTAINED	MALE			FEMALE		
AGE	AGGREGATE	SMOKER	NONSMOKER	AGGREGATE	SMOKER	NONSMOKER
70	28.61	51.49	22.79	16.58	25.23	13.94
71	31.52	56.04	25.55	18.36	27.53	15.65
72	34.69	60.91	28.64	20.35	30.04	17.56
73	38.17	66.08	32.07	22.57	32.76	19.72
74	41.97	71.59	35.87	25.04	35.73	22.13
75	46.12	77.44	40.07	27.79	38.96	24.84
76	50.66	83.65	44.71	30.85	42.47	27.86
77	55.61	90.22	49.80	34.26	46.31	31.24
78	61.01	97.16	55.39	38.06	50.48	35.02
79	66.90	104.50	61.05	42.28	55.04	39.23
80	73.31	112.23	68.17	46.98	60.01	43.92
81	80.30	120.39	75.43	52.21	65.43	49.14
82	87.89	128.97	83.32	58.02	71.37	54.94
83	96.15	138.01	91.87	64.47	77.86	61.38
84	105.11	147.51	101.12	71.62	84.97	68.53
85	114.84	157.51	111.13	79.56	92.76	76.45
86	125.38	168.02	121.91	88.36	101.31	85.22
87	136.78	179.07	133.54	98.09	110.70	94.91
88	149.12	190.71	146.05	108.85	121.02	105.60
89	162.43	202.98	159.49	120.73	132.37	117.39
90	176.78	215.92	173.93	133.83	144.86	130.36
91	192.23	229.59	189.41	148.25	158.60	144.62
92	208.83	244.06	205.98	164.10	173.72	160.26
93	226.63	259.40	223.71	181.47	190.34	177.39
94	245.68	275.70	242.63	200.48	208.59	196.10
95	266.03	293.06	262.81	221.22	228.61	216.50
96	287.69	311.58	284.27	243.79	250.52	238.67
97	310.71	331.38	307.04	268.28	274.44	262.69
98	335.08	352.57	331.15	294.74	300.46	288.64
99	360.81	375.28	356.58	323.22	328.65	316.56
100	390.00	401.79	385.43	355.89	361.22	348.57
101	433.21	443.01	428.14	403.46	409.04	395.18
102	503.12	511.77	497.24	478.80	485.12	468.97
103	612.44	620.86	605.27	594.76	602.43	582.55
104	773.84	783.19	764.78	764.21	774.00	748.52
105	1000.00	1000.00	1000.00	1000.00	1000.00	1000.00

Answers to Problems

Exercise 1.1

1. a) $2950   b) $1234   c) $10 263.01
2. a) 16.8%   b) 14.29%   c) 12%
3. 1328 days   4. $131.25; $129.45   5. 60%   6. $4880.38
7. $547.59   8. $1020.76; $1020.48   9. $1102.47   10. $80.14
11. 10.37%   12. $28.28; $19.63; $11.54; $1.78; total interest = $61.23
13. $164.38; $497.26; $427.41; $254.39; $57.33; total interest = $1400.77
14. 37.24%   15. 18.62%   16. 56.44%
17. a) 18.62%   b) $23.15

Exercise 1.2

1. February 6; $3000; 67 days; $2944.60
2. November 3; $1211.39; 27 days; $1207.15
3. June 27; $518.70; 185 days; $500.93
4. April 7; $4000; 37 days; $3965.83
5. a) $2038.11   b) 13.64%   c) 11.94%
6. $5024.85; $24.85; $128.03   7. $804.51   8. $1025.48
9. $5126.77   10. $1935.88; $435.88   11. $1960.09
12. a) $1517.62   b) 7.94%   c) 15.14%

Exercise 1.3

1. $1169.25   2. $238.60   3. $988.31   4. $532.94   5. $421.67
6. $739.93   7. a) $1128.97   b) $1190.44   c) $1255.92
8. $417.50   9. $529.41; $529.81   10. $1173.09
11. $161.87; $161.96   12. $2129.88; $1580.42
13. a) $543.05   b) $541.57   14. $125.30; $125.55; $125.47; $125.14

Exercise 1.4

1. $497.37; $503.54   2. $414.97; $421.48   3. $1133.33; $1136.68
4. $327.67; $334.30   5. $478.07

Exercise 1.5

1. a) $1941.23   b) $1939.45   **2.** a) $9.92%   b) 10.10%
3. $11.22; $688.78   **4.** 16.76%   **5.** $814.05
6. $1431.51; $51 611.99

Review Exercises 1.6

1. a) 20 months   b) 541 days
2. Ordinary interest: $1022.92; $977.60;
   Exact interest: $1022.60; $977.90
3. 123.87%   **4.** a) 37.63%   b) $163.46
5. a) 28.22%   b) $107.48   **6.** $691.87
7. $1395.35   **8.** $842.31   **9.** $957.50; $965.08
10. a) $1693.97   b) $1692.01   **11.** $131.70; $132.42
12. a) $926.67   b) $1079.14
13. a) $1523.29   b) 10.90%   c) 13.19%   d) 13.30%
14. a) $2039.40   b) 11.98%   c) 14.04%   **15.** $22.42; $814.05
16. 175 days   **17.** $26.94   **18.** $2.72
19. $1487.57   **20.** a) $1042.52   b) $1045.25

Exercise 2.1

Part A

1. $130.70; $30.70   **2.** $625.51; $125.51   **3.** $285.65; $65.65
4. $1695.88; $695.88   **5.** $80.61; $30.61   **6.** $1687.57; $887.57
7. $381.30; $81.30   **8.** $1221.37; $221.37
9. a) $541.50   b) $563.41   c) $586.14   **10.** $2851.52
11. a) $146.93   b) $148.02   c) $148.59   d) $148.98   e) $149.18
12. $8578.61   **13.** $35.236 billion; $163 000   **14.** a), b), c), d) $112.55

Part B

1. a) $61.83   b) $60.90   c) $61.69
4. $21 058.48; $22 071.36; $22 620.38; $23 003.87; $23 155.30; $23 194.62

Exercise 2.2

Part A

1. a) 7.12%   b) 16.99%   c) 8.24%   d) 12.75%   e) 19.56%
2. a) 5.91%   b) 8.71%   c) 9.57%   d) 15.70%   e) 7.70%
3. a) 7.92%   b) 6.05%   c) 18.27%   d) 9.96%
   e) 8.08%   f) 11.30%   g) 17.59%   h) 12.59%
4. 15.40%   **5.** 14.44%   **6.** 23.14%   **7.** $j_2 = 8.9$%
8. a) $j_2 = 15\frac{1}{2}$% is best; $j_{365} = 14.9$% is worst
   b) $j_2 = 6.5$% is best; $j_{365} = 5.9$% is worst
9. $m = 4$

Part B

5. a), b), c) $26 764.51; $26 878.33; $26 937.10; $26 977.00; $26 996.51
6. a) 6%   b) 6.70%   c) 6.53%
7. 6.14%   **8.** $j_2 = 12.37$%

Exercise 2.3

Part A

1. $83.64   **2.** $42.21   **3.** $655.56   **4.** $306.96   **5.** $558.17
6. $672.97   **7.** $1465.93   **8.** $978.85   **9.** $306.56   **10.** $1112.44
11. $442.94   **12.** $318.14   **13.** $1140.23
14. The cash plan is better by $10 657.22   **15.** $2735.69   **16.** $3258.08

Part B

1. $62.19   **2.** $746.03; $759.82; $766.99; $771.88; $773.78; $774.27
3. Select proposal A with higher net present value.

Exercise 2.4

Part A

1. $207.38; $207.45   **2.** $1150.16; $1150.21   **3.** $1680.84; $1681.55
4. $198.99; $199.21   **5.** $1046.53   **6.** $1757.77
7. $6810.12   **8.** $520.93   **9.** $1333.07; $275.79

Part B

3. $2111.09; $1224.25

Exercise 2.5

Part A

1. $j_4 = 10.96\%$ **2.** $j_{12} = 8.88\%$ **3.** $j_1 = 7.60\%$ **4.** $j_2 = 13.54\%$
5. 3 years 4 months 26 days   **6.** 2 years 11 months 23 days
7. 3 years 11 months 7 days   **8.** 2 years 6 months 12 days
9. $j_1 = 7.18\%$ **10.** $j_4 = 10.27\%$ **11.** 14.77%   **12.** $j_{365} = 13.52\%$
13. a) 15 years 199 days   b) 9 years 330 days
14. 6 years 208 days   **15.** 1039 days

Part B

1. a) $1542.21   b) $2378.41   **2.** 9 years 186 days
4. 8.5486117 years   **5.** $(1 + j_1)^2 - 1$ **6.** 21 years 31 days
7. 19 years 233 days   **8.** 7 years 128 days

Exercise 2.6

Part A

1. $1709.14   **2.** $809.40   **3.** a) $1678.58   b) $3370.45
4. $2468.20   **5.** a) $972.00   b) $1567.09   c) $2526.49
6. $186.14   **7.** $2185.25   **8.** $888.02   **9.** $232.13
10. a) $706.89   b) $459.58   **11.** $226.78   **12.** $193.61
13. $29 713.99   **14.** $884.98

Part B

2. $3951.33   **3.** $3611.27   **4.** $2504.12   **5.** $106.67

Exercise 2.7

Part A

1. $6273.74   **2.** $654.06   **3.** $646.28   **4.** $5264.27; $3264.27
5. $3612.72   **6.** $2067.29   **7.** $81 638.36
8. They should not accept the offer.
9. a) The payments option. b) The cash option. **10.** 7.57%

Part B

2. $686.76; $2.48   **3.** $2958.57   **4.** $15 903.69
5. $3494.79; 5.83%   **6.** 13.04%

7. 7.83%

Exercise 2.8

Part A

1. 87 645   **2.** 345   **3.** 6.50%
4. 6.6374573 years   **5.** $17 520.57   **6.** $217 324.45
7. a) 3.92%   b) 3.85%   c) 3.77%

Part B

1. a) 76 501   b) 0:47 a.m.
2. 36 886   **3.** $0.7968 U.S. = $1 Cdn.

Exercise 2.9

Part A

1. a) $1801.81   b) $1821.06   c) $1822.97
2. a) $5383.77   b) $5362.80   c) $5362.56
3. 13.52%; 14.47%   **4.** a), b) November 8, 2006
5. 7.9248125 years   **6.** She should accept offer c).

Part B

1. 8.38%   **2.** 14.624063 years   **3.** 4.0612599 years
4. 6.1250454 years   **5.** $328.59   **6.** 6.85%

Exercise 2.10

Part A

1. a) $1565.52   b) $1466.40
2. a) $3664.91   b) $3664.31
3. a) 7.95%   b) 5.82%
4. a) 9.72%   b) 8.42%   c) 7.84%   d) 7.23%
5. $4321.80   **6.** 9.06%   **7.** 95.600
8. a) $3908.30   b) $3916.26   c) $3911.06   d) $3913.04
9. a) 6.70%   b) 6.91%

Part B

2. 5.55%   **3.** .0142743

Review Exercises 2.11

1. $2080.76   **2.** $3996.16   **3.** $466.10
4. a) $2191.67; $2192.04   b) $1825.10; $1825.41
5. $827 323.97   **6.** $57.92   **7.** $228.04   **8.** j_2; j_4; j_{12}
9. $2276.06   **10.** 15 years 100 days   **11.** 11.14%
12. $10 017.03   **13.** September 3, 2003   **14.** $877.83
15. a) 11.04%   b), c) 10.99%   **16.** $1338.23
17. Cash option is better by $1709.88   **18.** $2199.72; $146.06
19. $980.21   **20.** $1213.12
21. a) $1349.87   b) $1348.85   c) $1349.86   d) $1350.88   e) $1362.58
22. $2584.47   **23.** a) $6253.15   b) $6253.11
24. $1618.57   **25.** a) $884.98   b) $885.53
26. a), b) 6 years 340 days

Exercise 3.2

Part A

1. a) $11 969.42   b) $12 832.52   **2.** $28 760.36   **3.** $3280.87
4. a) $610.51   b) $567.10   **5.** $612.12
6. a) $47 551.33   b) $78 806.67   **7.** $13 031.63   **8.** $30 400.45
9. $12 833.69   **10.** $29 477.01
11. a) $8111.34   b) $9281.38   **12.** $204.56
13. $5522.36   **14.** $278.03   **15.** $49 702.20

Part B

1. $25 732.82   **2.** $15 448.11   **4.** 14.2; 30

5. $3515.68   **6.** $\dfrac{(1 + i)^{20}s_{\overline{21}|i} - 21}{i}$ **9.** $n + \dfrac{n(n - 1)i}{2}$

11. a) $165.56   b) $195.18   **12.** $215.98
13. a) $2546.43   b) $3312.68   **14.** $103 343.97

Exercise 3.3

Part A

1. a) $3992.71   b) $3274.29   c) $3535.33
2. a) $12 126.49   b) $11 440.85   c) $11 711.81   **3.** $27 737.28
4. $15 569.49   **5.** $5478.92   **6.** $5760.04   **7.** $10 760.72
8. $5735.76   **9.** $10 894.14   **10.** $12 269.34   **11.** $12 132.54
12. a) $12 795.73   b) $1016.81   c) $8085.63   d) $7218.90
13. $84.39   **14.** $4824.70   **15.** $33.94
16. $135.27, $8116.20; $126.97, $7618.20; $118.95, $7137.00
17. a) $1012.63   b) $1217.12   **18.** $193.55

Part B

2. 50; 12   **4.** $17.\dot{7}$ **5.** 6.57
6. $(1 + i)^{-1} + (1 + 2i)^{-1} + \cdots + (1 + ni)^{-1}$ **7.** $a_{\overline{2n}|i} - a_{\overline{m}|i} = s_{\overline{m}|i} - a_{\overline{2n}|i} = \frac{1}{4i}$
8. Ratio $= (1 + i)^n$ **9.** $(1 + i)^n$
10. a) $6305.19   b) $1090.21   c) $6878.02
    d) $4149.10; smaller, since 18% > 15%
11. $13 559.93   **12.** The company should buy the drilling machine.
13. The company should purchase Machine A.
14. $35 632.60   **15.** $42 102.04   **16.** $13 589.73   **17.** $584.96

Exercise 3.4

Part A

1. $7066.97; $15 484.60   **2.** $508.79   **3.** $10.23   **4.** $525.47
5. $1080.31   **6.** $600.34   **7.** $82 534.24   **8.** $858.06   **9.** $4566.77
10. $4291.72   **11.** $1842.32   **12.** $40 416.40   **13.** $2338.80
14. $2532.43   **15.** $3081.69   **16.** $59.16   **17.** $276 963.65

Part B

1. $60 515.17   **2.** a) $4883.88   b) $5006.92   c) $2203.59
5. $17 531.17   **8.** $13 332.71
9. $a_{\overline{7}|i}(1+i)^{-3}$; $a_{\overline{7}|i}(1+i)$; $s_{\overline{4}|i}+a_{\overline{3}|i}$; $s_{\overline{7}|i}(1+i)$; $s_{\overline{7}|i}(1+i)^4$
10. a) $\ddot{s}_{\overline{3}|i}[3(1+i)^6+2(1+i)^3+1]$ b) $2s_{\overline{3}|i}(1+i)^2+3a_{\overline{3}|i}(1+i)$
11. $13 566.11   **12.** $228.91   **13.** $6461.63   **14.** $3096.92

Exercise 3.5

Part A

1. 15; $292.39   **2.** 185; $382.13; $133.11   **3.** 31; $474.56; $77.93
4. 7; $1410.28   **5.** 6; $192.99   **6.** 32; $345.58   **7.** 3; $1250.77
8. 31; $438.06 on July 1, 2011   **9.** a) 70; $1195.90   b) $7411.23
10. a) 21; $193.63 on January 1, 2016   b) $804.86
11. 15   **12.** $386.03

Part B

1. August 1, 2003; $16.18
2. a) 81   b) $1289.20   c) $1311.76
3. a) 43   b) $853.53   c) $938.88   **4.** $151 158.86
5. 12; $4456.81   **6.** $n=\frac{\log(\frac{S}{R}i+1)}{\log(1+i)}$ **7.** $n=-\frac{\log(1-\frac{A}{R}i)}{\log(1+i)}$ **8.** 20; $798.39   **9.** 48

Exercise 3.6

Part A

1. 7.98%   **2.** 18.40%   **3.** 8.70%   **4.** 19.61%; 21.47%
5. 35.07%   **6.** 19.48%   **7.** 10.75%

Part B

1. 21.20%   **2.** 35.07%   **3.** 31.72%   **4.** 26.62%; 30.12%
5. Option 1: $j = 22.73\%$; Option 2: $j = 19.98\%$ is better.
6. $24.19 a month to borrow and buy a TV set
7. 20.86%   **8.** 7.61%

Review Exercises 3.7

1. a) $96 757.14; $39 470.85   b) $333 943.43; $55 572.48
2. a) $1984.65   b) $2332.09   **3.** $3571.94
4. a) $2050.63   b) $8649.69   **5.** $10.75
6. $18 042.31; $6068.87   **7.** $71 580.38   **8.** $103 692.89
9. $74 402.93   **10.** $1973.95   **11.** $13 397.21   **12.** $2569.15
13. $28 630.72   **14.** 69; $2310.56; $323.66   **15.** 18; $627.75
16. April 1, 2005; $13.85   **17.** 65.66%   **18.** 21.55%   **19.** 17.97%
20. 70.04%   **21.** 14; $111.97 on January 1, 2007   **22.** $376.91

23. May 1, 2004; July 1, 2010   **24.** $14 796.40   **25.** $11 487.31
26. a) $7.38   b) $3719.28   c) 31 months; $97.97
27. 17; May 10, 2002; $61.08   **28.** 19.91%
29. a) $618.26   b) $694.53   c) $4551.60   **30.** $3263.56
31. a) $8904.62; $32 104.75   b) December 1, 2009; $475.78
32. 16.47%   **33.** $1630.72

Exercise 4.1

Part A

1. $1199.86   **2.** $4909.13   **3.** $2286.27
4. a) $11 469.92   b) $11 386.59   c) $11 629.86   d) $11 300.85
5. a) $2314.08   b) $2312.37   c) $2304.96   d) $2314.94
6. a) $74.56   b) $74.70   **7.** $1133.07
8. a) $221.85   b) $219.36   **9.** $172.97
10. a) $222.93   b) $223.25   c) $222.87
11. $330.67   **12.** $5843.61   **13.** a) $15 059.01   b) $15 058.46
14. a) $6781.37   b) $6782.71   **15.** Buying is cheaper.
16. $23 691.40   **17.** 14.86%   **18.** 16.62%   **19.** 112; $9.80   **20.** $239.89
21. a) $7078.18   b) $21 983.75   c) $7281.59   d) $22 615.51   e) $4498.31
22. a) $18 980.78   b) 1530.80   **23.** $123.49   **24.** 36; $71.19
25. a) $582.07   b) $582.08

Part B

4. a) $1019.62   b) $246.95   **5.** a) $328.93   b) $4199.26
8. $15 495.16   **9.** 11%   **10.** 29.70%   **11.** $3439.55
12. $111 188.52   **13.** a) $98 347.98   b) $104 376.31
14. a) $7804.35   b) $2058.21

Exercise 4.2

Part A

1. $687.48   **2.** $1177.89   **3.** $1534.67   **4.** $1579.05
5. a) $581.60   b) $700.42   c) $827.98
6. $1451.28; $1367.04; $1320.54
7. They would be better off with the government mortgage at $j_2 = 6\%$.
8. $219.47   **9.** 243; $240.13   **10.** They should take the seller's offer.
11. a) $885.42; $965.42; $1111.21   b) 19 years

Part B

1. $134 296.84   **2.** 8.00%   **3.** 15 years 8 months   **4.** $606.90
5. $945.56; $217.73   **6.** $179 085.59   **7.** 11 years; $286.81

Exercise 4.3

Part A

1. a) $6666.67   b) $5000   c) $4000   **2.** a) $5000   b) $3205.13
3. a) $10 714.29   b) $12 214.29   c) $6343.72
4. a) $2775.15   b) $3373.21   **5.** a) $2380.95   b) $2894.06
6. $8000   **7.** a) $32 970.56   b) $32 000   c) $43 618.37
8. $14 734.88   **9.** 12.5%; 12.89%   **10.** a) $131.16   b) $1508.88
11. $8646.29   **12.** 11%   **13.** $20 548.18   **14.** $1461.10

Part B

3. a) 12%   b) $200 000   c) 18   **4.** $34 457.12
5. 20 months; $182.88   **6.** $88 506.21   **7.** $1652.63
8. $100.98   **9.** 31 months; $98.33
10. a) $4912.66   b) $5764.96   **11.** a) $3920.63   b) $4600.83
12. $2222.49   **13.** $623.50   **14.** 14.654614 years **15.** $1989.36

Exercise 4.4

Part A

1. $7332.70   **2.** $1 167 898.49   **3.** $20 180.04   **4.** $17 326.12
5. $8323.06   **6.** $137 500   **7.** $8973.78   **8.** $113 990.19
9. $15 770.38   **10.** $1614.11   **11.** $7797.87   **12.** $43 627.25
13. $4780.77   **14.** $15 754.79

Part B

4. $4932.95   **5.** $16 090.80   **6.** She should choose fund A.
7. $\frac{1}{is_{\overline{2}|i}}(p + \frac{q}{is_{\overline{2}|i}})$ **9.** At the end of the 24th year.
10. $442.88; $2708.61   **11.** $147 928.85   **15.** b) $100 000
16. $97 489.02   **17.** $435.84   **18.** $8263.93

Review Exercises 4.5

1. a) $603.73   b) $98.86   **2.** $4892.00; $3281.39
3. a) $4046.12   b) $4045.61   **4.** a) 27; $88.29   b) 25; $56.12
5. $1273.45   **6.** 28; $38.03   **7.** 15.74%; 16.36%
8. $22 830.71   **9.** $8681.10; $4362.83
10. a) $9380.77   b) $7606.79   c) $15 705
11. Lease option is cheaper.   **12.** a) $41 884.02   b) $60 795.57
13. a) $610.31; $717.10; $1086.31   b) 15 years; $835.77
14. $115.72   **15.** $2256.98   **16.** $848.07
17. a) $16 435.51   b) $16 517.69   c) $14 654.25
18. $1352.74   **19.** $139.19
20. a) $6643.28   b) $8884.88   **21.** a) $187 500   b) $93 309.97
22. $14 012.21   **24.** $10 442.13; $100 727.85   **25.** $380 000
26. $245.36   **27.** $338 971.64   **28.** $449.21   **29.** $31 787.13
30. $358.01   **31.** $1118.60   **32.** $25 542.66

Exercise 5.1

Part A

1. a) $336.04   b) $335.53   **2.** a) $626.62   b) $623.85
3. $837.34   **4.** $155.30   **5.** $133.63
6. Last payment is $3226.91.   **7.** Last payment is $1842.92.
8. Debt at the end of 5 months is $2076.13; last payment is $117.38.
9. Last payment is $55.06.   **10.** $519.98   **11.** $424.91
12. $1190.71; $1297.74   **13.** $2072.15; $78.75

Part B

1. a) $43.75   b) $1639.45   **2.** $146.41
3. a) $1373.79   b) $343.45   c) 19 years 51 weeks
4. $1516.06   **5.** 16th   **6.** 8%   **7.** $8524.76

8. a) \$274.75   b) .0082071   c) 10%   d) \$29 738.47   e) 22 years
9. $100(\frac{n}{a_{\overline{m}|i}} - 1)$ **10.** U.S. uses j_{12}, bank statement is correct
11. a) \$1241.97   b) 24 years 39 weeks   c) \$209.54   **13.** 8.17%
14. 8%   **15.** 16.99%   **16.** \$1314.67   **17.** 43rd
18. Mortgage A: \$111 192.62; Mortgage B: \$97 256.92   **19.** 8.87%

Exercise 5.2

Part A

1. \$5607.30; \$5607.50   **2.** \$5200.37   **3.** \$3750.17; \$950.33
4. \$918.19   **5.** \$1281.01; \$245.39   **6.** \$1238.33; \$81.50
7. \$98 299.92   **8.** \$184 618.15; \$141 381.85
9. \$41 738.99; \$38 261.01   **10.** \$8334.94   **11.** \$1427.49
12. \$18 137.44   **13.** \$481.02   **14.** \$22 626.01

Part B

1. \$105 400.67   **2.** Take out \$125 000 mortgage at $j_2 = 11\%$
3. \$161 938.22   **4.** 17 years and 9 months; \$98.36   **5.** \$2126.40
6. a) 8 years and 5 months; \$236.76   b) 7 years and 8 months; \$150.40
7. a) \$251 038   b) \$147 230.31   c) \$94 366.49
9. November 30, 2002   **10.** \$3076.26
11. 11; \$93.46   **12.** \$123.54
13. a) \$915.86   b) \$154 754.41   c) i) 17 years and 2 months
     ii) \$99 091.38   iii) \$108 711.27   iv) \$202.91

Exercise 5.3

Part A

1. \$168.28; \$5.05   **2.** \$10.60   **3.** He should not refinance.
4. They should refinance.   **5.** \$1009.23   **6.** \$82 974.21
7. \$422.02   **8.** \$3787.21   **9.** \$479.18
10. a) \$264.13   b) \$7302.73
     c) They should refinance and save \$2.83 per month.

Part B

1. a) \$1114   b) \$871.41   c) \$15 682.65   d) \$97 554.57
2. \$125.26   **3.** \$2079.31
4. a) January 1, 2001 b) \$442.62   d) \$399.84   e) \$389.35
5. 1.836017314   **6.** a) \$1291.55; \$1281.44   b) \$146 666.52; \$140 349.12
     c) \$1331.73; it would not pay to refinance.   **7.** a) \$1339.65; \$1337.86
     b) \$590 010.50   c) \$180 910.94   d) \$12 402.25
8. \$2974.13   **9.** Choose Mortgage B.

Exercise 5.4

Part A

1. \$155.30   **2.** \$87.46   **3.** \$5469.48
4. \$11 928.33   **5.** \$169.53; \$3.80   **6.** \$6.21
7. a) At the end of 2 years: \$10 412.37; \$9584.29
     b) At the end of 5 years: \$10 023.71; \$8674.92
8. She will save money by refinancing the loan.

Part B
1. Take option 2.
2. b) $131.66; $90.11   c) $15 766.21; $15 489.54   d) Do not refinance.
3. $14 700.94

Exercise 5.5

Part A
1. $3428.87   **2.** $17 045.65   **3.** $241.38   **4.** $374.21; $6075.77
5. $1392.44   **6.** $184.27; $8711.54   **7.** $7958.92   **8.** $113.53
9. No final deposit needed; after 9.5 years the account will have $203 263.82
10. $41 971.91

Part B
1. $1290.02   **3.** $5168.85

Exercise 5.6

Part A
1. $706.63; $3919.24   **2.** $12 956.27; $36 706.27
3. a) $28 277.87   b) $164 182.71
4. a) $122 036.72   b) $1 070 154.78
5. $100.96   **6.** $1370.79
7. a) $325.12   b) $505.12   c) $1618.57   **8.** $5606.85

Part B
1. a) $24 394.59   b) $1 519 331.88   **2.** 3 years and 7 months; $95.61
3. $13 082.32   **4.** $4460.37   **5.** 6.5%   **6.** 14.74%
7. a) $846.42   b) $17 458.03   **8.** a) $6366.56   b) $789.34

Exercise 5.7

Part A
1. a) $7791   b) $7791   c) $8293.40
2. The sinking fund plan is cheaper by $302.25 per year.
3. Amortization is cheaper by $559.32 semi-annually.   **4.** 10.04%
5. 9.48%   **6.** 7.45%   **7.** a) $15 196.71   b) 8.92%
8. The amortization method is cheaper by $56.52 a month.

Part B
1. Amortization will save $117.61 a month.   **2.** 13.03%   **3.** $810.00
4. 7.25%   **5.** $1278.04 at the end of 12 years.   **6.** 8.35%

Review Exercises 5.8

1. a) $242.01   b) i) $12 540.76; $85.25; $156.76
ii) $13 423.23; $79.57; $162.44   **2.** $966.73; $243.08
3. $4736.06   **4.** 8.96%   **5.** $325.12
6. a) $536.70   b) $57 262.74   c) $69.16   d) $23 024.34
7. Do not refinance.
8. a) $703.79   b) $98 820.18   c) $818.59   d) $178.49; $640.10
9. $1198.27   **10.** Amortization will save $4.90 annually.

11. $3900   **12.** $14.36
13. a) $889.19   b) $41 870.31   c) i) $152.88; $736.31
   ii) −$51.93; $941.12   d) $992.59
14. a) $47.60   b) $63.54   **15.** a) $65.81   b) $3540.41
16. $429 476.31   **17.** $1129.42
18. a) $662.40   b) $642.66   c) $6754.97
   d) i) 22 months sooner   ii) $12 173.89
19. $7548.01
20. a) $2718.46   b) $14 718.46   c) 13.57%
21. a) $5489.74; $5489.52   b) $445.04   c) $37 277.73
22. a) Amortization: $14 144.72; sum-of-digits: $15 492.96
   b) Do not refinance.

Exercise 6.2

Part A

1. $549.48   **2.** $923.14   **3.** $1897.17   **4.** $5857.95   **5.** $792.96
6. $1918.48   **7.** $13 454.97   **8.** $5379.48   **9.** $1 068 973.16
10. $940.36; $1164.03   **11.** $107.25
12. a) $893.22   b) $1000   c) $1125.51

Part B

3. $6000   **4.** $1209.28   **5.** $1049   **6.** $846.28   **7.** $860.20
8. 10 years   **9.** $60.08   **10.** a) $957.88   b) 12 years
11. 7 years   **12.** $95

Exercise 6.3

Part A

1. Premium; $1025.79   **2.** Discount; $4866.79   **3.** Premium; $2046.13
4. Discount; $995.24   **5.** Premium; $2155.38   **6.** Discount; $10 599.34
7. $5.25; $1094.75

Part B

1. $16.32   **2.** $9.01   **3.** $1360   **4.** a) $1052.76 b) $1068.52
5. $909.91   **6.** $1019.14   **7.** $31.28   **8.** $881.53
9. a) $2216.30   **10.** $1801.07   **11.** $886.82
12. a) $925   b) Discount   c) $0.50; $0.53
13. b) $10 408.88

Exercise 6.4

Part A

1. a) $2345.84   b) $1699.07   **2.** a) $5627.57   b) $4466.12
3. a) $1098.96   b) $914.20

Part B

1. a) $1689.13   b) $2260.08   **2.** a) $1133.76 b) $883.17   **3.** $5419.36
4. a) $1092.77   b) i) Premium = $80
   ii) Premium = $12.77   c) 13 years.

Exercise 6.5

Part A

1. $857.47	**2.** $550.89	**3.** $1905.66	**4.** $9567.20
5. $1172.96	**6.** $1186.02	**7.** $1012.57	

Part B

3. a) $1003.67   b) $1167.14   **4.** October 3, 2003 **5.** $1885.95
6. a) $1087.15; $1074.14   b) $1094.28; $1081.28

Exercise 6.6

Part A

1. $1001.14	**2.** $1061.03	**3.** $1032.01	**4.** $1161.79
5. $128.87	**6.** 116.72	**7.** $97.30	**8.** 117.80

Part B

1. a) $4521.50   b) $4543.09   c) 121.17
3. a) $815.98   b) $823.34   c) i) $1109.16 ii) $634.48   iii) 90.57
4. a) $4581.84   b) 107.87   c) 94.29

Exercise 6.7

Part A

1. 11.42%	**2.** 6.11%	**3.** 10.64%	**4.** 7.82%
5. 11.48%	**6.** 6.09%	**7.** 10.55%	**8.** 7.97%
9. a) 7.44%	b) 7.50%	**10.** a) 9.87%	b) 9.86%
11. a) 10.96%	b) 10.50%	**12.** $j_2 = 10.20\%$	**13.** 6.76%
14. 8.62%	**15.** a) $913.54	b) $1074.39	c) 10.44%
16. 12.04%	**17.** a) Y2000: $80	b) Y2001: $120	

Part B

1. a) $802.71   b) 9.33%   **3.** 9.63%
4. a) $795.37   b) $1058.36   c) 12.32%
5. a) $944.61   b) 7.61%   **6.** a) $5\frac{1}{2}\%$ b) $4\frac{1}{2}$ years
7. $7\frac{1}{2}\%$ **8.** 28.50%   **9.** 8.69%
10. a) $1497.89   b) Discount; $2.11   c) 1.76%

Exercise 6.8

Part A

1. $11 926.56	**2.** $111 898.48	**3.** 8.99%	**4.** $301.23

Part B

2. $29 861 132.58	**3.** $242 773.02	**4.** a) $9614.46	b) $10 343.62
5. 9.20%	**6.** a) $47 167.03	b) $55 015.62	c) $46 707.38

Review Exercises 6.9

1. a) $11 155 738.60   b) $148 362.33   **2.** $10 691 791.51
3. a) $1200.00   b) 7.47%   **4.** a) $1007.98   b) Discount
5. a) $2156.30   b) 7.44%   c) 108.49

6. a) $907.99   b) $1093.68   c) 12.35%   d) $1100.13
7. $2334.76   8. a) $901.04   b) 20 years   c) 8.36%
9. a) 7.97%   b) 7.95%   10. a) 7%   b) 17.5 years
11. $1092.01   12. $8881.54   13. $4785.17
14. a) $1146.23   b) $34.27   c) $1272.38   d) 8.66%
15. a) Discount ($119.50)   b) Premium ($107.12)
16. $1023.41   17. $2240.50   18. $2224.46
19. $4693.99   20. 92.54   21. 13.12%

Exercise 7.1

Part A

1. +$11 700.02; proceed   2. +$15 461.25; proceed
3. +$43 293.16; proceed   4. +$25 170.46; proceed
5. At $j_1 = 4\%$: NPV(A) = $22 591.12; NPV(B) = $56 572.77; choose B
   At $j_1 = 7\%$: NPV(A) = $5009.87; NPV(B) = $29 217.93; choose B
6. NPV(A) = –$548.58; NPV(B) = +$2865.55; choose B
7. NPV at 16% = –$380 547; the company should not proceed with the project
8. NPV = +$9874.62; the fund should proceed with the investment
9. NPV at 7% = +$135.29; proceed

Part B

1. NPV = –$130.81; do not borrow
2. NPV of investment 1 = +$1478.15; NPV of investment 2 = +$5679.23;
   investment 2 provides the company with the highest profit
3.

Rate	2%	4%	6%	8%	10%	12%
NPV	$22 923.25	16 445.44	10 502.14	5037.15	1.24	–4648.79

Exercise 7.2

Part A

1. 15%   2. 16%   3. 27%
4. 29%   5. 11%   6. 12%

Part B

1. IRR(A) = 20%; IRR(B) = 20%. At 15%: NPV(A) = +$985.63
   NPV(B) = +$1609.89. At 25%: NPV(A) = –$757.12; NPV(B) = –$1217.92
2. a) 20%   b) 19%

Exercise 7.3

Part A

1. $23 824.08   2. K_1 = $266 601.48; K_2 = $259 867.53; build the cement
   block warehouse
3. K_1 = $238.88; K_2 = $223.21; the company should purchase the second battery
4. K_A = $48 297.09; K_Z = $54 042.88; machine A should be purchased
5. $29.41   6. Save $225.27 per year by buying the $700 000 roof
7. $11 679.69   8. $182 607.41   9. $23 996.81
10. a) $1666.67   b) $2650   11. $120 000

Part B

1. Purchasing the $25 000 car saves $4571.34 in capitalized costs
2. $1 905 412.60   5. $4433.49   6. $i = 2.26\%$ 7. 15 months
8. a) $570.42
   b) Total monthly costs = $279.52; the heat exchanger is cost effective
9. $55.72   10. $1 000 000

Exercise 7.4

Part A

1. a) $R_k = \$10\ 400$ b) $d = .331674938$ c) $R_k = \frac{n-k+1}{15}(52\ 000)$
   d) S.F. deposit = $9224.61
2. a) $R_k = \$4500$ b) $R_k = \frac{n-k+1}{21}(27\ 000)$ c) S.F. deposit = $3588.83
3. a) $19 200; $30 800   b) $36 202.70; $13 797.30
   c) $30 171.43; $19 828.57   d) $11 156.87; $38 843.13
4. a) $8000; $12 000   b) $12 500; $7500   c) $5942.43; $14 057.57
5. $B_5 = \$14\ 745.60$ 6. $B_7 = \$8827.66$; $R_8 = \$1713.88$ 7. 7 years
8. 6 years   9. Per-unit depreciation = $.50
10. Per-unit depreciation = $8.88 **11.** $70 000   **12.** $1 986 000; $2 780 400
13. a) $6375.95   b) Leasing the car is more economical. **14.** a) $d = 30\%$

Part B

7. $2000.04   8. $2216.09   9. $2580.39

Review Exercises 7.5

1. Accept all (D is best)
2. IRR of the project is 8%; for alternative projects: IRR (1) = 8%,
   IRR (2) = 8%, IRR (3) = 5%, IRR (4) = 20%, and IRR (5) = 18%
3. Buy machine 1   4. $23 996.82   5. $476.86
6. a) i) $2000   ii) $750   iii) $2433.31
   b) i) $20 000   ii) $13 750   iii) $22 200.64
7. a) $R_k = \$2666.67$ b) $d = .306638726$ c) Depreciation per km = 20¢
8. Depletion per truck load = $4.16
9. a) S.F. deposit = $15 650.16   b) $352 145.18   c) $31 693.07
10. $38 500   11. $23.43
12. a) $30 000   b) $18 114.47   c) $32 989.74   d) $17 500
13. $567.97   14. $27 747.57
15. a) $214 658.24   b) $d = .340246045$ c) S.F. deposit = $12 417.75
16. $i = .12$ 17. $14 167

Exercise 8.2

Part A

1. a) $\frac{2}{7}$ b) $\frac{3}{7}$ c) $\frac{2}{7}$ d) $\frac{1}{21}$ e) $\frac{4}{21}$ f) $\frac{1}{3}$
2. a) $\frac{1}{52}$ b) $\frac{2}{13}$ c) $\frac{1}{26}$ d) $1 - (\frac{51}{52})^2$ e) $\frac{1}{52}$ f) $\frac{8}{16\ 575}$
3. a) $\frac{1}{6}$ b) $\frac{1}{2}$ c) $\frac{1}{2}$ d) $\frac{1}{9}$ e) $\frac{13}{18}$
4. a) $\frac{1}{18}$ b) $\frac{1}{9}$ c) $\frac{13}{18}$ d) $\frac{1}{2}$ e) $\frac{5}{12}$ f) $\frac{1}{324}$
5. a) $\frac{3}{8}$ b) $\frac{1}{16}$ c) $\frac{1}{8}$ d) $\frac{1}{8}$
6. a) $\frac{3}{8}$ b) $\frac{1}{16}$ c) $\frac{1}{8}$ d) $\frac{1}{8}$

7. a) $\frac{64}{125}$ b) $\frac{61}{125}$ **8.** a) $\frac{18}{125}$ b) $\frac{54}{125}$ c) $\frac{117}{125}$

9. a) $\frac{3}{1600}$ b) $\frac{27}{80}$ c) $\frac{1597}{1600}$ d) $\frac{53}{80}$

10. a) .225   b) .225   **11.** $\frac{7}{20}$

Part B

1. $\frac{2}{17}$ **2.** $\frac{1}{6}$ **3.** $\frac{3}{5}$

Exercise 8.3

Part A

1. $8   **2.** $4.12   **3.** He should go on.
4. Students should take driver training.   **5.** $75
6. 54¢   **7.** $158   **8.** $3000

Part B

1. $$\frac{13}{27}$ **2.** $$\frac{1960}{64}$ **3.** $0
4. 21 units   **5.** She should stay with the present sales staff.

Exercise 8.4

Part A

1. 24.21%   **2.** $6575.34; 21.67%
3. a) $4631.58   b) $4642.11   **4.** 6.03%
5. Robert will receive $977 526; Tammy will receive $1 017 487
6. $16 865.74   **7.** $3402.87   **8.** $742.74   **9.** $794.57

Part B

1. a) $586.94   b) $619.12   c) 11.08%   **2.** 17.17%
3. $28 402.12; $29 294.14; $29 753.26   **4.** 1.18%   **5.** 23.60%

Review Exercises 8.5

1. a) $\frac{2}{13}$ b) $\frac{3}{26}$ **2.** a) .4479   b) .000064   c) .007873
3. a) $\frac{8}{27}$ b) $\frac{1}{9}$ c) $\frac{2}{27}$ **4.** $\frac{8}{16\ 575}$
5. $E(X) = 5.65$; $E(Y) = 5.50$ **6.** $10.69   **7.** 71 items; $354.25
8. 15.26%   **9.** $18 872.89   **10.** $5 507.09   **11.** $531.22

Exercise 9.2

Part A

1. a) .988450   b) .0028769   c) .0034737   d) .0071600   e) .0102629
2. 82.73   **3.** 999

Part B

3. a) .8187   b) .0952   c) .0086
4. a) .7333   b) .1266   c) .0143

Exercise 9.3

Part A

1. a) $74 274.47   b) $53 215.43   **2.** a) $4735.49   b) $3775.92
3. a) $7345.30   b) $6103.45   **4.** a) $4109.64   b) $4069.56
5. $1443.32

Part B

1. $94.50

Exercise 9.4

Part A

3. $1165.03   **4.** $4809.35   **5.** $359.78   **6.** $5813.84   **7.** $4156.59
8. a) $2322.01   b) $2509.68   **9.** $19 868.81   **10.** $331 194.46
11. $1388.12   **12.** Male $13 748.65; female $13 383.17 **13.** $129 961.52

Part B

1. $7553.85   **3.** $20 513.75   **6.** 4.1485   **7.** 5.57054

Exercise 9.5

Part A

1. $11.79   **2.** $380.39   **3.** $31 838.64   **4.** $6571.76
5. Male $83 951.04; female $82 196.93   **6.** $10 806.14

Part B

1. $2426.51   **3.** 0.04512841   **4.** 0.00785

Exercise 9.6

Part A

1. $2925.50   **2.** $19.94   **3.** $28 538.99
4. $2374.75   **6.** $100 101.76

Part B

1. $372.12   **2.** .07090

Review Exercises 9.7

1. $437.79   **2.** a) $1916.32   b) $3403.50   c) $15 522.81
4. 79.2   **5.** a) $\frac{29}{30}$ b) $\frac{19}{20}$ c) $\frac{1}{16}$
6. $p_{102} = .375; q_{102} = .625; p_{103} = \frac{1}{3}; l_{104} = 16; q_{104} = \frac{3}{4}; l_{105} = 4; p_{105} = 0$
7. $774.87   **8.** $666.75

Appendix Exercise 1

1. a) a^9 b) a^6 c) a^4 d) a^6
   e) a^2 f) a^4b^6 g) a^9/b^9 h) $(1.05)^{19}$
2. a) $a^{\frac{5}{6}}$ b) $a^{\frac{1}{6}}$ c) a^{-4} d) 1
   e) $\frac{1}{5}$ f) $a^{-\frac{3}{4}}/b^{\frac{1}{2}}$

3. a) $\sqrt[6]{a^5}$ b) $1/\sqrt[6]{a}$ c) $a^{-3}b^7$

4. a) .36036999   b) .830901089   c) 4175.88482   d) 24.18216933
    e) 367.6826338   f) 7.493367615   g) 4.817537887   h) 6.749870404

5. a) .045593188   b) .049565815   c) .163574046   d) .188869169
    e) .182487607   f) .005248373   g) .045971928   h) .24573094
    i) .030301   j) .008164846

6. a) 40.29758337   b) 4.816952038   c) 24.93728565   d) 21.22762776
    e) 21.52150537   f) 3.969362296   g) 4.988439193   h) 6.321928095
    i) 3.380238966   j) 5.209453366

7. a) 12.65507765   b) 7.598623985   c) 9.782407637
    d) 5.954792501   e) 9.453195311   f) 3.157282008

Appendix Exercise 2

1. a) Geometric; $t_4 = -\frac{1}{8}$; $t_8 = -\frac{1}{128}$; $S_{10} = .666015625$
    b) Arithmetic; $t_4 = 8$; $t_{15} = 41$; $S_{12} = 186$
    c) Arithmetic; $t_4 = 55$; $t_9 = 115$; $S_{10} = 730$
    d) Geometric; $t_4 = \frac{1080}{343}$; $t_7 = .24785591$; $S_{12} = 69.99731233$
    e) Arithmetic; $t_4 = -\frac{5}{12}$; $t_8 = -\frac{17}{12}$; $S_{10} = -\frac{95}{12}$
    f) Arithmetic; $t_4 = 5.6$; $t_{10} = -1.6$; $S_{15} = 12$

2. a) 45 150   b) 10 100   c) 12 051   d) 19 668

3. a) 20; 110   b) .00032; 781.24992
    c) 1.999004627; 14.48656247   d) .613913254; 7.721734929

4. a) 29; 155   b) 13; 7   c) 2; 132   d) 10; $\frac{70}{9}$ e) 8; −8

5. a) 10 240; 20 475   b) 6; 23.625   c) 12.8; 17.05
    d) 448; 895.125   e) 1.557967417; 19.1568813

6. $812.50   **7.** $155 000   **8.** 400 cm   **9.** $1073.74   **10.** 12.85%

11. a) 19.5 cm   b) 149.8046875 m   c) 150 m

12. a) $5 368 709.12   b) $10 737 418.23   c) $10.23; $10 475.52; $10 726 932.48

13. a) $22 616.89; $7052.05   b) $6637.64; $3653.68

14. a) .75   b) $3.\dot{3}$ c) 5   d) $\frac{1}{i}$

Appendix Exercise 3

1. a) 14.20   b) 54.06   c) 10.77   d) 15.74   e) 111.59   f) 16.03

2. a) 3.97%   b) 1.21%   c) 7.77%   d) 2.25%

3. 7.12%   **4.** 11.22%   **5.** 9.60%   **6.** 17.29%

Index